Series on Analysis, Applications and Computation – Vol. 5

ISAAC

Asymptotic Behavior of Generalized Functions

Series on Analysis, Applications and Computation

Series on Analysis, Applications and Computation – Vol. 5

ISAAC

Asymptotic Behavior of Generalized Functions

o Stevan Pilipović
University of Novi Sad, Serbia

o Bogoljub Stanković
University of Novi Sad, Serbia

o Jasson Vindas
Ghent University, Belgium

World Scientific

NEW JERSEY • LONDON • SINGAPORE • BEIJING • SHANGHAI • HONG KONG • TAIPEI • CHENNAI

Published by

World Scientific Publishing Co. Pte. Ltd.

5 Toh Tuck Link, Singapore 596224

USA office: 27 Warren Street, Suite 401-402, Hackensack, NJ 07601

UK office: 57 Shelton Street, Covent Garden, London WC2H 9HE

British Library Cataloguing-in-Publication Data
A catalogue record for this book is available from the British Library.

Series on Analysis, Applications and Computation — Vol. 5
ASYMPTOTIC BEHAVIOR OF GENERALIZED FUNCTIONS

Copyright © 2012 by World Scientific Publishing Co. Pte. Ltd.

ISBN-13 978-981-4366-84-7
ISBN-10 981-4366-84-6

Printed in Singapore by World Scientific Printers.

Preface

There are several approaches to generalized asymptotics within spaces of generalized functions. Probably the most developed approaches are that of Vladimirov, Drozhinov and Zavyalov [192], and of Kanwal and Estrada [56]. The first approach is followed by the authors of this book and extended in the direction of the S-asymptotics. The second approach is related to moment asymptotic expansions of generalized functions and the Cesàro behavior, and we refer to papers of Estrada, Kanwal and Vindas given in the references. We refer to [135] for the different definitions of the asymptotic behavior of distributions.

Asymptotic analysis is a very old and wide branch of mathematics. In general, the state of a system after certain period of time, or the behavior of a system in the frame of some limit procedure described by a suitable mathematical model, leads to the use and development of the methods of asymptotic analysis. With the general development of various branches of mathematical analysis and, especially, the theory of differential equations, the asymptotic analysis has obtained new impulses that resulted in new approaches and methods. In our analysis, such impulses came from the use of slowly varying functions and the introduction of asymptotic behavior of generalized functions.

In the first years of the twentieth century, the needs for a class of functions with growth order between the constant functions and power functions, in the scale of functions used in asymptotic analysis, appeared in many papers. Already Landau [87] used a term similar to "regular variation", but the theory of regularly varying and slowly varying functions (cf. **0.3**) belongs to Karamata and his pupils. Karamata's basic results were published in [76] and elaborated in [77]. Although the theory of regularly varying functions was invented for proving Tauberian type theorems for the Laplace-Stieltjes transform, such functions have been successfully employed in many other areas of mathematics: for Mercerian type theorems, in analytic number theory, the theory of entire functions and in the analysis of various classes of differential equations. Also, Karamata regularly varying functions were recognized as an important tool in probability theory [60].

At the present time the class of regularly varying functions is an integral part of asymptotic analysis. In the monographs [148], [64] and [9] one can find collected together the theory and applications of regular varying functions, the last one being certainly the most comprehensive treatise.

The class of regularly varying functions became more interesting and more applicable after the introduction of the notions of asymptotic behavior and expansion of generalized functions. We refer to [135] for a general overview of the development of the theory of asymptotic behavior of generalized functions up to 1989.

The notion of asymptotic behavior of generalized functions have had a very important role in quantum physics, [8], [10], [191], [192], where rigorous proofs of results in the foundation of this theory were obtained by the use of generalized asymptotic behavior. These papers motivated mathematicians to develop further the methods of asymptotic analysis on spaces of generalized functions.

The first definition of an asymptotic behavior for a generalized function is attributed to Lighthill [91]. His definition can be applied only to the so-called semi-regular distributions. Such a distribution T is equal to a locally integrable function f in a neighborhood of the observed point (in our case, infinity). The asymptotics of the function f at infinity gave the asymptotics of the distribution T at infinity. The same idea was adopted by Jones [72], Lavoine [88], Mangad [95], and Zemanian[203].

The next step forward was made by Silva in his axiomatic approach to distribution theory [147] and later by Lavoine and Misra [89], [90]. Silva introduced the growth order symbols small o and big O, in order to measure the asymptotic behavior of a distribution. Let $r \in C^\infty(a, \infty)$, $a \in \mathbf{R}$ and let $F \in C(a, \infty)$. If a distribution T satisfies

$$T = r \cdot D^m F \text{ in } (a, \infty), \ m \in \mathbf{N}_0, \text{ and } \lim_{x \to \infty} \frac{F(x)}{x^m} = 0,$$

then $T = o(r)$, $x \to \infty$. If there exists a constant M such that

$$\left| \frac{F(x)}{x^m} \right| \le M, \ x \in (b, \infty), \ b \ge a,$$

then $T = O(r)$, $x \to \infty$. Finally, if $T = D^m F$ in (a, ∞) and

$$\lim_{x \to \infty} \frac{F(x)}{x^m} = \frac{C}{m!},$$

then T behaves as the constant C in $\mathcal{D}'(\mathbf{R})$.

Lavoine and Misra defined the notion of *"equivalence at infinity"*. The distribution T is equivalent at infinity with Ax^α, $A \ne 0$, $\alpha \notin -\mathbf{N}$, if there exist $m \in \mathbf{N}_0$, $m + \alpha > 0$, and $F \in C(\mathbf{R})$, such that $T = D^m F$ on (a, ∞) and

$$F(x) \sim AC_{\alpha,m} x^{\alpha+m}, \ x \to \infty,$$

$$C_{\alpha,m} = \begin{cases} 1, & m = 0 \\ \Gamma(\alpha+1)/\Gamma(\alpha+m+1), & m > 0. \end{cases}$$

Later on, they replaced Ax^a by $Ax^a \log x$ in their definition of equivalence at infinity.

A natural generalization came with the introduction of regularly varying functions (cf. **0.3**) as the basic comparison functions for the asymptotics of generalized functions.

The notion of the quasi-asymptotic behavior (the quasi-asymptotics) of tempered distributions appeared as a qualitatively new important step in the asymptotic analysis of generalized functions. The first paper with the definition of the quasi-asymptotic behavior was the one of Zavyalov [200]. He introduced the quasi-asymptotics as automodel asymptotics "avtomodel'naya asimptotika". First, he defined a class M_+ of tempered distributions as follows: $T \in M_+$ if and only if there exists $a \in \mathbf{R}$ such that $T \in C^\infty(-\infty, a)$ and decreases faster than $|x|^r$, as $x \to -\infty$, for any positive r. For Zavyalov, $T \in M_+$ has the quasi-limit ($q - \lim$) equal to C if

$$T(kx) \to CH(x), \ k \to \infty \text{ in } \mathcal{S}'(\mathbf{R}),$$

where $H(x)$ is the Heaviside function. In short, $q - \lim_{x \to \infty} T(x) = C$. If $T \in M_+$ and $q - \lim T^{(\alpha)}(x) = C \ne 0$, $\alpha \in \mathbf{R}$, then T admits a quasi-asymptotic part of degree α at infinity. Zavyalov proved an equivalent statement: $T \in M_+$ admits a quasi-asymptotic part of degree α at infinity if and only if

$$k^{-\alpha} T(kx) \to g(x), \ x \to \infty \ \text{in} \ \mathcal{S}'(\mathbf{R})$$

for some distribution $g \neq 0$. This second statement has been used thereafter as the definition of the quasi-asymptotics.

Let us emphasize that the predominant role in the development of the quasi-asymptotics belongs to the Russian school with Vladimirov and his pupils and collaborators. Their results have given important impulses to the study of asymptotic behavior in different spaces of generalized functions and their applications. Especially, the robust work of Drozhzhinov and Zavyalov [38], [39], [40], [42], in the last ten years has been an extremely important contribution to the subject.

The notion of the shift asymptotics appeared for the first time in [146], Chapter VII, the second remark after Theorem VI:

$$\inf\{k \in \mathbf{R}; \ \{T(x+h)/(1+|h|^2)^{k/2}\}_{h \in \mathbf{R}^n} \ \text{is a bounded subset of} \ \mathcal{D}'(\mathbf{R}^n)\}$$

was called the degree of growth of the distribution T at infinity.

The shift asymptotics, defined in [15], has been later called S-asymptotics by the first two named authors of this book. It is said that $T \in \mathcal{S}'_+$ has shift asymptotics at infinity related to the regularly varying function $\rho(h) = h^\alpha L(h)$ if

$$T(x+h)/\rho(h) \to g(x), \ h \to \infty \ \text{in} \ \mathcal{S}'.$$

It was proved that if $\alpha > -1$, then T has quasi-asymptotics related to ρ at infinity. This result can give the impression that the S-asymptotics is a particular case of the quasi-asymptotics; however, this is not really the case. The relation between the S-asymptotics and the quasi-asymptotics is described in this book, Chapter I **2.13**. The first two authors of the book adapted and made precise the S-asymptotic behavior of generalized functions, and studied this type of asymptotic behavior for distributions ultradistributions and hyperfunctions, giving structural characterizations of comparison functions and the limit generalized function. It is worth to mention that the quasi-asymptotics is a natural notion in the analysis of tempered generalized functions, while the framework for the S-asymptotics is the space of exponential distributions $\mathcal{K}'_1$.

It should be mentioned that this book has a minor part in common with the previous books [135] and [192], which are also devoted to the asymptotic behavior of distributions. This was necessary in order to make the exposition complete and more transparent. Conceptually, we first treat the asymptotic behavior in a dual space $\mathcal{F}'_g$ of a suitable, but rather general, barrelled locally convex space of functions $\mathcal{F}_g$. Many important spaces of generalized functions are of this kind. The study is then specialized to a space $\mathcal{F}'$, a space of distribution, ultradistribution, or Fourier hyperfunction type. In this way we develop a general theory of asymptotics in $\mathcal{F}'$. We then specify the space $\mathcal{F}'$ in order to analyze special properties of asymptotic behaviors or expansions which are intrinsically connected with the nature of each space under consideration. We illustrate the theory by examples and point out some open problems which may serve as an encouragement for further investigations.

The book consists of two parts. The first one deals with the basic properties and analysis of the S-asymptotics and quasi-asymptotics in the space $\mathcal{F}'_g$ and in various special spaces of generalized functions. The asymptotic expansions and the Taylor expansion occupy also an important place in this part of the book.

In the second part one can find several applications of the asymptotic behavior of generalized functions. Abelian and Tauberian type theorems for integral transforms of convolution and Mellin-type convolution are studied for general kernels as well as for various special kernels such as the Laplace, Stieltjes, Weierstrass and Poisson transforms. This part also contains the asymptotic analysis of solutions to linear differential and partial differential equations, and to equations with ultradifferential or local operators. In addition, we give some applications of the quasi-asymptotic behavior of distributions to the study of summability of Fourier series and integrals.

We point out that our exposition, based on the work of the first two authors and their coauthors, has been extended with recent results of the third author and his collaborators. In this way, this book contains novel and strong results which give answers to some old problems in the area. Those results are presented in Chapter I **2.10–2.12** and Chapter II **5**, and make the presentation up-to-date.

Contents

Chapter I

Asymptotic Behavior of Generalized Functions

0 Preliminaries

0.1. We denote by $\mathbf{R}$ and $\mathbf{N}$ the sets of real and natural numbers; $\mathbf{N}_0 = \mathbf{N} \cup \{0\}$. The following notation will be used. If $x = (x_1, \ldots, x_n)$, $y = (y_1, \ldots, y_n) \in \mathbf{R}^n$, then $x \cdot y = x_1 y_1 + \cdots + x_n y_n$; $\|x\|^2 = x_1^2 + \cdots + x_n^2$; $x \geq 0 \iff x_i \geq 0$, $i = 1, \ldots, n$; $x \to \infty \iff x_i \to \infty$, $i = 1, \ldots, n$; $\mathbf{R}_+^n = \{x \in \mathbf{R}^n; \ x > 0\}$. If $x = (x_1, \ldots, x_n) \in \mathbf{R}^n$ and $k = (k_1, \ldots, k_n) \in \mathbf{N}_0^n$, then $|k| = k_1 + \cdots + k_n$, $k! = k_1! \ldots k_n!$, $x^k = x_1^{k_1} \ldots x_n^{k_n}$; $D^k = \partial^{k_1}/\partial x_1^{k_1} \ldots \partial^{k_n}/\partial x_n^{k_n}$, $f^{(k)} = D^k f$ $(f^{(0)} = f)$; $z = (z_1, \ldots, z_n) \in \mathbf{C}^n$; $D(a, r)$ denotes the *polydisk* $\{z \in \mathbf{C}^n; \ |z_i - a_i| < r_i, \ i = 1, \ldots, n\}$, where $|z_j|^2 = x_j^2 + y_j^2$, $z_j = x_j + iy_j \in \mathbf{C}$. $B(a, r)$ denotes the open ball $\{x \in \mathbf{R}^n, \ \|a - x\| < r\}$ and H is the *Heaviside function*: $H(t) = 0, t \leq 0$; $H(t) = 1, \ t > 0$.

0.2. A *cone* with vertex at zero in $\mathbf{R}^n$ is a non-empty set Γ such that $x \in \Gamma$ and $k > 0$ imply $kx \in \Gamma$. The cone Γ is called *solid* if int $\Gamma \neq \emptyset$. The *conjugate cone (dual cone)* Γ^* to the cone Γ is the set $\{\xi \in \mathbf{R}^n; \ x \cdot \xi \geq 0$ for each $x \in \Gamma\}$. It is obvious that Γ^* is also a cone which is convex and closed. The cone Γ is called *acute* if Γ^* is a solid cone.

0.3. A function $\rho : (a, \infty) \to \mathbf{R}$, $a \in \mathbf{R}_+$, is called *regularly varying at infinity* [76] if it is positive, measurable, and if there exists a real number α such that for each $x > 0$

$$\lim_{k \to \infty} \frac{\rho(kx)}{\rho(k)} = x^\alpha . \qquad (0.1)$$

The number α is called index of regular variation. If $\alpha = 0$, then ρ is called *slowly varying at infinity* and for such a function the letter "L" will be used. We then have that any regularly varying function can be written as $\rho(x) =$

$x^\alpha L(x)$, $x > a$, where L is slowly varying. It is known that the convergence of (0.1) is uniform on every fixed compact interval $[a', b']$, $a < a' < b' < \infty$, and that ρ is necessarily bounded (hence integrable) on it [9].

Let L be a slowly varying function at infinity. Then, for each $\varepsilon > 0$,

(i) there exist constants $C_1, C_2 > 0$ and $X > a$ such that

$$C_1 x^{-\varepsilon} \leq L(x) \leq C_2 x^\varepsilon, \quad x \geq X \, ; \tag{0.2}$$

(ii) $\lim\limits_{x \to \infty} x^\varepsilon L(x) = +\infty$, $\lim\limits_{x \to \infty} x^{-\varepsilon} L(x) = 0$.

We know ([9], p. 16) that if $L_2(x) \to \infty$, $x \to \infty$, and L_1, L_2 are slowly varying, then $L_1 \circ L_2 = L_1(L_2)$ is slowly varying, as well. Hence, for $x > -\infty$,

$$\lim\limits_{h \to \infty} \frac{L(x + h)}{L(h)} = \lim\limits_{u \to \infty} \frac{L(\log ut)}{L(\log u)} = 1 \, . \tag{0.3}$$

The definition of a regular varying function at zero is similar. For the definition of regularly generalized functions see [156] and [162].

0.4. The class of distributions f_α, $\alpha \in \mathbf{R}$, belonging to $\mathcal{S}'_+$ (see **0.5.1.**) is defined in the following way:

$$f_\alpha(t) = \begin{cases} H(t) t^{\alpha-1} / \Gamma(\alpha), & \alpha > 0, \\ \\ f_{\alpha+m}^{(m)}(t), & \alpha \leq 0, \ \alpha + m > 0, \end{cases}$$

where H is the Heaviside function, and the derivative $f^{(m)}$ is taken to be in the distributional sense (see [192], Chapter I, §1). We therefore have $f_0 = \delta$, the Dirac delta distribution; $f_{-m} = \delta^{(m)}$, $m \in \mathbf{N}$; and $f_p * f_q = f_{p+q}$. We also use the notations $t_+^\alpha = \Gamma(\alpha + 1) f_{\alpha+1}(t)$ and $t_-^\alpha = (-t)_+^\alpha$, $\alpha \notin -\mathbf{N}$.

Let $g \in \mathcal{S}'_+$. We denote $g^{(-\alpha)} = f_\alpha * g$, $\alpha \in \mathbf{R}$ ($*$ the is convolution symbol).

The sequence $\langle \delta_m \rangle_{m \in \mathbf{N}} \subset C^\infty(\mathbf{R}^n)$ is called a δ-*sequence* if: a) supp $\delta_m \subset [-\alpha_m, \alpha_m]$, $\alpha_m \to 0$, $m \to \infty$; b) $\delta_m \geq 0$, $m \in \mathbf{N}$; c) $\int\limits_{\mathbf{R}^n} \delta_m(t) dt = 1$, $m \in \mathbf{N}$.

If $\varphi \in \mathcal{D}$, then $\delta_m * \varphi \to \varphi$, $m \to \infty$ in $\mathcal{D}$, hence $\{\delta_m * \varphi; \ m \in \mathbf{N}\}$ is a bounded set in $\mathcal{D}$.

0.5. We will repeat definitions and some basic properties of generalized functions defined as elements of dual spaces.

0.5.1. The Schwartz spaces of test functions and distributions on $\mathbf{R}^n$ are denoted by $\mathcal{D} = \mathcal{D}(\mathbf{R}^n)$ and $\mathcal{D}' = \mathcal{D}'(\mathbf{R}^n)$, respectively ($\mathbf{R}^n$ will be omitted wherever n is not fixed). The space $\mathcal{D}$ is a locally convex, barrelled, Montel and complete space. If a filter with a countable basis is weakly convergent in $\mathcal{D}'$, then it is convergent in $\mathcal{D}'$ with the strong topology, as well (cf. [146]).

$\mathcal{S}$ is the *space of rapidly decreasing functions* and its dual $\mathcal{S}'$ is the *space of tempered distributions* [146]. For a closed cone $\Gamma \subset \mathbf{R}^n$, $\mathcal{S}'_\Gamma = \{f \in \mathcal{S}'; \operatorname{supp} f \subset \Gamma\}$. In the one-dimensional case $\mathcal{S}'_+ = \{f \in \mathcal{S}'(\mathbf{R}); \operatorname{supp} f \subset [0, \infty)\}$. Recall ([189]):

$$\mathcal{S}(\Gamma) = \{\varphi \in C^\infty(\Gamma); \|\varphi\|_p < \infty, \ p \in \mathbf{N}\},$$

where

$$\|\varphi\|_p = \sup_{x \in \Gamma, \ |\beta| \leq p} (1 + \|x\|^2)^{p/2} |\varphi^{(\beta)}(x)|.$$

By $\mathcal{S}_p(\Gamma)$ is denoted the completion of the set $\mathcal{S}(\Gamma)$ with respect to the norm $\|\varphi\|_p$. Note $\mathcal{S}(\Gamma) = \bigcap_{p \in \mathbf{N}_0} \mathcal{S}_p(\Gamma)$ and $\mathcal{S}'(\Gamma) = \bigcup_{p \in \mathbf{N}_0} \mathcal{S}'_p(\Gamma)$, where the intersection and the union have topological meaning. A sequence $\langle f_n \rangle_{n \in \mathbf{N}}$ in $\mathcal{S}'(\Gamma)$ converges to $f \in \mathcal{S}'(\Gamma)$ if and only if it belongs to some $\mathcal{S}'_q(\Gamma)$ and converges to $f \in \mathcal{S}'_q(\Gamma)$ in the norm of $\mathcal{S}'_q(\Gamma)$. *The space $\mathcal{S}'_\Gamma$ is isomorphic to $\mathcal{S}'(\Gamma)$* if Γ is closed convex solid cone ([189], [192]).

$\mathcal{E}'$ the *space of distributions with compact support*; it is isomorphic to the dual space of $\mathcal{E} = C^\infty(\mathbf{R}^n)$ (cf. [146]).

$\mathcal{D}_{L^p}$, $1 \leq p \leq \infty$, is the space of smooth functions with all derivatives belonging to L^p ([146]), $\mathcal{D}_{L^p} \subset \mathcal{D}_{L^q}$ if $p < q$.

$\dot{\mathcal{B}}$ is a subspace of $\mathcal{B} = \mathcal{D}_{L^\infty}$, defined as follows: $\varphi \in \dot{\mathcal{B}}$ if and only if $|\varphi^{(\alpha)}(x)| \to 0$ as $\|x\| \to \infty$ for every $\alpha \in \mathbf{N}_0^n$.

$\mathcal{D}'_{L^p}$, $1 < p \leq \infty$ is the dual space of $\mathcal{D}_{L^q}$, $1 \leq q < \infty$, $\frac{1}{p} + \frac{1}{q} = 1$. $\mathcal{D}'_{L^1}$ is the dual of $\dot{\mathcal{B}}$ and $\mathcal{D}'_{L^\infty}$ is denoted by $\dot{\mathcal{B}}'$.

$\mathcal{O}'_c$ *is the space of distributions with fast descent:*

$$\mathcal{O}'_c = \{T \in \mathcal{D}'; (1 + \|x\|^2)^m T \in \dot{\mathcal{B}}', \text{ for every } m \in \mathbf{N}\}.$$

$\mathcal{K}_p$, $p \geq 1$, is the spaces of functions $\varphi \in C^\infty$ with the property:

$$\nu_m(\varphi) = \sup_{x \in \mathbf{R}^n, \ |\alpha| \leq m} \exp(m \|x\|^p) |D^\alpha \varphi(x)| < \infty, \quad m = 1, 2, \dots$$

The elements of $\mathcal{K}_p$ are called *rapidly exponentially decreasing functions*. Then $\mathcal{K}'_p$ is the dual of $\mathcal{K}_p$.

The *convolution* in $\mathcal{D}'$ is defined as follows. Let $\langle \eta_m \rangle_{m \in \mathbf{N}}$ be a sequence in $\mathcal{D}(\mathbf{R}^{2n})$ such that for every compact set $K \subset \mathbf{R}^{2n}$ there exists $m_0(K)$ such that $\eta_m(x) = 1$, $x \in K$, $m \geq m_0(K)$ and

$$\sup_{x \in \mathbf{R}^{2n}} |\eta_m^{(\beta)}(x)| < C_\beta, \quad \beta \in \mathbf{N}_0^n.$$

The convolution of $T, S \in \mathcal{D}'$ is defined by

$$\langle T * S, \varphi \rangle = \lim_{m \to \infty} \langle T(x)S(y), \eta_m(x,y)\varphi(x+y) \rangle, \quad \varphi \in \mathcal{D},$$

if this limit exists for every $\langle \eta_m \rangle_m$ (then, it does not depend on $\langle \eta_m \rangle_m$). By the Banach–Steinhaus theorem we know that $T * S \in \mathcal{D}'$.

The spaces $\mathcal{D}'(\Gamma)$ and $\mathcal{S}'(\Gamma)$ with the operation $*$ are associative and commutative algebras. The convolution in this case is separately continuous.

We refer also to [63] and [3] for the theory of distributions.

0.5.2 *Ultradistribution spaces*

We follow the notation and definitions from [79], [81] and [86]. By $\langle M_p \rangle_p$ is denoted a sequence of positive numbers, $M_0 = M_1 = 1$, satisfying some of the following conditions:

(M.1) $M_p^2 \leq M_{p-1}M_{p+1}$, $p \in \mathbf{N}$;

(M.2) $M_p/(M_q M_{p-q}) \leq AB^p$, $0 \leq q \leq p$, $p \in \mathbf{N}$;

(M.2)$'$ $M_{p+1} \leq AB^p M_p$, $p \in \mathbf{N}_0$, $\mathbf{N}_0 = \mathbf{N} \cup \{0\}$;

(M.3) $\displaystyle\sum_{q=p+1}^{\infty} M_{q-1}/M_q \leq A_p M_p/M_{p+1}$, $p \in \mathbf{N}$;

(M.3)$'$ $\displaystyle\sum_{p=1}^{\infty} M_{p-1}/M_p < \infty$,

where A and B are constants independent of p.

In the sequel, we will always assume (M.1), (M.2)$'$ and (M.3)$'$.

Let h be a positive number and let $\langle h_p \rangle_p$ be a real positive sequence increasing to ∞. We denote

$$H_p = \begin{cases} h^p, & \text{for the ultradifferentiable functions of class } (M_p) \\ h_1 \ldots h_p, & \text{for the ultradifferentiable functions of class } \{M_p\}. \end{cases} \tag{0.4}$$

Let K be a compact set in an open set Ω of $\mathbf{R}^n$.

We denote by $\mathcal{E}_K^{H_p M_p}$ the space of smooth functions $\varphi \in \mathcal{E}$ such that

$$\sup_{x \in K, \alpha \in \mathbf{N}_0^n} |\varphi^{(\alpha)}(x)|/H_{|\alpha|} M_{|\alpha|} < \infty. \tag{0.5}$$

We set $\mathcal{D}_K^{H_p M_p} = \{\varphi \in \mathcal{E}_K^{H_p M_p}; \operatorname{supp} \varphi \subset K\}$, it is a Banach space with norm

$$q_{H_p M_p}(f) = \sup_{x \in K, \alpha \in \mathbf{N}_0^n} \frac{|\varphi^{(\alpha)}(x)|}{H_{|\alpha|} M_{|\alpha|}}. \tag{0.6}$$

Then, the basic spaces are defined by

$$\mathcal{D}_K^{(M_p)} = \operatorname{proj} \lim_{h \to 0} \mathcal{D}_K^{h^p M_p}, \quad \mathcal{D}_K^{\{M_p\}} = \operatorname{ind} \lim_{\{H_p = h_1 \dots h_p\}} \mathcal{D}_K^{H_p M_p}$$

$$\mathcal{D}^*(\Omega) \equiv \mathcal{D}_\Omega^* = \operatorname{ind} \lim_{K \subset\subset \Omega} \mathcal{D}_K^*,$$

where $*$ denotes either (M_p) or $\{M_p\}$.

The spaces with the upper index (M_p) are *the Beurling-type spaces of ultradifferentiable functions* and with the upper index $\{M_p\}$ *are the Roumieu-type spaces of ultradifferentiable functions*. Their strong duals are spaces of *Beurling and Roumieu-type ultradistributions, respectively.*

The space $\mathcal{E}'^* = \mathcal{E}'^*(\Omega)$ is the dual of

$$\mathcal{E}^{(M_p)} = \operatorname{proj} \lim_{K \subset\subset \Omega} \operatorname{proj} \lim_{h \to 0} \mathcal{E}_K^{h^p M_p};$$

$$\mathcal{E}^{\{M_p\}} = \operatorname{proj} \lim_{K \subset\subset \Omega} \operatorname{ind} \lim \mathcal{E}_K^{H_p M_p}.$$

Weighted ultradistribution spaces are defined by

$$\mathcal{D}_{L^1}^{(M_\alpha)}(\mathbf{R}^n) = \mathcal{D}_{L^1}^{(M_\alpha)} = \operatorname{proj} \lim_{h \to \infty} \mathcal{D}_{L^1,h}^{M_\alpha}, \quad \mathcal{D}_{L^1}^{\{M_\alpha\}} = \operatorname{ind} \lim_{h \to 0} \mathcal{D}_{L^1,h}^{M_\alpha},$$

where $\mathcal{D}_{L^1,h}^{M_\alpha}$, $h > 0$, is the Banach space of smooth functions φ on $\mathbf{R}^n$ with finite norm

$$\|\varphi\|_{L^1,h} = \sup_{\alpha \in \mathbf{N}_0^n} \frac{h^{|\alpha|}}{M_{|\alpha|}} \|\varphi^{(\alpha)}\|_{L^1}.$$

$\mathcal{D}^* = \mathcal{D}^*(\mathbf{R}^n)$ is dense in $\mathcal{D}_{L^1}^*$, and the inclusion mapping is continuous. The strong dual of $\mathcal{D}_{L^1}^*$ is denoted by $\mathcal{B}'^*$.

Spaces of tempered ultradistributions are defined as the strong duals of the following testing function spaces:

$$\mathcal{S}^{(M_\alpha)}(\mathbf{R}^n) = \mathcal{S}^{(M_\alpha)} = \operatorname{proj} \lim_{h \to \infty} \mathcal{S}_h^{M_\alpha},$$

$$\mathcal{S}^{\{M_\alpha\}}(\mathbf{R}^n) \equiv \mathcal{S}^{\{M_\alpha\}} = \text{ind} \lim_{h \to 0} \mathcal{S}_h^{M_\alpha},$$

where $\mathcal{S}_h^{M_\alpha}$, $h > 0$, is the Banach space of smooth functions φ on $\mathbf{R}$ with finite norm

$$\gamma_h(\varphi) = \sup_{\alpha,\beta \in \mathbf{N}_0^n} \frac{h^{|\alpha|+|\beta|}}{M_{|\alpha|}M_{|\beta|}} \|(1+|x|^2)^{|\alpha|/2}\varphi^{(\beta)}\|_{L^\infty}.$$

The Fourier transform is an isomorphism of $\mathcal{S}^*$ onto $\mathcal{S}^*$; $\mathcal{D}^* = \mathcal{D}^*(\mathbf{R}^n)$ is dense in $\mathcal{S}^*$, $\mathcal{S}^*$ is dense in $\mathcal{D}_{L^1}^*$ and the inclusion mappings are continuous.

The strong dual of $\mathcal{S}^*, \mathcal{S}'^*$, is *the space of tempered ultradistributions (of Beurling and Roumieu types)*. There holds $\mathcal{D}^* \hookrightarrow \mathcal{S}^* \hookrightarrow \mathcal{E}^*$, where $\hookrightarrow$ means that the left space is dense in the right one and that the inclusion mapping is continuous. Thus, $\mathcal{E}'^* \subset \mathcal{S}'^* \subset \mathcal{D}'^*$. We denote $\mathcal{S}'_+^* = \{f \in \mathcal{S}'^*(\mathbf{R}); \text{ supp } f \subset [0, \infty)\}$.

Let

$$\mathcal{S}_{[0,\infty)}^* = \{\psi \in C^\infty[0,\infty); \psi = \varphi|_{[0,\infty)} \text{ for some } \varphi \in \mathcal{S}^*\}$$

with the induced convergence structure from $\mathcal{S}^*$; its strong dual is in fact $\mathcal{S}'_+^*$.

An operator of the form: $P(D) = \sum\limits_{|\alpha|=0}^{\infty} a_\alpha D^\alpha$, $a_\alpha \in C$, $\alpha \in \mathbf{N}_0^n$, is called *ultradifferential operator* of class (M_p) (of class $\{M_p\}$) if there are constants $L > 0$ and $C > 0$ (for every $L > 0$ there is $C > 0$) such that $|a_\alpha| \leq CL^{|\alpha|}/M_{|\alpha|}$, $|\alpha| \in \mathbf{N}_0$.

0.5.3. *Fourier hyperfunctions.* There are many equivalent definitions of hyperfunctions, Laplace hyperfunctions and Fourier hyperfunctions (cf. [71], [80], [144], [145], [204]), but we will use definitions and results collected in [75]. Let I be a convex neighborhood of zero in $\mathbf{R}^n$ and let α be a non-negative constant. A function F, holomorphic on $\mathbf{R}^n + iI$, is said to *decrease exponentially* with type $(-\alpha), \alpha \geq 0$, if for every compact subset $K \subset\subset I$ and every $\varepsilon > 0$, there exists $C_{K,\varepsilon} > 0$ such that

$$|F(z)| \leq C_{K,\varepsilon} \exp(-(\alpha - \varepsilon)|\text{Re } z|), \quad z \in \mathbf{R}^n + iK. \tag{0.7}$$

The set of all such functions is denoted by $\tilde{\mathcal{O}}^{-\alpha}(\mathbf{D}^n + iI)$, $(\tilde{\mathcal{O}}(\mathbf{D}^n + iI)$ for $\alpha = 0$), where $\mathbf{D}^n$ denotes the directional compactification of $\mathbf{R}^n$: $\mathbf{D}^n = \mathbf{R}^n \cup S_\infty^{n-1}(S_\infty^{n-1}$ consists of all points at infinity in all direction). Space $\mathcal{P}_*$ is defined by

$$\mathcal{P}_* = \text{ind} \lim_{I \ni 0} \text{ ind} \lim_{\alpha \downarrow 0} \tilde{\mathcal{O}}^{-\alpha}(\mathbf{D}^n + iI). \tag{0.8}$$

The dual space of $\mathcal{P}_*$ *is the space* $\mathcal{Q}(\mathbf{D}^n)$ *of the Fourier hyperfunctions.* $\mathcal{Q}(\mathbf{D}^n)$ is a space of FS-type. We can give a representation of elements from $\mathcal{Q}(\mathbf{D}^n)$. Let $\mathcal{O}$ be the sheaf of holomorphic functions. Denote

$$U_j = (\mathbf{D}^n + iI) \cap \{\operatorname{Im} z_j \neq 0\}, \; j = 1, \ldots, n; \; (\mathbf{D}^n + iI)\#\mathbf{D}^n = U_1 \cap \cdots \cap U_n,$$

$$(\mathbf{D}^n + iI)\#_j\mathbf{D}^n = U_1 \cap \cdots \cap U_{j-1} \cap U_{j+1} \cap \cdots \cap U_n.$$

Then

$$\mathcal{Q}(\mathbf{D}^n) = \tilde{\mathcal{O}}((\mathbf{D}^n + iI)\#\mathbf{D}^n)/\sum_{j=1}^{n} \tilde{\mathcal{O}}((\mathbf{D}^n + iI)\#_j\mathbf{D}^n). \tag{0.9}$$

Thus, $f \in \mathcal{Q}(\mathbf{D}^n)$ is defined as the class $[F]$, where $F \in \tilde{\mathcal{O}}((\mathbf{D}^n + iI)\#\mathbf{D}^n)$. F is called a *defining function* of f and it is represented by 2^n functions F_σ, $F = \langle F_\sigma \rangle$, where $F_\sigma \in \tilde{\mathcal{O}}(\mathbf{D}^n + iI_\sigma)$; $\mathbf{D}^n + iI_\sigma = \mathbf{D}^n + i(I \cap \Gamma_\sigma)$ is an infinitesimal wedge of type $\mathbf{R}^n + i\Gamma_\sigma 0$, Γ_σ are open σ-th orthants in $\mathbf{R}^n$.

An $f \in L^1_{loc}(\mathbf{R}^n)$ is called a *function of infra-exponential type* if for every $\varepsilon > 0$, there exists $C_\varepsilon > 0$ such that $|f(x)| \leq C_\varepsilon \exp(\varepsilon|x|)$, $x \in \mathbf{R}^n$. Then by ℓf is denoted the hyperfunction defined by f.

We denote by Λ the set of n-th variations of $\{-1, 1\}$.

The boundary-value representation of $f \in \mathcal{Q}(\mathbf{D}^n)$ is:

$$f = [F] := \sum_{\sigma \in \Lambda} F_\sigma(x + i\Gamma_\sigma 0). \tag{0.10}$$

$F_\sigma(x + i\Gamma_\sigma 0)$ denotes the element of the quotient space given in (0.9); it is determined by F_σ.

The dual pairing between $\varphi \in \mathcal{P}_*$ and $g = [G] \in \mathcal{Q}(\mathbf{D}^n)$ is given by

$$\langle g, \varphi \rangle = \int_{\mathbf{R}^n} g(x)\varphi(x)dx = \sum_{\sigma \in \Lambda} \int_{\operatorname{Im} w = v_\sigma} G_\sigma(w)\varphi(w)dw, \; w = u + iv,$$

where $v_\sigma \in I_\sigma$.

Similarly, $\mathcal{Q}^{-\alpha}(\mathbf{D}^n)$, $\alpha > 0$, is defined using $\tilde{\mathcal{O}}^{-\alpha}$ instead of $\tilde{\mathcal{O}}$ (cf. Definition 8.2.5 in [75]).

An *infinite-order differential operator* $J(D)$ of the form

$$J(D) = \sum_{|a| \geq 0} b_a D^a \quad \text{with} \quad \lim_{|a| \to \infty} \sqrt[|a|]{|b_a|a!} = 0, b_a \in \mathbf{C},$$

is called a *local operator*. $J(D)$ maps continuously $\mathcal{O}(U)$ into $\mathcal{O}(U)$, U being an open set in $\mathbf{C}^n$, and also $\mathcal{Q}(\mathbf{D}^n)$ into $\mathcal{Q}(\mathbf{D}^n)$.

The *Fourier transform* on $\mathcal{Q}(\mathbf{D}^n)$ is defined by using the functions $\chi_\sigma = \chi_{\sigma_1}(z_1)\ldots\chi_{\sigma_n}(z_n)$, where $\sigma_k = \pm 1$, $k = 1,\ldots,n$, $\sigma = (\sigma_1,\ldots,\sigma_n)$ and $\chi_+(t) = e^t/(1+e^t)$, $\chi_-(t) = 1/(1+e^t)$, $t \in \mathbf{R}$. Let

$$u(x) \cong \sum_{\sigma\in\Lambda} U_\sigma(x+i\Gamma_\sigma 0) = \sum_{\sigma\in\Lambda}\sum_{\tilde\sigma\in\Lambda}(\chi_{\tilde\sigma}U_\sigma)(x+i\Gamma_\sigma 0),$$

where $\chi_{\tilde\sigma}U_\sigma \in \mathcal{Q}(\mathbf{D}^n + iI_\sigma)$, $\tilde\sigma \in \Lambda$ and $\chi_{\tilde\sigma}U_\sigma$ decreasing exponentially along the real axis outside the closed $\tilde\sigma$-th orthant. The Fourier transform of u is defined by

$$\mathcal{F}(u)(\xi) \cong \sum_{\sigma\in\Lambda}\sum_{\tilde\sigma\in\Lambda}\mathcal{F}(\chi_{\tilde\sigma}U_\sigma)(\xi - i\Gamma_{\tilde\sigma}0)$$

$$= \sum_{\sigma\in\Lambda}\sum_{\tilde\sigma\in\Lambda}\int_{\mathrm{Im}\,z=y^\sigma} e^{iz\zeta}(\chi_{\tilde\sigma}U_\sigma)(z)dz, \quad y^\sigma \in I_\sigma, \ \zeta = \xi + i\eta,$$

where $\mathcal{F}(\chi_{\tilde\sigma}U_\sigma) \in \tilde{\mathcal{O}}(\mathbf{D}^n - iI_{\tilde\sigma})$ and $\mathcal{F}(\chi_{\tilde\sigma}U_\sigma)(z) = O(e^{-\omega|x|})$ for a suitable $\omega > 0$ along the real axis outside the closed σ-orthant. $\mathcal{F}$ is an automorphism of $\mathcal{Q}(\mathbf{D}^n)$.

0.6. Let E be a locally convex topological vector space. An absolute convex closed absorbent subset of E is called a barrel. If every barrel is a neighborhood of zero in E, then E is called barreled. Throughout this book $\mathcal{F}_g$ stands for a locally convex barrelled complete Hausdorff space of smooth functions (the subscript g stands for "general"), $\mathcal{F}_g \hookrightarrow \mathcal{E} = \mathcal{E}(\mathbf{R}^n)$, and $\mathcal{F}'_g$ stands for the strong dual space of $\mathcal{F}_g$; observe $\mathcal{E}' \subset \mathcal{F}'_g$. If $T \in \mathcal{F}'_g$ and $\varphi \in \mathcal{F}_g$, then $\langle T, \varphi\rangle$ is the dual pairing between T and φ. We write $\mathcal{F}_0$ if all elements of $\mathcal{F}_g$ are compactly supported; $\mathcal{F}'_0$ denotes the dual space of $\mathcal{F}_0$; observe that a notion of support for elements of $\mathcal{F}'_0$ can be defined in the usual way. In $\mathcal{F}'_g$, a weakly bounded set is also a strongly bounded one (Mackey–Banach–Steinhaus theorem). The spaces of distributions, ultra-distributions, Fourier hyperfunctions, …, are of this kind. Furthermore, we shall always use the notation $\mathcal{F} = \mathcal{F}_g$ if $\mathcal{A} \hookrightarrow \mathcal{F}$, where $\mathcal{A} = \mathcal{D}, \mathcal{D}^*$, or $\mathcal{P}_*$; in such case $\mathcal{F}' \subset \mathcal{A}' = \mathcal{D}', \mathcal{D}'^*$, or $\mathcal{Q}(\mathbf{D}^n))$, respectively, and we say that $\mathcal{F}'$ is a distribution, ultradistribution, or Fourier hyperfunction space, respectively. We set $\mathcal{F}'_\Gamma = \{T \in \mathcal{F}'; \mathrm{supp}\,T \subset \Gamma\}$.

We suppose that the following operations are well defined on $\mathcal{F}'_g$:

Differentiation: We assume that ∂/∂_{x_i}, $\mathcal{F}_g \to \mathcal{F}_g$, are continuous operators. Let $k \in \mathbf{N}_0^n$. Then,

$$\left\langle \frac{\partial^k}{\partial_{x_1}^{k_1}\ldots\partial_{x_n}^{k_n}}T(x), \varphi(x)\right\rangle = \left\langle T(x), (-1)^{|k|}\frac{\partial^k}{\partial_{x_1}^{k_1}\ldots\partial_{x_n}^{k_n}}\varphi(x)\right\rangle, \quad \varphi \in \mathcal{F}_g.$$

Change of variables: If $T \in \mathcal{F}'_g$ and $k > 0$, then by definition

$$\langle T(kx), \varphi(x) \rangle = \left\langle T(x), \frac{1}{k^n} \varphi\left(\frac{x}{k}\right) \right\rangle, \quad \varphi \in \mathcal{F}_g.$$

If $T \in \mathcal{F}'_g$ and $h \in \mathbf{R}^n$,

$$\langle T(x+h), \varphi(x) \rangle = \langle T(x), \varphi(x-h) \rangle, \quad \varphi \in \mathcal{F}_g.$$

Furthermore, it is always assumed that $k \to \varphi(\cdot/k)$, $\mathbf{R}_+ \to \mathcal{F}_g$, and $h \to \varphi(\cdot + h)$, $\mathbf{R}^n \to \mathcal{F}_g$, are continuous. Consequently, by the mean value theorem, one readily verifies that both maps are indeed C^∞.

Let $\psi \in \mathcal{E}$, if $\varphi \mapsto \psi\varphi$ is a continuous mapping from $\mathcal{F}_g$ into $\mathcal{F}_g$, then we say that ψ is a *multiplier* of $\mathcal{F}'_g$. Then ψT is by definition $\langle \psi T, \varphi \rangle = \langle T, \psi\varphi \rangle$. The *set of multipliers* of $\mathcal{F}'_g$ is denoted by $M_{(\cdot)}$.

We shall say that $\theta \in \mathcal{E}$ is a *convolutor* of $\mathcal{F}'_g$ if the mapping $\varphi \mapsto \check{\theta} * \varphi$, $\mathcal{F}_g \to \mathcal{F}_g$, is well defined and continuous, where $\check{\theta}(t) = \theta(-t)$. We denote by $M_{(*)}$ the *set of convolutors* of $\mathcal{F}'_g$. If $\theta \in M_{(*)}$, then for $T \in \mathcal{F}'_g$, $(T * \theta)$ is defined by $\langle T * \theta, \psi \rangle = \langle T, \check{\theta} * \varphi \rangle$, $\varphi \in \mathcal{F}_g$.

$\langle \delta_m \rangle_{m \in \mathbf{N}}$, a sequence in $M_{(*)} \cap \mathcal{F}_g$, is called a δ-sequence in $\mathcal{F}_g$ if $\delta_m \geq 0$, $m \in \mathbf{N}$, and for every $\varphi \in \mathcal{F}_g$, $\delta_m * \varphi \to \varphi$ in $\mathcal{F}_g$, $m \to \infty$.

We shall say that the *convolution* with compactly supported elements is well defined in $\mathcal{F}'_g$ if a notion of support makes sense in $\mathcal{F}'_g$ and the following definition applies: given $S, T \in \mathcal{F}'_g$, where supp $S = K$ is compact in $\mathbf{R}^n$, their convolution is defined by

$$\langle S * T, \varphi \rangle = \langle S(x) \times T(y), \alpha(x)\varphi(x+y) \rangle = \langle S_x \times T_y, \alpha(x)\varphi(x+y) \rangle,$$

where the function $\alpha \in \mathcal{F}_g$ has compact support, supp $\alpha = \tilde{K}$, so that the set $K \subset \text{int } \tilde{K}$ and $\alpha(x) = 1$, $x \in K$. For $\mathcal{F}'$ it coincides with the usual definition of convolution.

We assume that $\mathcal{F}'_g$ contains *regular elements* $f \in L^1_{loc}$. They are identified with f itself:

$$f : \langle f, \varphi \rangle = \int_{\mathbf{R}^n} f(x)\varphi(x)dx, \quad \varphi \in \mathcal{F}_g,$$

if $f\varphi \in L^1$ for every $\varphi \in \mathcal{F}_g$ and if $\varphi_m \to 0$ in $\mathcal{F}$ implies $\langle f, \varphi_m \rangle \to 0$.

There is also another situation in which we shall identify locally integrable functions with elements $T \in \mathcal{F}'$. Let $\mathcal{A} = \mathcal{D}, \mathcal{D}^*$, or $\mathcal{P}_*$ and suppose $\mathcal{A} \hookrightarrow \mathcal{F}$. We identify T with $f \in L^1_{loc}$ if $\langle T, \varphi \rangle = \langle f, \varphi \rangle$, for all $\varphi \in \mathcal{A}$. In such a case we simply write $T = f$. For example, $T(x) = xe^{x^2}\sin(e^{x^2}) \in \mathcal{S}'(\mathbf{R})$ is defined in this way, and not as a regular element of $\mathcal{S}'(\mathbf{R})$ in the sense described above.

1 S-asymptotics in $\mathcal{F}'_g$

1.1 *Definition*

Definition 1.1. Let Γ be a cone with vertex at zero and let c be a positive real-valued function defined on Γ. It is said that $T \in \mathcal{F}'_g$ has S-asymptotic behavior related to c with limit U if $T(x+h)/c(h)$ converges weakly in $\mathcal{F}'_g$ to U when $h \in \Gamma$, $\|h\| \to \infty$, i.e. ($w. \lim$ = weak limit):

$$\underset{h\in\Gamma, \|h\|\to\infty}{w.\lim} T(x+h)/c(h) = U \quad \text{in} \quad \mathcal{F}'_g \tag{1.1}$$

or

$$\underset{h\in\Gamma, \|h\|\to\infty}{\lim} \langle T(x+h)/c(h), \varphi(x)\rangle = \langle U, \varphi\rangle, \quad \varphi \in \mathcal{F}_g. \tag{1.2}$$

If (1.1) is satisfied, it is also said that T has S-asymptotics and we write in short: $T(x+h) \overset{s}{\sim} c(h)U(x)$, $h \in \Gamma$.

Remarks. 1) If Γ is a convex cone we could use another limit in Γ. Let $h_1, h_2 \in \Gamma$. We say that $h_1 \geq h_2$ if and only if $h_1 \in h_2 + \Gamma$; Γ is now partially ordered.

For a real-valued function ρ defined on Γ, we write

$$\underset{h\in\Gamma, \, h\to\infty}{\lim} \rho(h) = A \in \mathbf{R}$$

if for any $\varepsilon > 0$ there exists $h(\varepsilon) \in \Gamma$ such that $\rho(h) \in (A - \varepsilon, A + \varepsilon)$ when $h \geq h(\varepsilon)$, $h \in \Gamma$.

If Γ is a convex cone, then the S-asymptotics with respect to this limit might be defined as:

$$\underset{h\in\Gamma, \, h\to\infty}{\lim} \langle T(x+h)/c(h), \varphi(t)\rangle = \langle u, \varphi\rangle, \quad \varphi \in \mathcal{F}_g. \tag{1.3}$$

In case $n = 1$, the limits (1.2) and (1.3) coincide. We will mostly use Definition 1.1 in this book, for S-asymptotics defined by (1.3) see also [135].

2) If $\mathcal{F}_g$ is a Montel space, then the strong and the weak topologies in $\mathcal{F}'_g$ are equivalent on a bounded set. If B is a filter with a countable basis and if $w. \underset{h\in B}{\lim} T(h) = U$ in $\mathcal{F}'_g$, then this limit exists in the sense of the strong topology. Hence, (1.1) is equivalent to

$$\underset{h\in\Gamma, \|h\|\to\infty}{s.\lim} T(x+h)/c(h) = U \quad \text{in} \quad \mathcal{F}'_g.$$

(Furthermore, from now on, we will omit the symbol *s.* for the strong convergence).

For the first ideas of the *S*-asymptotics see [3] and [146]. The starting point of the theory is [127].

1.2   *Characterization of comparison functions and limits*

Proposition 1.1. *Let Γ be a convex cone. Suppose $T \in \mathcal{F}'_g$ has the S-asymptotics $T(x + h) \stackrel{s}{\sim} c(h)U(x)$, $h \in \Gamma$. If $U \neq 0$, then:*

a) There exists a function d on Γ such that

$$\lim_{h \in \Gamma,\, \|h\| \to \infty} c(h + h_0)/c(h) = d(h_0), \quad \text{for every } h_0 \in \Gamma. \qquad (1.4)$$

b) The limit U satisfies the equation

$$U(\cdot + h) = d(h)U, \quad h \in \Gamma.$$

Proof. a) Since $U \neq 0$, there exists a $\tilde{\varphi} \in \mathcal{F}_g$ such that $\langle U, \tilde{\varphi} \rangle \neq 0$. For this $\tilde{\varphi}$ and a fixed $h_0 \in \Gamma$

$$\lim_{h \in \Gamma,\, \|h\| \to \infty} \frac{c(h + h_0)}{c(h)} \left\langle \frac{T(x + (h + h_0))}{c(h + h_0)}, \; \tilde{\varphi}(x) \right\rangle$$

$$= \lim_{h \in \Gamma,\, \|h\| \to \infty} \left\langle \frac{T((x + h_0) + h)}{c(h)}, \; \tilde{\varphi}(x) \right\rangle. \qquad (1.5)$$

Hence, for every $h_0 \in \Gamma$

$$\lim_{h \in \Gamma,\, \|h\| \to \infty} \frac{c(h + h_0)}{c(h)} = \frac{\langle U(x + h_0), \tilde{\varphi}(x) \rangle}{\langle U, \tilde{\varphi} \rangle} = d(h_0).$$

b) Now we can take in (1.5) any function $\varphi \in \mathcal{F}_g$ instead of $\tilde{\varphi}$. Then we have

$$d(h_0)\langle U, \varphi \rangle = \langle U(x + h_0), \varphi \rangle, \quad \varphi \in \mathcal{F}_g$$

which proves b). $\square$

We now restrict the space of generalized functions to a space of distribution, ultradistribution, or Fourier hyperfunction type. So, we have the following explicit characterization of the comparison function and limit.

Proposition 1.2. *Let Γ be a convex cone with $\operatorname{int}\Gamma \neq \emptyset$ ($\operatorname{int}\Gamma$ is the interior of Γ). Let $T \in \mathcal{F}'$ have S-asymptotics $T(x + h) \overset{s}{\sim} c(h)U(x)$, $h \in \Gamma$, where $U \neq 0$ and c is a positive function defined on $\mathbf{R}^n$. Then:*

a) For every $h_0 \in \mathbf{R}^n$ there exists

$$\lim_{h \in (h_0+\Gamma)\cap\Gamma, \|h\|\to\infty} c(h + h_0)/c(h) = \tilde{d}(h_0) \,.$$

b) There exists $\alpha \in \mathbf{R}^n$ such that $\tilde{d}(x) = \exp(\alpha \cdot x)$, $x \in \mathbf{R}^n$.

c) There exists $C \in \mathbf{R}$ such that $U(x) = C\exp(\alpha \cdot x)$.

Proof. a) Let $a \in \operatorname{int}\Gamma$. Then there exists $r > 0$ such that $B(a, r) \subset \Gamma$. Consequently, for every $\beta > 0$, $B(\beta a, \beta r) \subset \Gamma$, as well.

We shall prove that for every $h_0 \in \mathbf{R}^n$ and every $R > 0$ the set $(h_0 + \Gamma) \cap \Gamma \cap \{x \in \mathbf{R}^n; \|x\| > R\}$ is not empty. The first step is to prove that $(h_0 + \Gamma) \cap \Gamma$ is not empty.

Suppose that $y \in B(a, r/2) \subset \Gamma$. Then, for every $\beta \geq \beta_0 > 2\|h_0\|/r > 0$,

$$\|\beta a - (h_0 + \beta y)\| \leq \beta\|a - y\| + \|h_0\| \leq \beta r \,,$$

hence $h_0 + \beta y \in B(\beta a, \beta r) \subset \Gamma$.

For a fixed $R > 0$, we can choose β such that $\|h_0 + \beta y\| > R$. Then $h_0 + \beta y$ is a common element for $(h_0 + \Gamma)$, Γ and $\{x \in \mathbf{R}^n; \|x\| > R\}$.

Now we can use the limit (1.5) when $h \in (h_0 + \Gamma) \cap \Gamma$, and in the same way as in the proof of Proposition 1.1 a), we obtain

$$\lim_{h \in (h_0+\Gamma)\cap\Gamma,\ \|h\|\to\infty} \frac{c(h + h_0)}{c(h)} = \tilde{d}(h_0) = \frac{\langle U(t + h_0), \tilde{\varphi}\rangle}{\langle U, \tilde{\varphi}\rangle} \,.$$

From the existence of this limit, it follows:

1) $\tilde{d}$ extends d to the whole $\mathbf{R}^n$;

2) $\tilde{d}(0) = 1$; $\tilde{d} \in C^\infty(\mathbf{R}^n)$ and $\tilde{d}$ satisfies

$$\tilde{d}(h + h_0) = \tilde{d}(h)d(h_0), \quad h, h_0 \in \mathbf{R}^n \,. \tag{1.6}$$

We can take $h_0 = (0, \ldots 0, t_i, 0, \ldots,) \in \mathbf{R}^n$ in (1.6); then the limit

$$\lim_{t_i \to 0} \frac{\tilde{d}(h + h_0) - \tilde{d}(h)}{t_i} = \tilde{d}(h) \lim_{t_i \to 0} \frac{\tilde{d}(h_0) - \tilde{d}(0)}{t_i}$$

exists and gives

$$\frac{\partial}{\partial h_i}\tilde{d}(h) = \left(\frac{\partial}{\partial h_i}\tilde{d}(h)\right)_{h=0} \tilde{d}(h), \quad \text{for } i = 1, \ldots, n \,.$$

We introduce the function V given by

$$\tilde{d}(h) = e^{(\alpha \cdot h)}V(h), \quad \alpha_i = \left(\frac{\partial}{\partial h_i}\tilde{d}(h)\right)_{h=0}, \quad i = 1, \ldots, n.$$

Then $\dfrac{\partial}{\partial h_i}V(h) = 0$ for every $i = 1, \ldots, n$. Consequently, $V(h) = 1$, $h \in \mathbf{R}^n$.

For the proof of c), we now have $U(x + h) = \exp(\alpha \cdot h)U(x)$. Differentiating with respect to h and then setting $h = 0$, we obtain that U satisfies the differential equations $\dfrac{\partial}{\partial x_j}U = \alpha_j$, $1 \leq j \leq n$. Proceeding as in the proof of b), we obtain $U(x) = C \exp(\alpha \cdot x)$, for some $C \in \mathbf{R}$. $\qquad \square$

Remarks.

1. We only assumed that c is a positive function. But if we know that there exist $T \in \mathcal{F}'_g$ and $U \neq 0$ such that $T(x + h) \overset{s}{\sim} c(h)U(t)$, $h \in \Gamma$, then we can find a function $\tilde{c} \in C^\infty$ and with the property

$$\lim_{h \in \Gamma, \|h\| \to \infty} \tilde{c}(h)/c(h) = 1.$$

This function $\tilde{c}$ can be defined as follows: $\tilde{c}(h) = \langle T(x + h), \, \tilde{\varphi}(x)\rangle/\langle U, \tilde{\varphi}\rangle$, where $\tilde{\varphi}$ is chosen so that $\langle U, \tilde{\varphi}\rangle \neq 0$. In this sense we can suppose, whenever needed, that $c \in C^\infty$, and we do not loose generality.

Similarly, we have (see also Lemma 1.4 in 1.12) that

$$\lim_{h \in \Gamma', \|h\| \to \infty} c(h)/\tilde{c}(h + x) = \exp(-\alpha \cdot x) \quad \text{in } \mathcal{E}$$

if Γ is a convex cone, $\mathrm{int}\,\Gamma \neq \emptyset$, $\Gamma' \subset\subset \Gamma$ (Γ' is compact $\mathrm{int}\,\Gamma$).

2. If in Proposition 1.2 we replace $\mathcal{F}'$ for the general space $\mathcal{F}'_g$, then a) and b) still hold. On the other hand, c) will not be longer true, in general. We will show this fact in Remark 5 below by constructing explicit counterexamples.

3. In the one-dimensional case the cone Γ can be only $\mathbf{R}, \mathbf{R}_+$ or $\mathbf{R}_-$. In all these three cases $\mathrm{int}\,\Gamma \neq \emptyset$. Consequently, $\tilde{d}$ from Proposition 1.2 has always the form $\tilde{d}(x) = \exp(\alpha x)$, where $\alpha \in \mathbf{R}$.

Let us write $c(x) = L(e^x)\exp(\alpha x)$, $x \in \mathbf{R}$. We will show that L is a slowly varying function. Proposition 1.2 a) gives us the existence of the limit

$$\lim_{h \in \Gamma, \|h\| \to \infty} L(\exp(h + h_0))/L(\exp(h)) = 1, \quad h_0 \in \mathbf{R}.$$

If $\Gamma = \mathbf{R}_+$, then

$$\lim_{x \in \mathbf{R}_+, x \to \infty} L(xp)/L(x) = 1, \quad p \in \mathbf{R}_+$$

and this defines a slowly varying function (cf. **0.3**). Thus if $T \in \mathcal{F}'_g(\mathbf{R})$ and $T(x + h) \overset{s}{\sim} c(h)U(x)$, in $\mathbf{R}_+$, with $U \neq 0$, then it follows that c has the form $c(x) = \exp(\alpha x)L(\exp(x)), x \geq a > 0$, where L is a slowly varying function at infinity. Similarly, if $\Gamma = \mathbf{R}_-$, then L is slowly varying at the origin, while if $\Gamma = \mathbf{R}$, then L is slowly varying at both infinity and the origin.

4. The explicit form of the function c given in 3. is not known in the n-dimensional case, $n \geq 2$. This problem is related to the extension of the definition of a regularly varying function to the multi-dimensional case ([135], [162]) and with certain q-*admissible* and q-*strictly admissible* *functions* ([192]).

5. As mentioned before, c) in Proposition 1.2 does not have to hold for a general space $\mathcal{F}'_g$. From the proof of Proposition 1.2, we can still obtain the weaker conclusion $U(x + h) = \exp(\alpha \cdot h)U(x)$, $h \in \mathbf{R}^n$, which in turn implies the differential equations $(\partial/\partial x_j)U = \alpha_j$. For distribution, ultra-distribution, and Fourier hyperfunction spaces, these differential equations imply that U must have the form c) of Proposition 1.2. However, the latter fact is not true in general. We provide two related counterexamples below.

Let

$$\mathcal{A}_0(\mathbf{R}) = \{\varphi \in C^\infty(\mathbf{R}); \lim_{x \to -\infty} \varphi^{(m)}(x) \text{ exists and is finite}, m \in \mathbf{N}_0\},$$

it is a Frechet space with seminorms:

$$\beta_k(\varphi) = \sup_{x \in (-\infty, k], m \leq k} |\varphi^{(m)}(x)|.$$

Let us first observe that the definition of $\mathcal{A}_0(\mathbf{R})$ does not tell all the true about its elements. Notice that if $\varphi \in \mathcal{A}_0(\mathbf{R})$, then for $m = 1, 2, \ldots$, we have $\lim_{x \to -\infty} \varphi^{(m)}(x) = 0$, while $\lim_{x \to -\infty} \varphi(x)$ may not be zero. Indeed, the proof is easy, it is enough for $m = 1$, if $\lim_{x \to -\infty} \varphi'(x) = M$, then

$$\varphi(x) = \varphi(0) + \int_0^x \varphi'(t)dt \sim \varphi(0) + Mx, \quad x \to -\infty,$$

but since φ has limit at $-\infty$, then $M = 0$. So, we have

$$\mathcal{A}_0(\mathbf{R}) = \{\varphi \in C^\infty(\mathbf{R}); \lim_{x \to -\infty} \varphi(x) \text{ exists and } \lim_{x \to -\infty} \varphi^{(m)}(x) = 0, m \in \mathbf{N}\}.$$

Its dual space $\mathcal{A}'_0(\mathbf{R})$ contains a generalized function concentrated at $-\infty$ which contradicts c) of Proposition 1.2. Define the Dirac delta concentrated at $-\infty$ by

$$\langle \delta_{-\infty}, \varphi \rangle := \lim_{x \to -\infty} \varphi(x), \quad \varphi \in \mathcal{A}_0(\mathbf{R}).$$

Notice that the constant multiples of $\delta_{-\infty}$ are the only elements of $\mathcal{A}'_0(\mathbf{R})$ satisfying the differential equation $U' = 0$. This generalized function is translation invariant, i.e., $\delta_{-\infty}(\cdot + h) = \delta_{-\infty}$, for all $h \in \mathbf{R}$; in particular, it has the S-asymptotics

$$\delta_{-\infty}(x + h) \overset{s}{\sim} \delta_{-\infty}(x), \quad h \in \Gamma = \mathbf{R}.$$

Therefore, we have found an example of a non-constant limit for S-asymptotics related to the constant function $c(h) = 1$.

We can go beyond the previous example and give a counterexample for the failure of conclusion c) of Proposition 1.2 with a general c and S-asymptotics in $\mathcal{F}'_g$. Let $c(h) = \exp(\alpha h)L(\exp h)$, where L is slowly varying at infinity and $\alpha \in \mathbf{R}$. We assume that c is C^∞ and, for all $m \in \mathbf{N}$, $c^{(m)}(h) \sim \alpha^m c(h)$, $h \to \infty$ (otherwise replace c by $\tilde{c}$ given in Remark 1).

Next, we define

$$\mathcal{A}_c(\mathbf{R}) = \{\varphi \in C^\infty(\mathbf{R}); \lim_{x \to -\infty} \frac{\varphi^{(m)}(x)}{c(-x)} \text{ exists and is finite, } m \in \mathbf{N}_0\}.$$

It is a Frechet space with seminorms:

$$\beta_{k,c}(\varphi) = \sup_{x \in (-\infty, k], m \le k} \frac{|\varphi^{(m)}(x)|}{c(-x)}.$$

Note that $\mathcal{A}_c(\mathbf{R}) = c(-x) \cdot \mathcal{A}_0(\mathbf{R})$, consequently,

$$\varphi'(x) \sim -\alpha C_\varphi c(-x), \quad x \to -\infty,$$

where $C_\varphi = \lim_{x \to -\infty} \varphi(x)/c(-x)$. An inductive argument shows that for all m, $\varphi^{(m)}(x) \sim (-1)^m \alpha^m C_\varphi c(-x)$, $x \to -\infty$. Set now $g_{c,-\infty} = (1/c(-x)) \cdot \delta_{-\infty} \in \mathcal{A}'_c(\mathbf{R})$, a generalized function concentrated at infinity and given by

$$\langle g_{c,-\infty}, \varphi \rangle = \left\langle \delta_{-\infty}(x), \frac{\varphi(x)}{c(-x)} \right\rangle = \lim_{x \to -\infty} \frac{\varphi(x)}{c(-x)}, \quad \varphi \in \mathcal{A}_c(\mathbf{R}).$$

It is easy to show that the constant multiples of $g_{c,-\infty}$ are the only element of $\mathcal{A}'_c(\mathbf{R})$ satisfying the functional equation $U(x + h) = \exp(\alpha h)U(x)$ (and hence the differential equation $U' = \alpha U$); in particular,

$$g_{c,-\infty}(x + h) \overset{s}{\sim} e^{\alpha h} g_{c,-\infty}(x), \quad h \in \Gamma = \mathbf{R}.$$

In addition, there is an infinite number of elements of $\mathcal{A}'_c(\mathbf{R})$ having S-asymptotics in the cone $\Gamma = \mathbf{R}_+$ related to $c(h)$ with limit of the form $Cg_{c,-\infty}(x)$, $C \in \mathbf{R}$. In fact, consider $\delta^{(m)}$, the derivatives of the Dirac delta concentrated at the origin. We have that

$$\delta^{(m)}(x+h) \overset{s}{\sim} c(h)\alpha^m g_{c,-\infty}(x), \ \ h \in \Gamma = \mathbf{R}_+,$$

since

$$\lim_{h \to \infty} \frac{\langle \delta^{(m)}(x+h), \varphi(x) \rangle}{c(h)} = \lim_{h \to \infty} (-1)^m \frac{\varphi^{(m)}(-h)}{c(h)} = \alpha^m \langle g_{c,-\infty}(x), \varphi(x) \rangle \ .$$

1.3 *Equivalent definitions of the S-asymptotics in $\mathcal{F}'$*

Theorem 1.1. *Let $T \in \mathcal{F}'$ and let $\operatorname{int}\Gamma \neq \emptyset$. The following assertions are equivalent:*

a)
$$\underset{h \in \Gamma, \|h\| \to \infty}{w. \lim} \frac{T(x+h)}{c(h)} = U(x) = M \ \exp(\alpha x) \ \ in \ \mathcal{F}', \ M \neq 0. \quad (1.7)$$

b) For a δ-sequence $\langle \delta_m \rangle_m$ (cf. $\mathbf{0.6}$) there exists a sequence $\langle M_m \rangle_m$ in $\mathbf{R}$, such that $M_m \to M \neq 0$, $m \to \infty$, and

$$\underset{h \in \Gamma, \|h\| \to \infty}{w. \lim} \frac{(T * \delta_m)(x+h)}{c(h)} = M_m \exp(\alpha x), \ in \ \mathcal{F}', \ uniformly \ in \ m \in \mathbf{N}.$$
$$(1.8)$$

c) For a δ-sequence $\langle \delta_m \rangle_m$, (cf. $\mathbf{0.6}$),

$$\lim_{h \in \Gamma, \|h\| \to \infty} \frac{(T * \delta_m)(h)}{c(h)} = p_m, \ m \in \mathbf{N}, \quad (1.9)$$

where $p_m \neq 0$ for some m, and for every $\phi \in \mathcal{F}$,

$$\sup_{h \in \Gamma, \|h\| \geq 0} \left| \frac{(T * \phi)(h)}{c(h)} \right| < \infty. \quad (1.10)$$

If $\mathcal{F}' = \mathcal{D}'$ or $\mathcal{F}' = \mathcal{D}'^*$ the following assertion is also equivalent to a).

d) $(T(\cdot + h)/c(h)) * \varphi$ converges to $(U * \varphi)(h)$ in $\mathcal{E}(\mathbf{R}^n)$, for each $\varphi \in \mathcal{D}$ $(\varphi \in \mathcal{D}^*)$.

Proof. $a) \Rightarrow b)$. Let $\langle \delta_m \rangle_m$ be a δ-sequence. For any $\phi \in \mathcal{F}$, $\{\delta_m * \phi;$ $m \in \mathbf{N}\}$ is a compact set in $\mathcal{F}$. We have by the properties of the convolution (cf. **0.6**) and the Banach–Steinhaus theorem

$$\lim_{h\in\Gamma,\|h\|\to\infty} \left\langle \frac{(T*\delta_m)(x+h)}{c(h)}, \phi(x) \right\rangle = \lim_{h\in\Gamma,\|h\|\to\infty} \frac{\langle (T*\delta_m)(x), \phi(x-h) \rangle}{c(h)}$$

$$= \lim_{h\in\Gamma,\|h\|\to\infty} \frac{\langle T, (\check{\delta}_m * \phi)(\cdot - h) \rangle}{c(h)} = \lim_{h\in\Gamma,\|h\|\to\infty} \left\langle \frac{T(x+h)}{c(h)}, (\check{\delta}_m * \phi)(x) \right\rangle$$

$$= \langle Me^{\alpha\cdot x}, (\check{\delta}_m * \phi)(x) \rangle = \langle M_m e^{\alpha\cdot x}, \phi(x) \rangle, \tag{1.11}$$

uniformly in m. Now (1.11) implies (1.8) and b).

$b) \Rightarrow a)$. Let $\phi \in \mathcal{F}$ and

$$a_{m,h} = \left\langle \frac{(T*\delta_m)(x+h)}{c(h)}, \phi(x) \right\rangle = \frac{(T*(\check{\delta}_m * \phi))(h)}{c(h)}, \quad m \in \mathbf{N}, \ h \in \Gamma.$$

We have $a_{m,h} \to a_m$, $h \in \Gamma$, $\|h\| \to \infty$, uniformly for $m \in \mathbf{N}$, where

$$a_m = M_m \langle \exp(\alpha \cdot x), \phi(x) \rangle, \ m \in \mathbf{N},$$

$$a_m \to a = M \langle \exp(\alpha \cdot x), \phi(x) \rangle, \ m \to \infty.$$

Also $a_{m,h} \to a_h$, $m \to \infty$, where

$$a_h = \left\langle \frac{T(x+h)}{c(h)}, \phi(x) \right\rangle, \ h \in \Gamma.$$

This implies $a_h \to a$, $h \in \Gamma$, $\|h\| \to \infty$, what is in fact a).

$a) \Rightarrow c)$. From (1.7) it follows that $(T*\varphi)(h)/c(h)$ converges for every $\varphi \in \mathcal{F}$, when $h \in \Gamma$, $\|h\| \to \infty$. Hence, $(T*\varphi)(h)/c(h)$ is bounded, $h \in \Gamma$, $\|h\| \geq 0$; (1.9) follows directly from (1.7).

$c) \Rightarrow a)$. First, we shall prove that the set $G = \{\delta_m(\cdot + x), m \in \mathbf{N}, x \in \mathbf{R}^n\}$ is dense in $\mathcal{F}$. Suppose that $T \in \mathcal{F}'$ and that

$$\langle T, \delta_m(\cdot + x) \rangle = 0, \ m \in \mathbf{N}, \ x \in \mathbf{R}^n.$$

It follows that $(T*\check{\delta}_m)(-x) = 0$, $m \in \mathbf{N}$, $x \in \mathbf{R}^n$. Then, for any $\varphi \in \mathcal{F}$, $\langle T*\delta_m, \varphi \rangle = 0$, $m \in \mathbf{N}$, and consequently

$$\langle T, \varphi \rangle = \lim_{m\to\infty} \langle T, \delta_m * \varphi \rangle = \lim_{m\to\infty} \langle T*\check{\delta}_m, \varphi \rangle = 0.$$

This implies that $T = 0$ and hence, the set G is dense in $\mathcal{D}$ by the Hahn-Banach theorem. Thus, by (1.10) and the Banach–Steinhaus theorem, c) implies a).

a) $\Rightarrow$ d). Note that (1.1) implies the strong convergence of $T(\cdot + h)/c(h)$ to U in $\mathcal{D}'$ and $\mathcal{D}'^*$ respectively. Since the convolution in $\mathcal{D}'$ and $\mathcal{D}'^*$ is hypocontinuous, it follows the following equality in the sense of convergence in $\mathcal{E}(\mathbf{R}^n)$

$$\lim_{h \in \Gamma, \|h\| \to \infty} \left(\frac{T(\cdot + h)}{c(h)} * \varphi \right) = \left(\left(\lim_{h \in \Gamma, \|h\| \to \infty} \frac{T(\cdot + h)}{c(h)} \right) * \varphi \right) = U * \varphi$$

in both cases ($\varphi \in \mathcal{D}$, respectively $\varphi \in \mathcal{D}^*$). $\square$

The paper which can be consulted for related results is [128].

1.4 Basic properties of the S-asymptotics

Theorem 1.2. *Let $T \in \mathcal{F}'_g$.*

a) If $T(x+h) \overset{s}{\sim} c(h)U(x)$, $h \in \Gamma$, then for every $k \in \mathbf{N}_0^n$, $T^{(k)}(x+h) \overset{s}{\sim} c(h)U^{(k)}(x)$, $h \in \Gamma$.

b) Assume additionally that $\mathcal{F}_g$ is a Montel space. Let $g \in M_{(\cdot)}$ (set of multipliers of $\mathcal{F}'_g$ (cf. 0.6)); let c, c_1 be positive functions. If for every $\varphi \in \mathcal{F}_g$, $(g(x + h)/c_1(h))\varphi(x)$ converges to $G(x)\varphi(x)$ in $\mathcal{F}_g$ when $h \in \Gamma$, $\|h\| \to \infty$ and if $T(x + h) \overset{s}{\sim} c(h)U(x)$, $h \in \Gamma$, then $g(x+h)T(x+h) \overset{s}{\sim} c_1(h)c(h)G(x)U(x)$, $h \in \Gamma$.

c) If $T \in \mathcal{F}'_0$ and $\operatorname{supp} T$ is compact, then $T(x + h) \overset{s}{\sim} c(h) \cdot 0$, $h \in \Gamma$, for every positive function c.

d) Suppose that $\mathcal{F}'_g$ is a Montel spaces in which the convolution with compactly supported elements is well defined and hypocontinuous (cf. 0.6). Let $S \in \mathcal{F}'_g$, $\operatorname{supp} S$ being compact. If $T(x + h) \overset{s}{\sim} c(h)U(x)$, $h \in \Gamma$, then $(S * T)(x + h) \overset{s}{\sim} c(h)(S * U)(x)$, $h \in \Gamma$.

e) Let $T \in \mathcal{F}' = \mathcal{D}'^*$ and $T(x+h) \overset{s}{\sim} c(h) \cdot U(x)$, $h \in \Gamma$, in $\mathcal{D}'^*$. Assume that (M.2) holds. Let $P(D)$ be an ultradifferential operator of class $*$. Then

$$(P(D)T)(x + h) \overset{s}{\sim} c(h) \cdot (P(D)U)(x), \quad h \in \Gamma, \text{ in } \mathcal{D}'^*.$$

f) Let $T \in \mathcal{F}' = Q(\mathbf{D}^n)$. Let $P(D)$ be a local operator and let $T(x + h) \overset{s}{\sim} c(h) \cdot U(x)$, $h \in \Gamma$, $\|h\| \to \infty$ in $Q(\mathbf{D}^n)$, then $(P(D)T)(x + h) \overset{s}{\sim} c(h) \cdot (P(D)U)(x)$, $h \in \Gamma$, $\|h\| \to \infty$ in $Q(\mathbf{D}^n)$, as well.

Proof. a) The assertion is a consequence of the definition of the derivative of a generalized function. Namely,

$$\lim_{h \in \Gamma, \|h\| \to \infty} \left\langle \frac{T^{(k)}(x+h)}{c(h)}, \varphi(x) \right\rangle = \lim_{h \in \Gamma, \|h\| \to \infty} \left\langle \frac{T(x+h)}{c(h)}, (-1)^k \varphi^{(k)}(x) \right\rangle$$

$$= (-1)^k \langle U(x), \varphi^{(k)}(x) \rangle$$

$$= \langle U^{(k)}, \varphi \rangle, \quad \varphi \in \mathcal{F}_g.$$

b) Since $\mathcal{F}_g$ is a Montel space, we have that $\displaystyle\lim_{h \in \Gamma, \|h\| \to \infty} \frac{T(x+h)}{c(h)} = U(x)$ in $\mathcal{F}'_g$ with respect to strong topology. As $(g(x+h)/c_1(h) - G(x))\varphi(x)$, $h \in \Gamma, \|h\| \geq 0$, is a bounded set in $\mathcal{F}_g$, it follows:

$$\lim_{h \in \Gamma, \|h\| \to \infty} \langle g(x+h)T(x+h)/(c_1(h)c(h)), \varphi(x) \rangle$$

$$= \lim_{h \in \Gamma, \|h\| \to \infty} \left\langle T(x+h)/c(h), \left(\frac{g(x+h)}{c_1(h)} - G(x) \right) \varphi(x) \right\rangle$$

$$+ \lim_{h \in \Gamma, \|h\| \to \infty} \langle T(x+h)/c(h), G(x)\varphi(x) \rangle$$

$$= \left\langle U(x), \lim_{h \in \Gamma, \|h\| \to \infty} \left(\frac{g(x+h)}{c_1(h)} - G(x) \right) \varphi(x) \right\rangle + \langle U(x), G(x)\varphi(x) \rangle$$

$$= 0 + \langle U(x)G(x), \varphi(x) \rangle = \langle UG, \varphi \rangle, \quad \varphi \in \mathcal{F}_g.$$

c) For each $\varphi \in \mathcal{F}_0$ there exists $r_\varphi > 0$ such that $\operatorname{supp} \varphi \subset B(0, r_\varphi) = \{x \in \mathbf{R}^n; \|x\| < r_\varphi\}$. The support of $T(x+h)$ is $(\operatorname{supp} T - h)$. Thus, by our assumption, there exists β_{r_φ} such that for all $h \in \Gamma$, $\|h\| > \beta_{r_\varphi}$ the set $(\operatorname{supp} T - h) \cap B(0, r_\varphi)$ is empty and consequently $\langle T(x+h), \varphi(x) \rangle = 0, h \in \Gamma, \|h\| \geq \beta_{r_\varphi}$.

d) By definition of the convolution

$$\langle (S * T)(x+h), \varphi(x) \rangle = \langle (S * T)(x), \varphi(x - h) \rangle$$

$$= \langle S_t \times T_y, \alpha(t)\varphi(t + y - h) \rangle \qquad (1.12)$$

$$= \langle S_t \times T_y(\cdot + h), \alpha(t)\varphi(t + y) \rangle.$$

Hence, $(S * T)(x+h) = (S * T(\cdot + h))(x)$.

Since in a Montel space the weak and the strong convergence are equivalent, we have by (1.12)

$$\lim_{h\in\Gamma,\|h\|\to\infty}(S*T)(x+h)/c(h) = \lim_{h\in\Gamma,\|h\|\to\infty}\left(S*\frac{T(\cdot+h)}{c(h)}\right)(x) = (S*U)(x).$$

For the proof of e) and f), we note that an ultradifferential operator $P(D)$ maps continuously $\mathcal{D}'^*$ into $\mathcal{D}'^*$ and a local operator maps continuously $\mathcal{Q}(\mathbf{D}^n)$ into $\mathcal{Q}(\mathbf{D}^n)$.   □

Remark. From assertion a) of Theorem 1.2, a natural question arises for spaces $\mathcal{F}'$: The limit U can be a constant generalized function, hence $U' = 0$. *Is there a positive function $\tilde{c}$ such that T' has S-asymptotics related to this $\tilde{c}$, but with a limit different from zero?*

In general the answer is negative as shown by the following example: Let T be defined by $x^2 + \sin(\exp x^2)$, $x \in \mathbf{R}$. Then, $T(x+h) \overset{s}{\sim} h^2 \cdot 1$, $h \in \mathbf{R}_+$. But $T'(x) = 2x(1 + \exp(x^2)\cos(\exp x^2))$. The same situation is obtained with the distribution $f(x) = x^2 + x\sin x$.

We can now formulate an open problem: *Suppose that $T(x + h) \overset{s}{\sim} c(h)U(x)$, $h \in \Gamma$. If $U(x)$ is a constant generalized function, then the problem is to find some additional conditions on T which guarantee the existence of a function $\tilde{c}$ such that T' has the S-asymptotics in Γ related to $\tilde{c}$.*

More generally, let $S \in \mathcal{F}'$ and $T = (\partial/\partial x_k)S$. If $T(x + h) \overset{s}{\sim} c(h)U(x)$, $h \in \Gamma$. The question is what we can say about the S-asymptotics of S.

Recall the well known result: Let h be a *real-valued function which has the first derivative $h'(x) \neq 0$, $x \geq x_0$ and $h(x) \to \infty$, $x \to \infty$. If a function F has its first derivative on (x_0, ∞) such that there exists $\lim_{x\to\infty} F'(x)/h'(x) = A$, then there exists $\lim_{x\to\infty} F(x)/h(x) = A$, as well.*

We know that an opposite assertion does not hold, and this is at the basis of the open problem quoted above.

We give a theorem to illustrate the relation between the S-asymptotics of a distribution and the S-asymptotics of its primitive. We refer to [114] for the proof.

Theorem 1.3. *1) Let $f, g \in \mathcal{D}'(\mathbf{R})$ and for some $m \in \mathbf{N}$, $g^{(m)} = f$.*

a) If $f(x+h) \overset{s}{\sim} h^\nu L(h) \cdot 1$, $h \in \mathbf{R}_+$, where $\nu > -1$, then $g(x+h) \overset{s}{\sim} h^{\nu+m} L(h) \cdot 1$, $h \in \mathbf{R}_+$.

b) If $f(x+h) \overset{s}{\sim} \exp(\alpha h) L(\exp h) \exp(\alpha x)$, $h \in \mathbf{R}_+, \alpha \in \mathbf{R}$, and $\int_0^x \exp(\alpha h) L(\exp h) dh \to \infty$, when $x \to \infty$, then

$$g(x+h) \overset{s}{\sim} \left(\int_0^h \int_0^{h_{m-1}} \cdots \int_0^{h_1} \exp(\alpha t) L(\exp t) dt dh_1 \ldots dh_{m-1} \right) \exp(\alpha x), h \in \mathbf{R}_+.$$

2) Let $\phi_0 \in \mathcal{D}'(\mathbf{R})$ such that $\int \phi_0(t) dt = 1$. If

$$\lim_{h \to \infty} \left\langle \frac{g^{(i)}(x+h)}{\exp(\alpha h) L(\exp h)}, \phi_0(x) \right\rangle = \alpha^i \langle \exp(\alpha x), \phi_0(x) \rangle, i = 0, 1, \ldots, m-1$$

and

$$f(x+h) \overset{s}{\sim} \exp(\alpha h) L(\exp h) \alpha^m \exp(\alpha x), \ h \in \mathbf{R}_+,$$

then

$$g(x+h) \overset{s}{\sim} \exp(\alpha h) L(\exp h) \exp(\alpha x), \ h \in \mathbf{R}_+.$$

3) Suppose that $T \in \mathcal{D}'$, $\Gamma = \{x \in \mathbf{R}^n; \ x = (0, \ldots, x_k, 0, \ldots, 0)\}$ and $T = (\partial/\partial x_k) S$. If $T(x+h) \overset{s}{\sim} c(h) U(x)$, $h \in \Gamma$ and $c(h)$ is locally integrable in h_k such that

$$c_1(h_k) = \int_{h_k^0}^{h_k} c(v) dv_k \to \infty \quad as \ \ h_k \to \infty, \ h_k^0 \geq 0,$$

then $S(x+h) \overset{s}{\sim} c_1(h) U(x)$, $h \in \Gamma$.

4) Suppose that $S \in \mathcal{D}'$ and that for an $m \in \{1, 2, \ldots, n\}$,

$$(D_{t_m} S)(x+h) \overset{s}{\sim} c(h) \cdot U(x), \ h \in \Gamma.$$

Let $V \in \mathcal{D}'$, $D_{t_m} V = U$ and $\phi_0 \in \mathcal{D}(\mathbf{R})$, $\int_{\mathbf{R}} \phi_0(\tau) d\tau = 1$. Let

$$\lim_{h \in \Gamma, \|h\| \to \infty} \langle S(x+h)/c(h), \phi_0(x_m) \lambda_m(\tilde{x}) \rangle = \langle V, \phi_0 \lambda_m \rangle,$$

where $\tilde{x} = (x_1, \ldots, x_{m-1}, x_{m+1}, \ldots, x_n)$ and

$$\lambda_m(\tilde{x}) = \int_{\mathbf{R}} \psi(x_1, \ldots, x_m, \ldots, x_n) dx_m, \quad \psi \in \mathcal{D}.$$

Then $S(x+h) \overset{s}{\sim} c(h) V(x), h \in \Gamma$.

The following theorem asserts that the S-asymptotics is a local property if the elements of $\mathcal{F}_g$ are compactly supported.

Theorem 1.4. *Let $T_1, T_2 \in \mathcal{F}_0'$. Let the open set $\Omega \subset \mathbf{R}^n$ have the following property: for every $r > 0$ there exists a $\beta_r > 0$ such that the ball $B(0,r) = \{x \in \mathbf{R}^n; \|x\| < r\}$ is in $\{\Omega - h; h \in \Gamma, \|h\| \geq \beta_r\}$. If $T_1 = T_2$ on Ω and $T_1(x + h) \overset{s}{\sim} c(h)U(x), h \in \Gamma$, then $T_2(x + h) \overset{s}{\sim} c(h)U(x), h \in \Gamma$, as well.*

Proof. Let $\varphi \in \mathcal{F}_0$ with $\operatorname{supp} \varphi \subset B(0,r)$. We shall prove that

$$\lim_{h \in \Gamma, \|h\| \to \infty} \left\langle \frac{T_1(x+h) - T_2(x+h)}{c(h)}, \varphi(x) \right\rangle = 0. \qquad (1.13)$$

The complement of the set $\operatorname{supp}(T_1(x+h) - T_2(x+h))$ contains the set $\{\Omega - h, h \in \Gamma\}$. By our supposition the number β_r is fixed in such a way that the sets $\{\Omega - h; h \in \Gamma, \|h\| \geq \beta_r\}$ contain $B(0,r)$ and consequently $\operatorname{supp} \varphi$. Since T_1 has the S-asymptotics related to c and with the limit U, (1.13) implies

$$\lim_{h \in \Gamma, \|h\| \to \infty} \frac{\langle T_2(x+h), \varphi(x) \rangle}{c(h)} = \lim_{h \in \Gamma, \|h\| \to \infty} \frac{\langle T_1(x+h), \varphi(x) \rangle}{c(h)} = \langle U, \varphi \rangle.$$

$\square$

Remark. The open set Ω, by its property, has to contain a set $\Gamma^0 \cap \{x \in \mathbf{R}^n; \|x\| > R\}$ where Γ^0 is an open acute cone such that $\overline{\Gamma} \subset \Gamma^0$ and R is a positive number.

1.5 S-asymptotic behavior of some special classes of generalized functions

1.5.1 Examples with regular distributions

1. $\exp(a \cdot (x + h)) \overset{s}{\sim} \exp(a \cdot h) \exp(a \cdot x), h \in \mathbf{R}^n$.

2. $\exp(\sqrt{(x+h)^2 + (x+h)}) \overset{s}{\sim} \exp h \exp\left(x + \frac{1}{2}\right), h \in \mathbf{R}_+$.

3. Let $w \in S^{n-1} = \{x \in \mathbf{R}^n; \|x\| = 1\}$, $p = (p_1, \ldots, p_n) \in \overline{\mathbf{R}}_+^n$ and $\Gamma = \{qw; q \in \mathbf{R}_+\}$. Denote, $J = \{k \in \{1, \ldots, n\}; w_k \neq 0\}$ and $\beta = \sum_{i \in J} p_i$. Then

$$(x + h)^p \overset{s}{\sim} q^\beta \prod_{i \in J} w_i^{p_i} \cdot 1, \quad h \in \Gamma,$$

and

$$(1 + x + h)^{-p} \overset{s}{\sim} q^{-\beta} \prod_{i \in J} w_i^{-p_i} \cdot 1, \quad h \in \Gamma,$$

$$((x + h)^p = (x_1 + h_1)^{p_1} \ldots (x_n + h_n)^{p_n}).$$

4. *For a slowly varying function* $L(t), t \geq \alpha > 0$ *we have*

$$L(t + h) \overset{s}{\sim} L(h) \cdot 1, \quad h \in \mathbf{R}_+ .$$

Namely,

$$\lim_{h \to \infty} \langle L(t + h)/L(h), \varphi(t) \rangle = \lim_{h \to \infty} \int_{-r}^{r} \varphi(t) L(t + h)/L(h) dt$$

$$= \lim_{q \to \infty} \int_{e^{-r}}^{e^r} \varphi(\log y) L(\log(yq))/L(\log q) \frac{dy}{y} = \int_{\mathbf{R}} \varphi(t) dt, \quad \varphi \in \mathcal{D}(\mathbf{R}) .$$

We used above that $L(\log y)$ is also a slowly varying function (cf. **0.3**) and that $L(\log(uh))/L(\log h)$ converges uniformly to 1 as $h \to \infty$ if u stays in compact intervals $[\alpha_1, \alpha_2]$, $0 < \alpha_1 < \alpha_2 < \infty$.

5. *Let* $g \in L^1(\mathbf{R})$. *A distribution defined by* g *has the S-asymptotic behavior related to* $c = 1$ *and with limit* $U = 0$. To show this, let $\alpha \in \mathbf{R}$ and

$$T(t) = \int_{\alpha}^{t} g(x) dx, \quad t \in \mathbf{R} .$$

Then, $T(t + h) \overset{s}{\sim} 1 \cdot \int_{\alpha}^{\infty} g(x) dx, h \in \mathbf{R}_+$. By Theorem 1.2 a), $g(t + h) \overset{s}{\sim} 1 \cdot 0, h \in \mathbf{R}_+$.

1.5.2 *Examples with distributions in subspaces of* $\mathcal{D}'$

6. Let f_α, $\alpha \in \mathbf{R}$, be given as in **0.4**. It defines a distribution in $\mathcal{D}'_{[0,\infty)}$.

For $\alpha = -k$, $k = 0, 1, \ldots, f_{-k} = \delta^{(k)}$. Since $\delta^{(k)}$ is supported by $\{0\}$, it has the S-asymptotics zero related to every $c > 0$ (cf. Theorem 1.2 c)). For $\alpha > 0$,

$$f_\alpha(x + h) \overset{s}{\sim} (1/\Gamma(\alpha)) h^{\alpha-1} \cdot 1, \quad h \in \mathbf{R}_+ .$$

In the case $\alpha < 0$, $\alpha \neq -1, -2, \ldots$, $f_\alpha(x) = (1/\Gamma(\alpha)) \cdot \mathrm{Pf}(H(x)x^{\alpha-1})$, where Pf is the finite part or "partie finie" (see [146], pp. 41–43).

Note, $\operatorname{supp}\phi(x - h) = \operatorname{supp}\phi + h$, $\phi \in \mathcal{D}(\mathbf{R})$. Thus for $\phi \in \mathcal{D}(\mathbf{R})$, we can find h_0 such that $\operatorname{supp}\phi(x - h) \subset (a, \infty)$, $a > 0$, $h \geq h_0$. Then

$$\lim_{h\to\infty} \langle f_\alpha(x)/h^{\alpha-1}, \phi(x - h)\rangle = \lim_{h\to\infty} \int_a^\infty \frac{1}{\Gamma(\alpha)}\left(\frac{x}{h}\right)^{\alpha-1}\phi(x - h)dx$$

$$= \lim_{h\to\infty} \int_R \frac{1}{\Gamma(\alpha)}\left(\frac{u+h}{h}\right)^{\alpha-1} \cdot \phi(u)du = \int_R \frac{1}{\Gamma(\alpha)}\phi(u)du.$$

Hence, $f_\alpha(x + h) \overset{s}{\sim} h^{\alpha-1} \cdot \dfrac{1}{\Gamma(\alpha)}$, $h \in \mathbf{R}_+$, $\alpha < 0$, $\alpha \neq -1, -2, \ldots$

7. If $T \in \mathcal{S}'$, then there exists a real number k_0 such that T has S-asymptotic behavior related to $c(h)\|h\|^{k_0}$, where $c(h)$ tends to infinity as $|h| \to \infty$, $h \in \mathbf{R}^n$ and with limit $U = 0$.

By Theorem VI, chapter VII in [146], there exists a number k_0 such that the set of distributions $\{T(\cdot + h)/(1 + \|h\|^2)^{k_0/2}; \ h \in \mathbf{R}^n\}$ is bounded in $\mathcal{D}'$. Hence, this set is weakly bounded and

$$\left\langle T(x + h)/(c(h)\|h\|^{k_0}), \varphi(x)\right\rangle = \frac{(1 + \|h\|^2)^{k_0/2}}{c(h)\|h\|^{k_0}}\left\langle \frac{T(x+h)}{(1+\|h\|^2)^{k_0/2}}, \varphi(x)\right\rangle$$

tends to zero as $\|h\| \to \infty$.

8. Let $T \in \mathcal{D}'(\mathbf{R})$ have the following property:

For a δ-sequence $\langle\delta_m\rangle_m$ there is a sequence $\langle p_m\rangle_m$ in $\mathbf{R}$, such that $p_m \to p \neq 0$, $m \to \infty$, and

$$\lim_{h\to\infty} \frac{(T * \delta_m)(h)}{c(h)} = p_m, \quad m \in \mathbf{N}, \tag{$*$}$$

where the limit is uniform for $m \in \mathbf{N}$.

Then T has S-asymptotics related to c.

We will prove that b) in Theorem 1.1 is satisfied, which is equivalent to a) in the same theorem. For every compact set $K \subset \mathbf{R}$

$$\frac{(T * \delta_m)(x + h)}{c(h)} = \frac{(T * \delta_m)(x + h)}{c(x + h)}\frac{c(x + h)}{c(h)} \to p_m \exp(\alpha x), h \in \Gamma, \|h\| \to \infty,$$

uniformly for $x \in K$, because of $\dfrac{c(x + h)}{c(h)} \to \exp(\alpha x)$, $\|h\| \to \infty$, uniformly for $x \in K$ (cf. Remarks after Proposition 1.2). Then, T has the S-asymptotics related to c.

Property (*) is not equivalent to the existence of the S-asymptotics. The next example illustrates that this condition is not necessary.

Assume that $S \in C(\mathbf{R}) \cap L^1(\mathbf{R})$ but not being bounded on $\mathbf{R}$. This function has S-asymptotics equal to zero related to $c = 1$ (see **5.**). For T we take $1 + S(x)$. Then, $T(x + h) \overset{s}{\sim} 1 \cdot 1$, $h \in \mathbf{R}_+$ and

$$\lim_{h \to \infty} [(1 + S) * \delta_m](h) = \lim_{h \to \infty} \langle 1 + S(x + h), \check{\delta}_m(x) \rangle = \langle 1, \check{\delta}_m(x) \rangle = p_m \,.$$

This limit is not uniform in $m \in \mathbf{N}$ because $\lim_{m \to \infty} [(1 + S) * \delta_m](h)$ does not necessarily exist.

9. Every distribution in $\mathcal{D}'_{L^p}, 1 \leq p < \infty$ has S-asymptotic behavior related to $c = 1$, and $\Gamma = \mathbf{R}^n$, with limit $U = 0$. Let us show this.

By Theorem XXV, Chapter VI in [146] it follows that $(T * \varphi) \in L^p(\mathbf{R}^n)$ for every $\varphi \in \mathcal{D}$. Every derivative of $(T * \varphi)$, $\dfrac{\partial}{\partial h_k}(T * \varphi)(h) = (T * \dfrac{\partial}{\partial x_k}\varphi)(h)$, $h \in \mathbf{R}^n$, is also in $L^p(\mathbf{R}^n)$. Hence $(T * \check{\varphi}) \in \mathcal{D}_{L^p}$. We know that every element of $\mathcal{D}_{L^p}$, $1 \leq p < \infty$ is bounded over $\mathbf{R}^n$ and tends to zero when $\|h\| \to \infty$ ([146], p. 199).

It has been examined in [153] how slowly the function $(T * \varphi) \in \mathcal{D}_{L_p}$, $\varphi \in \mathcal{D}$, tends to zero as $\|h\| \to \infty$.

In fact, the question is:

Let $p \geq 1$. *Is it possible to find a positive function* c *such that* $c(x) \to 0$, $\|x\| \to \infty$ *and* $|\varphi(x)/c(x)| \leq C_\varphi$, $\|x\| \geq R_\varphi$, *for every* $\varphi \in \mathcal{D}_{L^p}$? C_φ *and* R_φ *are positive constants depending on* φ.

The answer is negative (see [153]).

A similar question can be asked for $T \in \mathcal{D}'_{L^p}$, $1 \leq p < \infty$, but related to the S-asymptotics. Precisely, *whether there exists* $c(h) > 0, c(h) \to 0, \|h\| \to \infty$, *such that*

$$|\langle T(x + h)/c(h), \phi(x) \rangle| \leq M_\phi, \|h\| > \beta_\phi \quad \phi \in \mathcal{D},$$

where β_ϕ *and* M_ϕ *are positive constants depending on* ϕ.

The answer to this question is also negative (cf. [153]).

10. Let $T \in \mathcal{K}'_p$. Then there exists $k_0 \in \mathbf{N}_0$ such that T has S-asymptotic behavior with limit $U = 0$ related to $c(h) \exp(k_0 \|h\|^p)$, where $c(h)$ tends to infinity as $\|h\| \to \infty$.

First, we prove that there exists a positive integer k, such that the set $\{T(\cdot + h)\exp(-k\|h\|^p),\ h \in \mathbf{R}^n\}$ is bounded in $\mathcal{D}'$.

We start by giving a bound for the seminorms $\nu_k(\varphi(\cdot - h)),\ \varphi \in \mathcal{K}_p$:

$$\nu_k[\varphi(\cdot - h)] = \sup_{x \in \mathbf{R}^n, |a| \leq k} \exp(k\|x\|^p)|D^a\varphi(x - h)|$$

$$= \sup_{x \in \mathbf{R}^n, |a| \leq k} \exp(k\|x + h\|^p)|D^a\varphi(x)|$$

$$\leq \exp(2^p k\|h\|^p) \sup_{x \in \mathbf{R}^n, |a| \leq 2pk} \exp(2^p k\|x\|^p)|D^a\varphi(x)|$$

$$\leq \exp(2^p k\|h\|^p)\nu_{2^p k}(\varphi).$$

By assumption, T is a continuous linear functional on $\mathcal{K}_p$. Note, the sequence of norms $\langle\nu_k\rangle_k$ is increasing. Thus, there exist $\varepsilon > 0$ and $k_0 \in \mathbb{N}_0$ such that

$$|\langle T, \varphi\rangle| \leq 1 \quad \text{for} \quad \varphi \in \mathcal{K}_p,\ \nu_{k_0}(\varphi) \leq \varepsilon.$$

This inequality holds for all $k \geq k_0$. Hence

$$|\langle T, \varphi\rangle| \leq \varepsilon^{-1}\nu_k(\varphi),\ k \geq k_0 \quad \text{for every} \quad \varphi \in \mathcal{K}_p.$$

We know that $\mathcal{D} \subset \mathcal{K}_p$ and that the inclusion is continuous. Let $\varphi \in \mathcal{D}$, Then

$$|\langle \exp(-2^p k\|h\|^p)T(x + h), \varphi(x)\rangle| = |\langle T(x), \exp(-2^p k\|h\|^p)\varphi(x - h)\rangle|$$

$$\leq \varepsilon^{-1}\exp(-2^p k\|h\|^p)\nu_k[\varphi(x - h)] \leq \varepsilon^{-1}\nu_{2^p k}(\varphi),\ k > k_0.$$

We can choose $k_0 \geq 2^p k$. The set $\{\exp(-k_0\|h\|^p)T(x + h);\ h \in \mathbf{R}^n\}$ is bounded in $\mathcal{D}'$ (and weakly bounded in $\mathcal{D}'$, as well).

Now, for every $\varphi \in \mathcal{D}$:

$$\lim_{\|h\| \to \infty} \langle \exp(-k_0\|h\|^p)T(x + h)/c(h), \varphi(x)\rangle$$

$$= \lim_{\|h\| \to \infty} \frac{1}{c(h)}\langle \exp(-k_0\|h\|^p)T(x + h), \varphi(x)\rangle = 0.$$

1.5.3 S-asymptotics of ultradistributions and Fourier hyperfunctions — Comparisons with the S-asymptotics of distributions

11. If a distribution has S-asymptotics in $\mathcal{D}'$ (see Definition 1.1), it has the same S-asymptotics in $\mathcal{D}'^*$, as well. But the opposite is not true. This is illustrated by the following example:

$$T(x) = 1 + \sum_{n=1}^{\infty} \delta^{(n)}(x-n)/M_n, \ x \in \mathbf{R},$$

has S-asymptotics in $\mathcal{D}'^{(M_p)}(\mathbf{R})$, but it does not have S-asymptotics in $\mathcal{D}'(\mathbf{R})$. We will show this. Let $\phi \in \mathcal{D}^{(M_p)}(\mathbf{R})$. Then,

$$\left\langle \sum_{n=1}^{\infty} \delta^{(n)}(x+h-n)/M_n, \ \phi(x) \right\rangle = \sum_{n=1}^{\infty} (-1)^n \phi^{(n)}(n-h)/M_n \to 0, \ h \to \infty.$$

This is a consequence of the property

$$\sup_{x \in \mathbb{R}} |\phi^{(n)}(x)|/k^n M_n \to 0, \ n \to \infty \text{ for every } k > 0.$$

Suppose that there exists a function c such that for every $\varphi \in \mathcal{D}(\mathbf{R})$

$$\sum_{n=1}^{\infty} (-1)^n \varphi^{(n)}(n-h)/M_n c(h),$$

converges, as $h \to \infty$. Taking $h = n$ this implies that $\varphi^{(n)}(0)/M_n c(n)$ converges to zero, as $n \to \infty$, for every $\varphi \in \mathcal{D}(\mathbf{R})$. However, such a function c does not exist (Borel's theorem). Namely, for every given sequence $M_n c(n)$, $n \in \mathbb{N}$, there exists $\varphi \in C_0^{\infty}(\mathbb{R})$ such that $\varphi^{(n)}(0)/M_n c(n) \to \infty$, as $n \to \infty$.

This example is the motivation for the following assertion.

12. Let $T \in \mathcal{D}'(\mathbf{R})$ be such that $\lim_{h \to \infty} T(x+h)/c(h)$ exists in $\mathcal{D}'^*(\mathbf{R})$. Assume that for some $s \in \mathbf{N}$ and $\omega \in \mathcal{D}^*(\mathbf{R})$ with the property

$$\int_{\mathbf{R}} \omega(t)dt = 1, \ \int_{\mathbf{R}} t^j \omega(t)dt = 0, \ j = 1, \ldots, s,$$

the following limit

$$\lim_{h \to \infty} \left\langle \frac{T(t+h)}{c(h)}, \frac{t^s}{p^{s+1}} \omega \left(\frac{t}{p} \right) \right\rangle, \ p \in (0, 1],$$

exists uniformly in p. Then, $\lim_{h \to \infty} T(x+h)/c(h)$ exists in $\mathcal{D}'(\mathbf{R})$ as well.

Proof. Let $\phi \in \mathcal{D}(\mathbf{R})$, $h \geq h_0 > 0$. The function $F_h(t) = \langle T(x + h + t), \phi(x) \rangle$, $t \in \mathbf{R}$, is smooth and by the Taylor formula, we have:

$$F_h(t) = F_h(0) + t\frac{d}{dt}F_h(0) + \cdots + \frac{t^{s-1}}{(s-1)!}\frac{d^{s-1}}{dt^{s-1}}F_h(0)$$

$$+ \frac{t^s}{(s-1)!}\int_0^1 (1-p)^{s-1}F_h^{(s)}(pt)dp, \quad s \geq 1, \ 0 < p \leq 1.$$

This implies

$$\langle T(x+h), \phi(x) \rangle = \langle T(x+h+t), \phi(x) \rangle - \langle T(x+h), \phi'(x) \rangle$$

$$- \cdots - \frac{(-1)^{s-1}t^{s-1}}{(s-1)!}\langle T(x+h), \phi^{(s-1)}(x) \rangle$$

$$- \frac{(-1)^s t^s}{(s-1)!}\int_0^1 (1-p)^{s-1}\langle T(x+h+pt), \phi^{(s)}(x) \rangle dp,$$

$$h \geq h_0, \ t \in \mathbf{R}.$$

Multiplying both sides of the last equality by $\omega(t)$, integrating with respect to t, and using Fubinni's theorem, we obtain

$$\langle T(x+h), \phi(x) \rangle = \langle\langle T(x+h+t), \omega(t) \rangle, \phi(x) \rangle - \cdots - \frac{(-1)^s}{(s-1)!}$$

$$\int_0^1 (1-p)^{s-1}\langle\langle T(x+h+pt), t^s\omega(t) \rangle, \ \phi^{(s)}(x) \rangle dp.$$

Set

$$G(x,h,p) = \langle T(x+h+pt), t^s\omega(t) \rangle, \quad x \in \mathbf{R}, \ h \geq h_0, \ p \in [0,1].$$

We have $G(x,h,0) = 0$, $x \in \mathbf{R}$, $h \geq h_0$ and

$$\lim_{h\to\infty} G(x,h,p)/c(h) = Ce^{ax}\langle e^{apt}, t^s\omega(t) \rangle, \quad x \in \mathbf{R}, \ p \in (0,1],$$

for some $a \in \mathbf{R}$ (cf. Proposition 1.2 a)), where the limit is uniform in $p \in (0,1]$ and $x \in \text{supp}\,\phi$. The limit function is continuous in $x \in \mathbf{R}$ and $p \in [0,1]$, because of $\langle e^{apt}, t^s\omega(t) \rangle \to 0$, as $p \to 0$. Because of that, we obtain

$$\lim_{h\to\infty} \frac{1}{c(h)}\langle\langle T(x+h+pt), t^s\omega(t) \rangle, \ \phi^{(s)}(x) \rangle$$

$$= \left\langle \lim_{h\to\infty} \frac{1}{c(h)}\langle T(x+h+pt), t^s\omega(t) \rangle, \phi^{(s)}(x) \right\rangle$$

$$= \langle\langle Ce^{ax+apt}, t^s\omega(t) \rangle, \phi^{(s)}(x) \rangle = C\langle e^{ax}, \phi^{(s)}(x) \rangle\langle e^{apt}, t^s\omega(t) \rangle.$$

This implies

$$\lim_{h\to\infty} \left\langle \frac{1}{c(h)} \langle T(x+h), \phi(x)\rangle \right\rangle = \left\langle \left\langle \lim_{h\to\infty} \frac{T(x+h+t)}{c(h)}, \omega(t) \right\rangle, \phi(x) \right\rangle$$

$$- \cdots - \frac{(-1)^s}{(s-1)!} \int\limits_0^1 (1-p)^{s-1} \langle e^{apt}, t^s\omega(t)\rangle dp \langle Ce^{ax}, \phi^{(s)}(x)\rangle,$$

which proves the assertion.   □

Such an assertion can be proved in the multi-dimensional case by adjusting the previous argument. But **an open problem** *is to find necessary and sufficient conditions for a distribution T, which has S-asymptotics in $\mathcal{D}'^*$, to have the S-asymptotics in $\mathcal{D}'$ as well.*

13. We shall construct an ultradistribution out of the space of Schwartz distributions and having S-asymptotics.

Assume (M.2) holds. Let $P(D)$ be an ultradifferential operator of class $*$ of infinite order ($a_\alpha \neq 0$ for infinitely α) (see **0.5.2**). Then, $P(D)\delta$ is an element of $\mathcal{D}'^*$ which is not a distribution and which has S-asymptotics in $\mathcal{D}'^*$ equal to zero related to any c.

If $T \in \mathcal{D}'$ and $T(x+h) \overset{s}{\sim} 1 \cdot 1$, $h \in \Gamma$ in $\mathcal{D}'$, the ultradistribution $T + P(D)\delta$ is not a distribution, but $(T + P(D)\delta)(x+h) \overset{s}{\sim} 1 \cdot 1$, $h \in \Gamma$ in $\mathcal{D}'^*$:

$$\lim_{h\in\Gamma, \|h\|\to\infty} (\langle\langle T(x+h), \phi(x)\rangle + \langle P(D)\delta(x+h), \phi(x)\rangle)$$

$$= \langle 1, \phi\rangle + \lim_{h\in\Gamma, \|h\|\to\infty} \langle \delta(x+h), \sum(-1)^m a_m \phi^{(m)}(x)\rangle = \langle 1, \phi\rangle, \quad \phi \in \mathcal{D}^*.$$

14. Let $P(D)$ be a local operator $\sum_{|\alpha|\geq 0} b^\alpha D^\alpha$, $b^\alpha \neq 0$, $|\alpha|\geq 0$ (see **0.5.3.**). The Fourier hyperfunction $f = 1 + P(D)\delta$ has S-asymptotics related to $c = 1$ in any cone Γ and with limit $U = 1$, but f is not a distribution. For the S-asymptotics of f it is enough to prove that

$$\lim_{h\in\Gamma, \|h\|\to\infty} \langle P(D)\delta(x+h), \varphi(x)\rangle = 0, \quad \varphi \in \mathcal{P}_*.$$

Since $P(D)$ maps $\mathcal{P}_*$ into $\mathcal{P}_*$,

$$\langle P(D)\delta(x+h), \varphi(x)\rangle = \langle \delta(x+h), P(-D)\varphi(x)\rangle = \psi(h),$$

where $\psi = P(-D)\varphi$. By the property of the elements of $\mathcal{P}_*$ (see **0.5.3**), we have $\lim\limits_{h\in\Gamma,\|h\|\to\infty} \psi(h) = 0$, for every cone Γ.

A hyperfunction g supported by the origin is uniquely expressible as $g = \tilde{P}(D)\delta$, where $\tilde{P}(D)$ is a local operator. In such a way, the above proof implies that every Fourier hyperfunction with support at $\{0\}$ has S-asymptotics with limit equal zero.

Since $P(D)\delta = \sum\limits_{|\alpha|\geq 0} b_\alpha D^\alpha\delta$ is a distribution if and only if $b_\alpha \neq 0$ for a finite number of α, we note that $1 + P(D)\delta$ is not a distribution, but it has the S-asymptotics related to $c = 1$.

We can also find coefficients b_α of a local operator $P(D)$ such that $f = 1 + P(D)\delta$ is not defined by an ultradistribution belonging to the Gevrey class $\mathcal{D}^{(s)'}$ or $\mathcal{D}^{\{s\}'}$, $s > 1$ (see [81]). For the sake of simplicity, we shall consider the one-dimensional case. Choose $P(D)$ such that the coefficients of $P(D)$ are: $b_n = (n!)^{-(1+c_n)}$, $n \in N$, where $c_n = (\log\log n)^{-1}$. With these coefficients, $P(D)$ is a local operator. Namely,

$$\lim_{n\to\infty} \sqrt[n]{b_n n!} = \lim_{n\to\infty} (n!)^{-\frac{1}{n\log\log n}} = 0.$$

Also, any ultradistribution in the Gevrey class $s > 1$, supported by $\{0\}$, is of the form

$$J(D)\delta = \sum_{n=0}^{\infty} a_n D^n\delta, \text{ necessarily with } |a_n| \leq Ck^n/(n!)^s$$

for some constants k and C (Beurling's type) or for any $k > 0$ with a constant C (Roumieu's type). But the coefficients $b_n = (n!)^{-(1+c_n)}$ do not satisfy these conditions, therefore $P(D)\delta$ cannot represent an ultradistribution. Namely, since $c_n \to 0$ when $n \to \infty$, for any $s > 1$, there exists n_0 such that $1 + c_n < s$, $n \geq n_0$. Thus,

$$(n!)^{-(1+c_n)} > Ck^n/(n!)^s, \quad n \geq n_0, \ k > 0.$$

Consequently, $P(D)\delta$ does not represent an ultradistribution of Gevrey type.

On the other hand, if we suppose that $g = P(D)\delta$ is an ultradistribution with support $\{0\}$ in the Gevrey class $s > 1$, then, we would have an ultradifferential operator $J_1(D)$ such that

$$g = J_1(D)\delta = \sum_{n=0}^{\infty} e_n D^n\delta, \ |e_n| \leq Ck^n/(n!)^s.$$

But in this case $J_1(D)$ is also a local operator and $J_1(D) \neq P(D)$. This contradicts the fact that a hyperfunction with support at $\{0\}$ is given by a unique local operator.

1.6 *S-asymptotics and the asymptotics of a function*

We suppose in this subsection that elements of $\mathcal{F}$ are compactly supported and, as usual, that the topology in $\mathcal{F}$ is stronger than the topology in $\mathcal{E}$. Recall, we use the notation $\mathcal{F}_0$ in this case.

Every locally integrable function f defines an element of $\mathcal{F}'_0$ (regular generalized functions).

We shall compare the asymptotic behavior of a locally integrable function f and the S-asymptotic behavior of the generalized function generated by it.

A function f has asymptotics at infinity if there exists a positive function c such that $\lim\limits_{x\to\infty} f(x)/c(x) = A \neq 0$, *(in short $f(x) \sim Ac(x)$, $x \to \infty$) .*

1. The following example points out that a continuous and L^1-integrable function can have S-asymptotics as a distribution without having an ordinary asymptotics. Suppose that $g \in L^1(\mathbf{R}) \cap C(\mathbf{R})$ has the property that $g(n) = n$, $n \in \mathbf{N}$ and that it is equal to zero outside suitable small intervals $I_n \ni n$, $n \in \mathbf{N}$. Denote by $f(t) = e^t \int_0^t g(x)dx$, $t \in \mathbf{R}$. It is easy to see that

$$f(t+h) \overset{s}{\sim} e^h \cdot e^t \int_0^\infty g(x)dx, \ h \in \mathbf{R}_+ .$$

By Theorem 1.2 a) $f'(t)$ has S-asymptotics related to e^h and with the same limit. But, in view of the properties of g, $f'(t) = f(t) + e^t g(t)$ has not the same asymptotics (in the ordinary sense). Moreover, g can be chosen so that f' has no asymptotics at all.

2. The following example shows that a function f can have asymptotic behavior without having S-asymptotics with limit U different from zero. An example is $x \mapsto \exp(x^2), x \in \mathbf{R}$. Suppose that $\exp(x^2)$ has S-asymptotics related to a $c(h) > 0$, $h \in \mathbf{R}_+$ with a limit U different from zero. By Proposition 1.2 c), U has the form $U(x) = C\exp(ax), C > 0$. Then, for every $\varphi \in \mathcal{F}_0$ such that $\varphi > 0$ we have

$$\lim_{h \to \infty} \frac{1}{c(h)} \int \exp[(x+h+h_0)^2]\varphi(x)dx = e^{ah_0}\langle Ce^{ax}, \varphi(x)\rangle .$$

Therefore,

$$e^{ah_0}\langle U, \varphi \rangle = \exp(h_0^2) \lim_{h \to \infty} \frac{1}{c(h)} \int e^{(x+h)^2} e^{2h_0(x+h)} \varphi(x)dx$$

$$\geq \exp(h_0^2)\langle U, \varphi \rangle, \quad \text{for every } h_0 > 0 .$$

But this inequality is absurd. Consequently, $\exp(x^2)$ cannot have such an S-asymptotic behavior.

One can prove a more general assertion.

Proposition 1.3. *Let $f \in L^1_{loc}(\mathbf{R}) \subset \mathcal{F}'_0(\mathbf{R})$ have one of the four properties for $\alpha > 1, \beta > 0, x \geq x_0, h > 0, M > 0$ and $N > 0$:*

a) $f(x+h) \geq M \exp(\beta h^\alpha)f(x) \geq 0$,

a') $-f(x+h) \geq -M \exp(\beta h^\alpha)f(x) \geq 0$,

b) $0 \leq f(x+h) \leq N \exp(-\beta h^\alpha)f(x)$,

b') $0 \leq -f(x+h) \leq -N \exp(-\beta h^\alpha)f(x)$.

Then f cannot have S-asymptotics with limit $U \neq 0$, but the function f can have asymptotics.

For the proof see ([135], p. 89).

It is easy to show that for some classes of real functions f on $\mathbf{R}$ the asymptotic behavior at infinity implies the S-asymptotics.

Proposition 1.4. *a) Let c be a positive function and let $T \in L^1_{loc}(\mathbf{R}^n)$. Suppose that there exist locally integrable functions $U(x)$ and $V(x), x \in \mathbf{R}^n$, such that for every compact set $K \subset \mathbf{R}^n$*

$$|T(x+h)/c(h)| \leq V(x), \quad x \in K, \ \|h\| > r_K ,$$

$$\lim_{h \in \Gamma, \|h\| \to \infty} T(x+h)/c(h) = U(x), \quad x \in K .$$

Then, $T(x+h) \overset{s}{\sim} c(h)U(x), \ h \in \Gamma$ in $\mathcal{F}'_0$.

b) Let $T \in L^1_{loc}(\mathbf{R})$ have the ordinary asymptotic behavior

$$T(x) \sim \exp(\alpha x)L(\exp x), \quad x \to \infty, \ \alpha \in \mathbf{R} ,$$

where L is a slowly varying function. Then,

$$T(x+h) \overset{s}{\sim} \exp(\alpha h)L(\exp h)\exp(\alpha x), \quad h \in \mathbf{R}_+, \ \text{in } \mathcal{F}'_0(\mathbf{R}) .$$

Proof. a) For every $\varphi \in \mathcal{F}_0$

$$\lim_{h \in \Gamma, \|h\| \to \infty} \left\langle \frac{T(x+h)}{c(h)}, \varphi(x) \right\rangle = \lim_{h \in \Gamma, \|h\| \to \infty} \int_{\mathbf{R}^n} \frac{T(x+h)}{c(h)} \varphi(x) dx. \quad (1.14)$$

Since $\operatorname{supp} \varphi \subset K \subset \mathbf{R}^n$ and T has all the listed properties, Lebesgue's theorem implies the result.

b) It is enough to use in (1.14) that $L(yt)/L(t) \to 1$, $t \to \infty$ uniformly in y, when y stays in compact interval contained in $\mathbf{R}_+$. $\qquad\square$

A more general result is the following one ([135], pp. 89–90).

Proposition 1.5. *Let Γ be a cone and let $\Omega \subset \mathbf{R}^n$ be an open set such that for every $r > 0$ there exists a β_r such that $B(0,r) \subset \{\Omega - h;\ h \in \Gamma,\ \|h\| \geq \beta_r\}$. Suppose that $G \in L^1_{loc}(\Omega)$ and it has the following properties: There exist locally integrable functions U and V in $\mathbf{R}^n$ such that for every $r > 0$ we have*

$$|G(x+h)/c(h)| \leq V(x), \quad x \in B(0,r),\ h \in \Gamma,\ \|h\| \geq \beta_r;$$

$$\lim_{h \in \Gamma, \|h\| \to \infty} G(x+h)/c(h) = U(x), \quad x \in B(0,r).$$

If $G_0 \in \mathcal{F}'_0$ coincides with G on Ω, then

$$G_0(x+h) \overset{s}{\sim} c(h)U(x), \quad h \in \Gamma.$$

Proof. By Theorem 1.4, it is enough to proof that

$$\lim_{h \in \Gamma, \|h\| \to \infty} \int_{\Omega} \frac{G(x+h)}{c(h)} \varphi(x) dx = \int_{\mathbf{R}^n} U(x)\varphi(x) dx;$$

but as in the proof of Proposition 1.4 a), the exchange of the limit and the integral sign is justified by our assumptions and Lebesgue's theorem. $\qquad\square$

The following proposition gives a sufficient condition under which the S-asymptotics of $f \in L^1_{loc}(\mathbf{R})$, in $\mathcal{D}'(\mathbf{R})$, implies the ordinary asymptotic behavior of f.

Proposition 1.6. *Let $f \in L^1_{loc}(\mathbf{R})$, $c(h) = h^\beta L(h)$, where $\beta > -1$ and L be a slowly varying function. If for some $m \in N, x^m f(x), x > 0$, is monotonous and $f(x+h) \overset{s}{\sim} c(h) \cdot 1, h \in \mathbf{R}_{+,}$, in $\mathcal{D}'(\mathbf{R})$, then $\lim_{h \to \infty} f(h)/c(h) = 1$. If we suppose that L is monotonous, then we can omit the hypothesis $\beta > -1$.*

For the proof see [115].

1.7   *Characterization of the support of $T \in \mathcal{F}_0'$*

We suppose in this subsection that the topology in $\mathcal{F}_0$ is defined in such a way that a sequence $\{\varphi_n\}$ in $\mathcal{F}_0$ converges if and only if there exists a compact set $K \subset \mathbf{R}^n$ and $\varphi_m^{(k)}$ converges to $\varphi^{(k)}$ uniformly on K for every $k \in \mathbf{N}_0^n$ as $m \to \infty$.

We already proved in Theorem 1.4 a relation between the support of a distribution and its S-asymptotics. Now, we shall complete this result.

We need a property of the S-asymptotics given in the next lemma.

Lemma 1.1. *Let Γ be a cone and let $\tilde{\Gamma}$ be a convex cone (it is partially ordered). A necessary and sufficient condition that for every $c(h) > 0$, $h \in \Gamma$*

a) $\underset{\|h\| \to \infty, h \in \Gamma}{w. \lim} T(x+h)/c(h) = 0$ in $\mathcal{F}'_0$,

b) $\underset{h \to \infty, h \in \tilde{\Gamma}}{w. \lim} T(x+h)/c(h) = 0$ in $\mathcal{F}'_0$,

is that for every $\phi \in \mathcal{F}_0$ the following holds:

In case a): There exists $\beta(\phi) > 0$ such that

$$\langle T(x+h), \phi(x) \rangle = 0, \quad \|h\| \geq \beta(\phi), \ h \in \Gamma. \tag{1.15}$$

In case b): There exists $h_\phi \in \tilde{\Gamma}$ such that

$$\langle T(x+h), \phi(x) \rangle = 0, \quad h \geq h_\phi, \ h \in \tilde{\Gamma}. \tag{1.16}$$

Proof. We have to prove only that the condition is necessary in both cases. It is obvious that the condition is sufficient. Let us suppose the opposite, that is, the condition is not necessary. Then, we could find a sequence $\langle a_m \rangle_m$ in Γ such that in case a) $\|a_m\| \to \infty$ in Γ, and in case b) $a_m \to \infty$ in $\tilde{\Gamma}$, $(m \to \infty)$ and that

$$\langle T(x+a_m), \phi(x) \rangle = \alpha_m \neq 0, \ m \in \mathbf{N}.$$

Let find c such that $c(h) = \alpha_m$ for $h = a_m, m \in \mathbf{N}$. Clearly, for such a $c(h)$ the function $h \mapsto \langle T(x+h)/c(h), \phi(x) \rangle$ cannot converge to zero as $\|h\| \to \infty$, $h \in \Gamma$, or $h \to \infty$, $h \in \tilde{\Gamma}$. This contradicts our supposition that the S-asymptotics equals zero in both cases. $\qquad\square$

Theorem 1.5. *Let Γ be a cone and let $T \in \mathcal{F}'_0$. A necessary and sufficient condition that for every $r > 0$ there exists β_r such that the sets*

$$\operatorname{supp} T \cap B(h,r), \quad h \in \Gamma, \|h\| \geq \beta_r \quad \text{are empty}$$

is that $T(x+h) \overset{s}{\sim} c(h) \cdot 0, h \in \Gamma$ for every positive function c on Γ.

Proof. Theorem 1.2 c) and Theorem 1.4 assert that the condition in Theorem 1.5 is necessary. We have to prove only that this condition is also sufficient.

Let us suppose that

$$\underset{\|h\|\to\infty, h\in\Gamma}{w.\lim} T(x+h)/c(h) = 0 \quad \text{in } \mathcal{F}'_0.$$

Let $\phi \in \mathcal{F}_0$. By Lemma 1.1 a), we know that there exists $\beta_0(\phi) = \inf \beta(\phi)$, where $\beta(\phi)$ is such that (1.15) holds. We shall prove that the set $\{\beta_0(\phi); \phi \in \mathcal{F}_0, \operatorname{supp}\varphi \subset K\}$ is bounded for every compact set $K \subset \mathbf{R}^n$. Let us suppose the opposite. Then we could find a sequence $\langle \phi_k \rangle_k$ in $\mathcal{F}_0, \operatorname{supp}\varphi_k \subset K, k \in \mathbf{N}$, and a sequence $\langle h_k \rangle_k$ in $\Gamma, \|h_k\| \to \infty$ such that

$$\langle T(t+h_k), \phi_p(t) \rangle = A_{k,p} = \begin{cases} a_k \neq 0, & p = k \\ 0, & p < k. \end{cases}$$

We give the construction of sequences $\{\phi_k\}$ and $\{h_k\}$. Let $\phi_k \in \mathcal{F}_0, \operatorname{supp}\phi_k \subset K, k \in \mathbf{N}$, be such that $\langle \beta_0(\phi_k) \rangle_k$ is a strictly increasing sequence which tends to infinity. Then, there exist $\langle h_k \rangle_k$ in Γ and $\varepsilon_k > 0$, $k \in \mathbf{N}$, such that $\beta_0(\phi_{k-1}) + \varepsilon_k \leq \|h_k\| \leq \beta_0(\phi_k) - \varepsilon_k$. Now, we shall construct the sequence $\langle \psi_p(t) \rangle_p$ in $\mathcal{F}_0, \operatorname{supp}\psi_p \subset K, p \in \mathbf{N}$, for which we have

$$\langle T(t+h_k), \psi_p(t) \rangle = \begin{cases} 0, & p \neq k \\ a_k, & p = k. \end{cases} \tag{$*$}$$

Put

$$\psi_p(t) = \phi_p(t) - \lambda_1^p \phi_1(t) - \cdots - \lambda_{p-1}^p \phi_{p-1}(t), \quad p > 1, t \in \mathbf{R}^n.$$

The sequence $\langle \lambda_i^p \rangle_i$ will be determined in such a way that $\psi_p(t)$ satisfies $(*)$ the sought property, $p \in \mathbf{N}$.

It is easy to see that $\langle T(t+h_k), \psi_k(t) \rangle = a_k$ and $\langle T(t+h_k), \psi_p(t) \rangle = 0, k > p$. For a fixed p and $k < p$ we can find λ_i^p, $i = 1, \ldots, p-1$, such that for $k = 1, \ldots, p-1$,

$$0 = \langle T(t+h_k), \psi_p(t) \rangle = A_{k,p} - \lambda_1^p A_{k,1} - \cdots - \lambda_{p-1}^p A_{k,p-1}.$$

Hence

$$\lambda_1^p A_{k,1} + \cdots + \lambda_{p-1}^p A_{k,p-1} = A_{k,p}, \ k = 1, \ldots, p-1, \ p > 1.$$

Since $A_{k,k} \neq 0$ for every k, the system always has a solution.

We introduce a sequence of numbers $\langle b_k \rangle_k$,

$$b_k = \sup\{2^k |\psi_k^{(i)}(t)|; \ i \leq k\}, \ k \in \mathbb{N}.$$

Then, the function

$$\psi(t) = \sum_{p=1}^{\infty} \psi_p(t)/b_p, \ t \in \mathbf{R}^n, \ \text{is in } \mathcal{F}_0 \ \text{and} \ \text{supp}\psi \subset K.$$

Thus, we have

$$\langle T(t + h_k), \psi(t) \rangle = \sum_{p=1}^{\infty} \langle T(t + h_k), \psi_p(t)/b_p \rangle = a_k/b_k.$$

Now, if we choose $c(h)$ such that $c(h_k) = a_k/b_k$, $k \in \mathbf{N}$, then $\left\langle \dfrac{T(t+h), \psi(t)}{c(h)} \right\rangle$ does not converge to zero when $\|h\| \to \infty, h \in \Gamma$.

This proves that for every compact set K there exists a $\beta_0(K)$ such that

$$\langle T(t + h), \phi(t) \rangle = 0, \ \|h\| \geq \beta_0(K), \ h \in \Gamma, \ \phi \in \mathcal{F}_0, \ \text{supp}\phi \subset K.$$

It follows that $T(t + h) = 0$ over $B(0, r), \|h\| \leq \beta(r), h \in \Gamma$ and $T(t) = 0$ over $B(h, r), \|h\| \geq \beta(r), h \in \Gamma$. $\qquad\square$

Remark. The condition of Theorem 1.5 implies that the support of T has the following property: The distance from supp T to a point $h \in \Gamma$, $d(\text{supp}\, T, h)$, tends to infinity when $\|h\| \to \infty$, $h \in \Gamma$.

The next proposition shows that if in Definition 1.1 we take the limit (1.3) instead of the limit (1.2), then a more precise result is obtained.

Theorem 1.6. *Let $T \in \mathcal{F}'_0$ and $\tilde{\Gamma}$ be an acute, open and convex cone (partially ordered, see Remark 1 after Definition 1.1). A necessary and sufficient condition for*

$$\text{supp}\, T \in C_{\mathbf{R}^n}(a + \tilde{\Gamma}) \ \text{for some } a \in \tilde{\Gamma} \tag{1.17}$$

is that

$$\lim_{h \in \tilde{\Gamma}, h \to \infty} T(x + h)/c(h) = 0 \ \text{in } \mathcal{F}'_0 \ \text{for every } c(h) \tag{1.18}$$

$(C_{\mathbf{R}^n} A = \mathbf{R}^n \setminus A).$

Proof. If (1.17) holds, then for any ball $B(0,1)$, there exists a $h_r \in \tilde{\Gamma}$ such that $B(h,r) \subset (a + \tilde{\Gamma})$ for $h \geq h_r, h \in \tilde{\Gamma}$. This implies (1.18), and by Lemma 1.1, (1.17) follows.

Let us suppose now that (1.16) and consequently (1.18) hold, but for any $a \in \tilde{\Gamma}$, $\operatorname{supp} T \not\subset C_{\mathbf{R}^n}(a + \tilde{\Gamma})$. We fix such an $a = a_0 > 0$. There exists an $a_1 \in (a_0 + \tilde{\Gamma}) \cap \operatorname{supp} T$. Since $\operatorname{supp} T \not\subset C_{\mathbf{R}^n}(2a_1 + \tilde{\Gamma})$, there exists an $a_2 \in (2a_1 + \tilde{\Gamma} \cap \operatorname{supp} T)$. In such a way, we construct a sequence $\langle a_k \rangle_k$ in $\tilde{\Gamma}$ such that $a_k \in (ka_{k-1} + \tilde{\Gamma}) \cap \operatorname{supp} T$ and $a_k \geq ka_0, k \in \mathbf{N}$.

Since $a_k \in \operatorname{supp} T, k \in \mathbf{N}$, it follows that there exists a sequence $\langle \phi_k \rangle_k$ in $\mathcal{F}_0$ such that $\operatorname{supp} \phi_k \subset B(0,1), k \in N$, and

$$\langle T(x + a_k), \phi_k(x) \rangle \neq 0, k \in \mathbf{N}.$$

We put now:

$$c_{k,i} = \langle T(x + a_i), \phi_k(x) \rangle, \ k, i \in \mathbf{N};$$

$$b_k = \operatorname{supp}\{2^k |\phi^{(j)}(x)|; j \leq k, x \in \mathbf{R}^n\}, k \in \mathbf{N}.$$

We have to prove that there exists a sequence $\langle c_k \rangle_k$ such that $c_k \geq 1, k \in \mathbf{N}$ and

$$\sum_{k \geq 1} c_{k,i}/(b_k c_k) \neq 0, i \in \mathbf{N}. \tag{1.19}$$

First, we notice that $c_{k,k} \neq 0$. If $\sum_{k=1}^{\infty} c_{k,1}/b_k = 0$, then we take $c_1 > 1$, and if this series is different from 0, then we put $c_1 = 1$. Let $i = 2$. If

$$c_{1,2}/(b_1 c_1) + \sum_{k \geq 2} c_{k,2}/b_k = 0 \ (\neq 0), \text{ we take } c_2 > 1 \ (c_2 = 1)$$

such that

$$c_{1,1}/(c_1 b_1) + c_{2,1}/(c_2 b_2) + \sum_{k \geq 3} c_{k,1}/b_k \neq 0.$$

Let $i = 3$,

$$c_{1,3}/(b_1 c_1) + c_{2,3}/(b_2 c_2) + \sum_{k=3}^{\infty} c_{k,3}/b_k = 0 \text{ or } \neq 0.$$

Then we take $c_3 > 1$ and $c_3 = 1$ respectively such that

$$c_{1,1}/(b_1 c_1) + c_{2,1}/(b_2 c_2) + c_{3,1}/(b_3 c_3) + \sum_{k=4}^{\infty} c_{k,1}/b_k \neq 0,$$

$$c_{1,2}/(b_1c_1) + c_{2,2}/(b_2c_2) + c_{3,2}/(b_3c_3) + \sum_{k=4}^{\infty} c_{k,2}/b_k \neq 0.$$

Continuing in this way, we construct a sequence $\langle c_k \rangle_k$ for which (1.19) holds.

Let us put

$$\psi_k(x) = \phi_k(x)/(b_kc_k), k \in \mathbf{N} \text{ and } \psi = \sum_{k \geq i} \psi_k.$$

From the properties of sequences $\langle b_k \rangle_k$ and $\langle c_k \rangle_k$, we can easily show that

$$\sum_{k=1}^{N} \psi_k \to \psi \text{ in } \mathcal{F}_0 \text{ when } N \to \infty.$$

Relation (1.18) implies

$$\langle T(x + a_i), \psi(x) \rangle = \sum_{k=1}^{\infty} c_{k,1}/(b_kc_k) \neq 0, \ i \in \mathbf{N}.$$

We obtain that (1.18) does not hold for ψ. This completes the proof. $\square$

In Theorem 1.6 the support of T can be just $C_{\mathbf{R}^n}(a + \tilde{\Gamma})$. The question is: *Is it possible to obtain a similar proposition for the S-asymptotics given by Definition 1.1?* This question is analyzed in the next examples. We take $\mathcal{F}_0 = \mathcal{D}$ in which the S-asymptotics by the weak and strong convergence are equivalent.

Examples. i) Let $T(x,y) = \sum_{m \geq 1} m\delta(x - m, y)$. The given series converges in $\mathcal{D}'(\mathbf{R}^2)$. Since for a $\phi \in \mathcal{D}(\mathbf{R}^2)$, supp $\phi \subset B(0, r)$, we have

$$\lim_{n \to \infty} \langle \sum_{m=1}^{n} m\delta(x - m, y), \phi(x, y) \rangle = \sum_{1 \leq m \leq r} m\phi(m, 0).$$

It follows that T is a distribution on $\mathbf{R}^2$ (see Theorem XIII, Chapter 3 in [146]).

Let us remark that the support of T lies on the half line $\gamma \equiv \{(p, 0) \in \mathbf{R}^2; p > 0\}$. We can take for Γ the cone $\mathbf{R}^2_+ \equiv \{(\alpha, \beta) \in \mathbf{R}^2; \alpha > 0, \beta > 0\}$. It is a convex, open and acute cone in $\mathbf{R}^2$. We shall show that the limit

$$\lim_{h \in \Gamma, \|h\| \to \infty} \langle T(u + h), \phi(u) \rangle$$

does not exist. To do this, it is enough to take the limit over the half line $\gamma' \equiv \{(0, \alpha_0) + \gamma\}, \alpha_0 > 0$, which belongs to Γ. If we choose ϕ such that $\phi > 0$ and $\phi(0, -\alpha_0) = 1$, then for $h = (p, \alpha_0) \in \gamma'$

$$\langle T(u+h), \phi(u) \rangle = \sum_{m \geq 1} m\phi(m-p, -\alpha_0) \geq p.$$

Note, if $\|h\| \to \infty$, then $p \to \infty$, as well. Consequently, the answer to the posed question is negative.

ii) The following example shows that if

$$\lim_{h \in \tilde{\gamma}, \|h\| \to \infty} T(x+h)/c(h) = 0 \text{ in } \mathcal{D}'$$

for every positive $c(h)$, and every $\tilde{\gamma} = \{pw, p > 0\}, w \in \Gamma$, then this does not imply that

$$\lim_{h \in \Gamma, \|h\| \to \infty} T(x+h)/c(h) = 0 \text{ in } \mathcal{D}'.$$

Let us remark that both limits on a $\tilde{\gamma}$, when $h \to \infty$ or $\|h\| \to \infty$, are equal.

Let T be given by $T(x, y) = \sum_{m \geq 1} m\delta(x - m, y - \frac{1}{m})$ on $\mathbb{R}^2$. The support of T lies on the curve $\{(x, 1/x); x > 0\}$. Let $\Gamma = \mathbb{R}^2_+, w = (\cos \alpha, \sin \alpha), 0 < \alpha < \frac{\pi}{2}$ and $\tilde{\gamma} \equiv \{pw; p > 0\}$. Then, for a $\phi \in \mathcal{D}(\mathbb{R}^2)$

$$\langle T(x+h_1, y+h_2), \phi(x, y) \rangle = \sum_{m \geq 1} m\phi\left(m - h_1, \frac{1}{m} - h_2\right) \to 0, \|h\| \to \infty, h \in \tilde{\gamma}.$$

In order to show that

$$\lim_{h \in \Gamma, \|h\| \to \infty} \langle T(x+h_1, y+h_2), \phi(x, y) \rangle$$

does not exist, take h to belong to $\gamma = \{(x, a); x > 0\}$ for a fixed $a > 0$, as we did in example i).

Remarks. As a consequence of Theorem 1.5 and Theorem 1.6, we have some results concerning the convolution product (see [135], p. 102).

Let

$$G_1 \equiv \{f \in C^\infty; \text{supp} f \subset C_{\mathbb{R}^n}\{h \in \Gamma; \|h\| \geq \beta_f\}\},$$

$$G_2 \equiv \{f \in C^\infty; \text{supp} f \subset C_{\mathbb{R}^n}\{h \in \tilde{\Gamma}; h \geq h_f\}\}.$$

Corollary 1.1. *For a fixed $T \in \mathcal{D}'$, the convolution $T * \phi$ maps $\mathcal{D}$ into G_1 if and only if the support of T has the property given in Theorem 1.5.*

Corollary 1.2. *For a fixed $T \in \mathcal{D}'$ and for a convex, open and partially ordered cone $\tilde{\Gamma}$ the convolution $T * \phi$ maps $\mathcal{D}$ into G_2 if and only if $\text{supp} T \subset C_{\mathbb{R}^n}(a + \tilde{\Gamma})$ for some $a \in \tilde{\Gamma}$.*

1.8 *Characterization of some generalized function spaces*

Theorem 1.7. *A necessary and sufficient condition for a distribution T to belong to:*

a) $\mathcal{E}'$ is that $T(x + h) \stackrel{s}{\sim} c(h) \cdot 0, h \in \mathbf{R}^n$ for every positive c.

b) $\mathcal{O}'_c$ is that T has S-asymptotic behavior related to every $c(h) = \|h\|^{-\alpha}, \alpha \in \mathbf{R}_+$ and with limit $U = 0$.

c) $\mathcal{B}'$ is that T has S-asymptotic behavior related to every positive $c, c(h) \to \infty, \|h\| \to \infty$, and with limit $U = 0$.

Proof. a) It is a direct consequence of Theorem 1.5.

b) It is enough to apply Theorem IX Chapter VII in [146] which states: *A necessary and sufficient condition that a distribution T belongs to $\mathcal{O}'_c$ is that for every $\varphi \in \mathcal{D}$ the function $(T * \varphi)(h)$ is continuous and of fast descent at infinity.* (see **0.5.1**).

c) By Theorem XXV, Chapter VI in [146] a distribution $T \in \mathcal{B}'$ if and only if $T * \varphi \in L^\infty(\mathbf{R}^n)$ for every $\varphi \in \mathcal{D}$. Suppose that $T \in \mathcal{B}'$. Then for every $\varphi \in \mathcal{D}$ and $c(h) \to \infty, \|h\| \to \infty$

$$\lim_{h \in \Gamma, \|h\| \to \infty} \left\langle \frac{T(x + h)}{c(h)}, \varphi(x) \right\rangle = \lim_{h \in \Gamma, \|h\| \to \infty} \frac{(T * \check{\varphi})(h)}{c(h)} = 0 \,.$$

Suppose that $(T * \check{\varphi})(h)/c(h) \to 0, \|h\| \to \infty$, for every $\varphi \in \mathcal{D}$ and for every $c, c(h) \to \infty$ as $\|h\| \to \infty$. We will show that $(T * \check{\varphi})(h) \in L^\infty(\mathbf{R}^n)$ for every $\varphi \in \mathcal{D}$. Then, by the same theorem, it follows that $T \in \mathcal{B}'$.

Let us assume the contrary, i.e., that $(T * \check{\varphi}_0)(h)$ is not bounded for a $\varphi_o \in \mathcal{D}$. Then, we could find two sequences $\langle h_m \rangle_m$ in $\mathbf{R}^n$ and $\langle c_m \rangle_m$ in $\mathbf{R}$ such that $|c_m| \to \infty$ as $m \to \infty, \|h_m\| \geq m$ and $(T * \check{\varphi}_0)(h_m) = c_m$. Now, for $c_o(h)$ such that $c_o(h_m) = \sqrt{|c_m|}, m \in \mathbb{N}$, the limit $< T(x + h)/c_o(h), \varphi_o(x) >$ would not exist, as $\|h\| \to \infty$. This is in a contradiction with our assumption that T has the S-asymptotics related to every $c(h)$ which tends to infinity as $\|h\| \to \infty$. $\qquad\square$

Proposition 1.7. *a) If for every rapidly decreasing function c, T has the S-asymptotic behavior related to c^{-1} and with limit U_c ($U_c = 0$ is included), then $T \in \mathcal{S}'$.*

b) If for every rapidly exponentially decreasing function c (for every $k > 0, c(h) \exp(k\|h\|) \to 0, \|h\| \to \infty$) a distribution T has the S-asymptotic behavior related to c^{-1} with limit U_c ($U_c = 0$ is included), then $T \in \mathcal{K}'_1$.

Proof. a) Let c be given. There exists β_0 such that for every h, $\|h\| \geq \beta_0$, and for every $\varphi \in \mathcal{D}$

$$|< T(x + h) \cdot c(h), \varphi(x) >| \leq |< U_c, \varphi >| + \varepsilon_\varphi \leq M_\varphi + \varepsilon_\varphi.$$

Therefore, the set $\{T(x + h) \cdot c(h); h \geq \beta_0\}$ is weakly bounded and thus bounded in $\mathcal{D}'$. By Theorem VI 4°, Chapter VII in [146] if $\{c(h)T(x + h); \|h\| > \beta_0\}$ is bounded in $\mathcal{D}'$, for every c of fast descent, then $T \in \mathcal{S}'$.

b) The proof is similar to that of a), if we use the following theorem proved which will be showed below (cf. Theorem 1.15):

Let $T \in \mathcal{D}'$. If for every rapidly exponentially decreasing function r on $\mathbf{R}^n$ the set $\{r(h)T(x + h); h \in \mathbf{R}^n\}$ is bounded in $\mathcal{D}'$, then $T \in \mathcal{K}'_1$. $\square$

1.9 *Structural theorems for S-asymptotics in $\mathcal{F}'$*

In the analysis of the S-asymptotics and its applications, it is useful to know analytical expression for generalized functions having S-asymptotics, especially if it is given via continuous functions and their derivatives. The next theorems are of this kind. These types of theorems are usually referred as structural theorems.

Our first result concerns the one-dimensional case, a refinement will be obtained in Theorem 1.9 below.

Theorem 1.8. *Let $\alpha \in \mathbf{R}$ and let L be a slowly varying function. Suppose that $f \in \mathcal{D}'(\mathbf{R})$. If $f(x + h) \overset{s}{\sim} c(h) \cdot e^{\alpha x}, h \in \mathbf{R}_+$, then there is an $m_0 \in \mathbf{N}$ such that for every $m \geq m_0$ the following holds:*

(1) Let $\alpha \neq 0$ and $c(h) = e^{\alpha h}L(e^h)$, $h > 0$. Then there are $g_{m,i} \in C(1, \infty), i = 0, 1, \ldots, m$ such that

$$f(x) = \sum_{i=0}^{m} g_{m,i}^{(i)}(x), \quad x \in (1, \infty),$$

and

$$g_{m,i}(x) \sim C_i x^m \exp(\alpha x)L(\exp x), \quad x \to \infty, \ i = 0, \ldots, m,$$

where C_i are suitable constants.

(2) Let $\alpha = 0$ and $c(h) = h^\nu L(h)$, $h > 0$. Then,

a) If $\nu > -1$, then there is $F_m \in C(1, \infty)$ such that $f = F_m^{(m)}$ and $F_m(x) \sim x^{m+\nu} L(x)$, $x \to \infty$;

b) If $\nu \le -1$, then there are $f_{m,i} \in C(1, \infty)$ and $A_{m,i} \ne 0$, $i = 0, 1, \ldots, m$, such that

$$f_{m,i}(x) \sim A_{m,i} x^{m+\nu-i} L(x), i = 0, 1, \ldots, m$$

and

$$f(x) = \sum_{i=0}^{m} f_{m,i}^{(m-i)}(x), \quad x \in (1, \infty).$$

Proof. Case $\alpha \ne 0$. By the remark 1. after Proposition 1.2, there exists a function $\tilde{c} \in C^\infty(\mathbf{R})$ such that $c(h)/\tilde{c}(x+h) \to \exp(-\alpha x)$, $h \to \infty$ in $\mathbf{R}$ and $\tilde{c}^{(i)}(x+h)/c(h) \to \alpha^i \exp(\alpha x)$, $i \in \mathbf{N}_0$.

Since $f(x+h)/c(h) \to \exp(\alpha x)$ with strong convergence and the set $\{c(h)\phi(x)/\tilde{c}(x+h); h \ge A\}$ is bounded in $\mathcal{D}(\mathbf{R})$, we can then use Theorem XI, Chapter III in [146] to obtain

$$\lim_{h \to \infty} \langle \frac{f(x+h)}{\tilde{c}(x+h)}, \phi(x) \rangle = \lim_{h \to \infty} \langle \frac{f(x+h)}{c(h)}, \frac{c(h)}{\tilde{c}(c+h)}\phi(x) \rangle$$

$$= \langle 1, \phi(x) \rangle, \quad \phi \in \mathcal{D}(\mathbf{R}).$$

Let $\theta \in C^\infty, \theta(x) = 0$ for $x < 0$ and $\theta(x) = 1$ for $x > 1$. We have

$$\frac{\theta(x+h)f(x+h)}{\tilde{c}(x+h)} \to 1 \quad \text{in} \quad \mathcal{D}'(\mathbf{R}) \text{ as } h \to \infty.$$

Thus, $\{\theta(x+h)f(x+h)/\tilde{c}(x+h); h > 0\}$ is a bounded subset of $\mathcal{D}'(\mathbf{R})$. This implies that this set is bounded in $\mathcal{S}'(\mathbf{R})$ as well (cf. Theorem XXV, Chapter VI in [146]). Since $\mathcal{D}(\mathbf{R})$ is dense in $\mathcal{S}(\mathbf{R})$, by the Banach–Steinhaus theorem, we obtain

$$\frac{\theta(x+h)f(x+h)}{\tilde{c}(x+h)} \to 1 \quad \text{in} \quad \mathcal{S}'(\mathbf{R}) \text{ as } h \to \infty, \text{ i.e.,}$$

$$\lim_{h \to \infty} \left\langle \frac{(\theta f/\tilde{c})(x+h)}{d(h)}, \phi(x) \right\rangle = \langle 1, \phi \rangle, \quad \text{for every } \phi \in \mathcal{S}(\mathbf{R}),$$

where $d(h) = 1, h > A$. We shall now borrow two theorems about quasi-asymptotics (see Definition 2.2) which will be shown later. Since the S-asymptotics in $\mathcal{S}'_+$ with $\nu > -1$ implies the quasi-asymptotics of $\theta f/\tilde{c}$

(see Theorem 2.47), the structural theorem for the quasi-asymptotics (see Theorem 2.2) implies that there is $m_0 \in \mathbf{N}$ such that for every $m > m_0$ there is $F_m \in C(\mathbf{R})$ such that

$$(\theta f / \tilde{c})(x) = F_m^{(m)}(x), \quad x \in \mathbf{R},$$

and $F_m(x) \sim x^m$ as $x \to \infty$. Thus, we obtain

$$f(x) = \tilde{c}(x) F_m^{(m)}(x), \quad x \in (1, \infty).$$

The Leibnitz formula implies

$$f(x) = \sum_{i=0}^{m} \binom{m}{i} (-1)^i (\tilde{c}^{(i)}(x) F_m(x))^{(m-i)}, \quad x \in (1, \infty).$$

Since

$$\frac{\tilde{c}^{(i)}(x+h)}{c(h)} \to \alpha^i e^{\alpha x}, h \to \infty, x \in \mathbf{R},$$

we obtain

$$\tilde{c}^{(i)}(h) \sim \alpha^i c(h), \quad h \to \infty.$$

This implies the result when $\alpha \neq 0$ and $c(h) = e^{\alpha h} L(e^h), h > 0$.

In case $\alpha = 0$ and $c(h) = h^\nu L(h)$, $h > 0$, $\nu > -1$, the S-asymptotics of θf related to $c(h) = h^\nu L(h), h > 0$, $\nu > -1$, implies the quasi-asymptotics of θf related to the same $c(h)$ (see Theorem 2.47). The assertion follows now from Theorem 2.2.

If $\nu \leq -1$, then we take $k > 0$ such that $k + \nu > -1$. With θ as in the preceding proof and by Theorem 1.2 b), we have

$$(1 + (x+h)^2)^{k/2} \theta(x+h) f(x+h) \overset{s}{\sim} h^{k+\nu} L(h) \cdot 1, \quad h \in \mathbf{R}_+.$$

By the same arguments as in the preceding proof, we have that there is $m_0 \in \mathbf{N}$ such that for every $m > m_0$ there is an $F_m \in C(\mathbf{R})$, $\operatorname{supp} F_m \subset (0, \infty)$

$$F_m(x) \sim x^{\nu+k+m} L(x), \quad x \to \infty$$

$$(1 + x^2)^{k/2} \theta(x) f(x) = F_m^{(m)}(x), \quad x \in \mathbf{R}.$$

Thus, for $x \in (1, \infty)$,

$$f(x) = \sum_{i=0}^{m} (-1)^i \binom{m}{i} \left(\left(\frac{1}{(1+x^2)^{k/2}} \right)^{(i)} F_m(x) \right)^{(m-i)}.$$

The result now follows from the fact

$$\left(\frac{1}{(1+x^2)^{k/2}}\right)^{(i)} \sim C_i x^{-k-i}, x \to \infty,$$

where $C_i \neq 0$ are suitable constants, $i = 0, \ldots, m$. $\square$

We can show a more general result which describes the precise structure of a distribution having S-asymptotics.

Theorem 1.9. *If $T \in \mathcal{D}'$ has S-asymptotics related to the open cone Γ and the continuous and positive function $c(h), h \in \Gamma$, then for the ball $B(0,r)$ there exist continuous functions $F_i, |i| \leq m$, such that $F_i(x+h)/c(h)$ converges uniformly for $x \in B(0,r)$ when $h \in \Gamma, \|h\| \to \infty$, and the restriction of the distribution T on $B(0,r) + \Gamma$ can be given in the form $T = \sum\limits_{|i| \leq m} D^i F_i$.*

We need the following lemma for the proof of the theorem:

Lemma 1.2. *Let $T \in \mathcal{D}', T(x+h) \overset{s}{\sim} c(h) \cdot U(x), h \in \Gamma$. Then, for an open ball $B(0,r)$ and a relatively compact open neighborhood Ω of zero in $\mathbf{R}^n$, there exists an $m \in \mathbf{N}_0$ such that for every $\varphi, \psi \in \mathcal{D}_\Omega^m$ the function $(T*\varphi*\psi)(x)$ is continuous for $x \in B(0,r) + \Gamma$. Moreover, the set of functions $\{(T_h*\varphi*\psi)(x); h \in \Gamma\}$ converges uniformly for $x \in B(0,r)$ to $(U*\varphi*\psi)(x)$, as $h \in \Gamma, \|h\| \to \infty; T_h = T(x+h)/c(h)$.*

Proof. Suppose that T has S-asymptotics related to $c(h)$. Then, the set $\{T(x+h)/c(h) \equiv T_h; h \in \Gamma\}$ is weakly bounded in $\mathcal{D}'$ and consequently, bounded in $\mathcal{D}'$. A necessary and sufficient condition that a set $\mathcal{B}' \subset \mathcal{D}'$ is bounded in $\mathcal{D}'$ is: for every $\alpha \in \mathcal{D}$ the set of functions $\{T * \alpha; T \in \mathcal{B}'\}$ is bounded on every compact set $K \subset \mathbf{R}^n$ (see §7, Chapter VI in [146]). Moreover $\{T * \alpha; T \in \mathcal{B}'\}$ defines a bounded set of regular distributions.

Let $Cl(\Omega) = K$ ($Cl(\Omega)$ is the closure of Ω); K is a compact set. For a fixed $\alpha \in \mathbf{C}_0^\infty, \operatorname{supp} \alpha \subset K$, the linear mappings $\beta \mapsto (T_h * \alpha) * \beta, \, h \in \Gamma$, are continuous mappings of $\mathcal{D}_K$ into $\mathcal{E}$ because of the separate continuity of the convolution. Since the set $\{T_h * \alpha; h \in \Gamma\}$ is a bounded set in $\mathcal{D}'$, for every ball $B(0,r)$ the set of mappings $\beta \to \{(T_h * \alpha) * \beta; h \in \Gamma\}$ is the set of equicontinuous mappings of $\mathcal{D}_K$ into L_B^∞, where $B = B(0,r)$. Now there exists an $m \geq \mathbf{N}_0$ such that the linear mappings $(\alpha, \beta) \to T_h * \alpha * \beta$ which map $\mathcal{D}_K \times \mathcal{D}_K$ into L_B^∞ can be extended to $\mathcal{D}_\Omega^m \times \mathcal{D}_\Omega^m$ in such a way that $(\alpha, \beta) \mapsto T_h * \alpha * \beta, h \in \Gamma$, are equicontinuous mappings of $\mathcal{D}_\Omega^m \times \mathcal{D}_\Omega^m$ into L_B^∞ (see for example the proof of Theorem XXII, Chapter VI in [146]).

We proved that for every $\varphi, \psi \in \mathcal{D}_\Omega^m$ and every $h \in \Gamma$ the functions $x \mapsto (T_h * \varphi * \psi)(x)$ are continuous in $x \in B(0, r)$. From the relation $(T_h * \varphi * \psi)(x) = (T * \varphi * \psi)(x + h)/c(h)$ and from the properties of c it follows that $y \mapsto (T * \varphi * \psi)(y)$ is a continuous function for $y \in B(0, r) + \Gamma$ and $\varphi, \psi \in \mathcal{D}_\Omega^m$.

It remains to prove that $T_h * \varphi * \phi$ converges to $U * \varphi * \phi$ as $\|h\| \to \infty$, $h \in \Gamma$, in L_B^∞ for $\varphi, \psi \in D_\Omega^m$. We know that $\mathcal{D}$ is a dense subset of $\mathcal{D}^m, m \ge 0$. We can construct a subset A of $\mathcal{D}_K$ to be dense in $\mathcal{D}_\Omega^m$. The set of functions $T_h * \alpha * \beta$ converges in L_B^∞ for $\alpha, \beta \in A$, when $h \in \Gamma, \|h\| \to \infty$. Taking care of the equicontinuity of the mappings $\mathcal{D}_\Omega^m \times \mathcal{D}_\Omega^m$ into L_B^∞, defined by $T_h * \varphi * \psi$, we can use the Banach–Steinhaus theorem to prove that $T_h * \varphi * \psi$ converges in L_B^∞ when $h \in \Gamma, \|h\| \to \infty$.

Proof of the Theorem. We shall use (VI, 6; 23) from [146].

$$\Delta^{2k} * (\gamma E * \gamma E * T) - 2\Delta^k * (\gamma E * \xi * T) + (\xi * \xi * T) = T, \qquad (1.20)$$

where E is a solution of the iterated Laplace equation $\Delta^k E = \delta; \gamma, \xi \in \mathcal{D}_\Omega$. We have only to choose the natural number k large enough so that E belongs to $\mathcal{D}_\Omega^m$. Now, it is possible to take $F_1 = \gamma E * \gamma E * T, F_2 = \gamma E * \xi * T$ and $F_3 = \xi * \xi * T$. All of these functions are of the form $F_i = T * \varphi_i * \psi, \varphi_i, \psi, \in \mathcal{D}_\Omega^m, i = 1, 2, 3$.

The following holds:

$$F_i(x + h)/c(h) = (F_i(x)/c(h)) * \tau_{-h} = (T * \varphi_i * \psi_i,)(x) * \tau_{-h}/c(h)$$
$$= ((T(x) * \tau_{-h})/c(h) * (\varphi_i * \psi_i) = (T_h * \varphi_i * \psi_i)(x), \quad x \in B(0, r), \ h \in \Gamma.$$

Hence, by Lemma 1.2 it follows that $F_i(x + h)/c(h)$ converges uniformly for $x \in B(0, r)$ when $h \in \Gamma, \|h\| \to \infty$. $\qquad\square$

Consequences of Theorem 1.9

a) If the functions $F_i, |i| \le m$, have the property given in Theorem 1.9 and if $\Gamma = \mathbf{R}^n$, then the regular distributions defined by the functions F_i/c have the S-asymptotic behavior related to $c_1(h) \equiv 1$.

b) If $T \in \mathcal{S}'$, then all functions $F_i, |i| \le m$, are continuous for $x \in B(0, r) + \Gamma$, and of slow growth ($F_i = (1 + r^2)^q f_i, |i| \le m, q \in \mathbf{R}$, where $r = \|x\|$ and $f_i, |i| \le m$, are continuous and bounded functions).

c) The converse of Theorem 1.9 is also true. Therefore, it completely characterizes those distributions having S-asymptotics.

Proof. a) Suppose that $\Gamma = \mathbf{R}^n$. Then by the properties of F_i, $|i| \leq m$, the functions F_i/c are continuous and $F_i(x)/c(x)$ converge to the numbers C_i as $\|x\| \to \infty$. Now, for $|i| \leq m$ we have

$$\lim_{\|h\| \to \infty} \langle F_i(x+h)/c(x+h), \varphi(x) \rangle$$

$$= \lim_{\|h\| \to \infty} \int_{\mathbf{R}^n} (F_i(x+h)/c(x+h))\varphi(x)dx = \langle C_i, \varphi \rangle, \quad \varphi \in \mathcal{D}.$$

b) If $T \in \mathcal{S}'$, then there exists a $q \in \mathbf{R}$ such that the set of distributions $\{T(x+h)/(1+\|h\|^2)^q; h \in \mathbf{R}^n\} = W$ is bounded in $\mathcal{D}'$ (Theorem VI, Chapter VII in [146]). We can now repeat the first part of the proof of Lemma 1.2 but with $c(h) = (1+\|h\|^2)^q$, $h \in \Gamma = \mathbf{R}^n$. In this may we obtain that there exists a $p \in \mathbf{N}_0$ such that for $\varphi, \psi \in \mathcal{D}_\Omega^p$ and $x \in \mathbf{R}^n$ the function $x \mapsto (T*\varphi*\psi)(x)$ is continuous and $(T*\varphi*\psi)(x)/(1+\|x\|^2)^q$, $x \in \mathbf{R}$ is bounded. It remains only to choose the number k in (1.20) large enough so that $\gamma E \in \mathcal{D}_\Omega^{\max(m,p)}$.

c) It follows directly from Theorem 1.2 a) and Proposition 1.5. $\qquad \square$

We can also characterize the structure of ultradistributions having S-asymptotics.

Theorem 1.10. ([132]). *Let $T \in \mathcal{D}_\Omega'^*$. Suppose: (M.1), (M.2) and (M.3) are satisfied by M_p; $\Omega = \Omega_1 + \Gamma_1$, where $\Omega_1 \subset \mathbf{R}^n$ is open set and $\Gamma_1 \subset \mathbf{R}^n$ is convex cone; Γ is a subcone of Γ_1. Then T has the S-asymptotics related to c and Γ if and only if for a given open and relatively compact set $A(\overline{A} \subset \Omega)$, there exist an ultradifferential operator $P(D)$ of class $*$ and continuous functions f_1 and f_2 on $A + \Gamma$ such that*

$$\lim_{h \in \Gamma, \|h\| \to \infty} f_i(x+h)/c(h), \quad i = 1, 2,$$

exist uniformly in $x \in A$, $i = 1, 2$, and $T = P(D)f_1 + f_2$ on $A + \Gamma$.

Proof. One can easily prove that the condition is sufficient.

The condition is necessary. Suppose that $T(x+h) \overset{s}{\sim} c(h) \cdot U(x)$, $h \in \Gamma$ in $\mathcal{D}_\Omega'^*$. Denote by $T_h = \tau_h T/c(h)$ and by $(CB)_A$ the space of continuous and bounded functions on A. By Theorem 6.10 in [79], $F_h : \varphi \mapsto T_h * \varphi$, $h \in \Gamma$, are continuous mappings: $\mathcal{D}_{B_{r+\varepsilon}}^* \to (CB)_A$. Consequently, F_h are the continuous mappings: $\mathcal{D}_K^* \to (CB)_A$, where $K = \overline{B}_r \subset \Omega$. The set $\{T_h; h \in \Gamma, \|h\| \geq \gamma\}$, $\gamma > 0$, is bounded in $\mathcal{D}_\Omega'^*$ because of the S-asymptotics of T.

Thus, for a fixed $\varphi \in \mathcal{D}_K^*$, the set $\{F_h(\varphi);\ h \in \Gamma,\ \|h\| \geq \gamma\}$ is bounded in $(CB)_A$. Since $\mathcal{D}_K^*$ is a barrelled space, by the Banach theorem it follows that the family of functions $\{F_h\} = \{F_h;\ h \in \Gamma,\ \|h\| \geq \gamma\}$ is equicontinuous. Therefore, there exists H_p, $p \in \mathbf{N}$ of the form (0.1) and $\beta > 0$ such that any function in $\{F_h\}$ maps the neighborhood of zero

$$V_\beta = \left\{ \phi \in \mathcal{D}_K^*;\ \sup_{x \in K, |\alpha| \in \mathbf{N}_0} |\phi^{(\alpha)}(x)| / H_{|\alpha|} M_{|\alpha|} < \beta \right\} \tag{1.21}$$

into the unit ball $B(0,1)$ in $(CB)_A$. Denote by $\tilde{\mathcal{D}}_K^{H_p M_p}$ the completion of $\mathcal{D}_K^*$ under the norm $q_{H_p M_p}$ (see (0.5)).

Using the extension of a function through its continuity (see [12]), we shall show that the family $\{F_h\}$ can be extended on $\tilde{\mathcal{D}}_K^{H_p M_p}$, keeping the uniform continuity; let us denote it by $\{\overline{F}_h\}$.

Let $\phi \in \tilde{\mathcal{D}}_K^{H_p M_p}$ and let $\langle \phi_j \rangle_j$ be a sequence in $\mathcal{D}_K^*$ which converges to ϕ being a Cauchy sequence in the norm $q_{H_p M_p}$. We shall prove that $T_h * \psi_j$ converges, as $j \to \infty$, in $(CB)_A$ uniformly in $h \in \Gamma$, $\|h\| \geq \gamma$. It is enough to prove that $\langle T_h * \psi_j \rangle_j$ is a Cauchy sequence in $(CB)_A$.

Every neighborhood of zero W in $(CB)_A$ contains the ball $B(0, \delta)$ for some $\delta > 0$. The neighborhood of zero $V_{\beta\delta}$, given by (1.21), satisfies $T_h * V_{\beta\delta} \subset W$, when $h \in \Gamma$, $\|h\| \geq \gamma$.

Let $j_0 \in \mathbf{N}_0$ be such that $\psi_i - \psi_j \in V_{\beta\delta}$ if $i, j \geq j_0$. Then,

$$T_h * \psi_i - T_h * \psi_j = T_h * (\psi_i - \psi_j) \in W, \quad h \in \Gamma, \ \|h\| \geq \gamma.$$

This proves the existence of the family $\{\overline{F}_h\}$ and that $\lim_{j \to \infty} T_h * \psi_j = g_h \in (CB)_A$, $\|h\| \geq \gamma$. We shall prove that $g_h = T_h * \phi$. The sequence $\langle \psi_j \rangle_j$ converges to ϕ also in $\mathcal{E}'^*_{B_{r+\varepsilon}}$, as well. So, $T_h * \psi_j$ converges to $T_h * \phi$ as $j \to \infty$, in $\mathcal{D}'^*_\omega$, where $\Omega = \omega - B_{r+\varepsilon}$ (see [79], p. 73). Thus, $(T_h * \phi)|_A$ must be g_h.

It remains to prove that for $\phi \in \tilde{\mathcal{D}}_K^{H_p M_p}$, $T_h * \phi$ converges in $(CB)_A$ as $h \in \Gamma$, $\|h\| \to \infty$. Since $\{\overline{F}_h\}$ is an equicontinuous family of functions, for every $\phi \in \tilde{\mathcal{D}}_K^{H_p M_p}$ the set $\{T_h * \phi;\ h \in \Gamma,\ \|h\| \geq \gamma\}$ is bounded in $(CB)_A$. By the Banach–Steinhaus theorem $T_h * \phi$ converges in $(CB)_A$.

Now we will give the analytic form of T. Let $\phi \in \mathcal{D}_Q$ such that Q is a compact subset of the interior of K and $q_{H_p M_p}(\phi) < \infty$. One can easily prove that $\phi \in \tilde{\mathcal{D}}_K^{H_p M_p}$. Now, by Theorem 2.11 in [81], $T = P(D)(\phi * T) - w * T$, where we can take that $\phi \in \tilde{\mathcal{D}}_K^{H_p M_p}$ and $w \in \mathcal{D}_K^*$. Let us denote $\phi * T$

by f_1 and $w * T$ by f_2. The properties of these two functions follow by the previous part of the proof.   □

We pass now to the case of Fourier hyperfunctions.

Theorem 1.11. *Let* $f = [F'] \in \mathcal{Q}(\mathbf{D}^n), F' \in \tilde{\mathcal{O}}((\mathbf{D}^n + iI \# \mathbf{D}^n)$. *If* f *has the S-asymptotics related to* c *and* Γ, *then there exist an elliptic local operator* $J(D)$ *and functions* $q_s \in C^\infty(\mathbf{R}^n)$, $s \in \Lambda$, *is the set of n-vectors with entry* $\{-1, 1\})$, *of infra exponential type such that:*

1. $q_s(z) \in \tilde{\mathcal{O}}(\mathbf{D}^n + iI_s)$, *where* $\mathbf{D}^n + iI_s$ *is an infinitesimal wedge of type* $\mathcal{D}^n + i\Gamma_s 0$, $s \in \Lambda$.

2. $f = J(D) \sum_{s \in \Lambda} q_s(x)$.

3. For every $s \in \Lambda$ *there exists* $\varepsilon_0 > 0$ *such that* $q_s(x + i\varepsilon s)/c(x)$, $s \in \Lambda$ *converge as* $\|x\| \to \infty$, $x \in \Gamma$, *for every fixed* ε, $0 < \varepsilon < \varepsilon_0$.

Proof. Let

$$f \cong \sum_{\sigma \in \Lambda} F_\sigma(x + i\Gamma_\sigma), \quad F_\sigma = \mathrm{sgn}\sigma F'_\sigma$$

and

$$\mathcal{F}(f) = [R] \cong \sum_{\sigma \in \Lambda} \sum_{\tilde{\sigma} \in \Lambda} \mathcal{F}(\chi_\sigma F_\sigma)(\xi - i\Gamma_{\tilde{\sigma}} 0).$$

Then there exists a monotone increasing continuous positive valued function $\varepsilon(r)$, $r \geq 0$, which satisfies $\varepsilon(0) = 1$, $\varepsilon(r) \to \infty$, $r \to \infty$ and such that

$$|R(\zeta)| \leq C_k \exp(|\zeta|/\varepsilon(|\zeta|), \frac{1}{k} \leq |\mathrm{Im}\,\zeta_j| \leq 1,$$

where $j = 1, \ldots, n$ and $k \in \mathbf{N}$ (cf. [74], p. 652).

By Lemma 1.2 in [73], we can choose an entire function J of infra exponential type on $\mathbf{C}^n$ which satisfies the estimate:

$$|J(\zeta)| \geq C \exp(|\zeta|/\varepsilon(|\zeta|)), \quad |\mathrm{Im}\,\zeta| \leq 1.$$

Then $J^{-2}(\zeta) \in \tilde{\mathcal{O}}(\mathbf{D}^n + i\{|\mu| < 1\})$ and

$$|R(\zeta)/J^2(\zeta)| \leq C_k \exp(-|\zeta|/\varepsilon(|\zeta|)) \leq C_k \exp(-|\zeta|^\gamma),$$

where $\frac{1}{k} \leq |\mathrm{Im}\,\zeta_j| \leq 1$, $j = 1, \ldots, n$, and $0 < \gamma < 1$.

Denote by $g = \mathcal{F}^{-1}(1/J^2)$. By Theorem 8.2.6 in [75], $g \in \mathcal{Q}^{-1}(\mathbf{D}^n)$. By Proposition 8.4.3. in [75]

$$f * g = \int_{\mathbf{R}} f(\cdot - t)g(t)dt \in \mathcal{Q}(\mathbf{D}^n)$$

and

$$\mathcal{F}(f * g) = \mathcal{F}(f)\mathcal{F}(g).$$

By the Corollaries of Proposition 2 in [165]

$$f = J_0(D)(g * f), \quad J_0 = J^2. \tag{1.22}$$

We can always assume that there exists $\alpha \in \mathbf{R}_+$ such that $I = (-\alpha, \alpha)^n$. Then we denote by $I_\sigma = I \cap \Gamma_\sigma$, $\sigma \in \Lambda$.

By Proposition 8.3.2 in [75] the following assertions are true:

$F_\sigma \chi_{\tilde{\sigma}} \in \tilde{\mathcal{O}}(x + iI_\sigma)$ and it decreases exponentially outside any cone containing $\overline{\Gamma}_{\tilde{\sigma}}$ as a proper subcone.

$\mathcal{F}(F_\sigma \chi_{\tilde{\sigma}}) \in \tilde{\mathcal{O}}(x - iI_\sigma)$ and it decreases exponentially outside any cone containing $\overline{\Gamma}_{\tilde{\sigma}}$ as a proper subcone.

$\mathcal{F}(F_\sigma \chi_{\tilde{\sigma}})J^{-2}$ has the same cited properties as $\mathcal{F}(F_\sigma \chi_{\tilde{\sigma}})$.

$\mathcal{F}(F_\sigma \chi_{\tilde{\sigma}})J^{-2}\chi_s \in \tilde{\mathcal{O}}(x - iI_s)$ and it decreases exponentially outside any cone containing $\overline{\Gamma}_\sigma$ and $\overline{\Gamma}_s$ proper subcone.

$\mathcal{F}^{-1}(\mathcal{F}(F_\sigma \chi_{\tilde{\sigma}})J^{-2}\chi_s) \in \tilde{\mathcal{O}}(x + i(I_\sigma \cup I_s))$ and it decreases exponentially outside any cone containing $\overline{\Gamma}_s$ as a proper subcone.

Consider now the Fourier hyperfunction $f * g$ given in (1.22):

$$f * g = \mathcal{F}^{-1}(\mathcal{F}(f)\mathcal{F}(g))$$
$$\cong \frac{1}{(2\pi)^n} \sum_{\sigma \in \Lambda} \sum_{\tilde{\sigma} \in \Lambda} \int_{\mathbf{R}^n} e^{iz_\sigma \zeta_\sigma} \mathcal{F}(\chi_{\tilde{\sigma}} F_\sigma)(\zeta_{\tilde{\sigma}})/J^2(\zeta_{\tilde{\sigma}})d\xi,$$

where $\eta_{\tilde{\sigma}} \in -I_{\tilde{\sigma}}$ and $z_\sigma \in \mathbf{R}^n + iI_\sigma$.

Let $\sigma \in \Lambda$ be fixed. Then for $\tilde{\sigma} \in \Lambda$ and $z_\sigma \in \mathbf{R}^n + iI_\sigma$, we have

$$S_{\sigma,\tilde{\sigma}}(z_\sigma) = \frac{1}{(2\pi)^n} \int_{\mathbf{R}^n} e^{iz_\sigma \zeta_{\tilde{\sigma}}} \mathcal{F}(\chi_{\tilde{\sigma}} F_\sigma)(\zeta_{\tilde{\sigma}})/J^2(\zeta_{\tilde{\sigma}})d\xi;$$

$$|S_{\sigma,\tilde{\sigma}}(z_\sigma)| \leq \frac{1}{(2\pi)^n} \int_{\mathbf{R}^n} e^{-x\eta_{\tilde{\sigma}} - y_\sigma \xi} \mathcal{F}(\chi_{\tilde{\sigma}} F_\sigma)(\zeta_{\tilde{\sigma}})/J^2(\zeta_{\tilde{\sigma}})d\xi.$$

One can see that $S_{\sigma,\tilde{\sigma}}(z_\sigma), z_\sigma \in \mathbf{R}^n + I_\sigma$ are continuable to the real axis. The corresponding functions $x \mapsto S_{\sigma,\tilde{\sigma}}(x)$ are continuous and of infra exponential type on $\mathbf{R}^n$. By Lemma 8.4.7 in [75], for $x \in \mathbf{R}^n$,

$$S_{\sigma,\tilde{\sigma}}(x) \cong S_{\sigma\tilde{\sigma}}(x + i\Gamma_\sigma 0), \quad \tilde{\sigma} \in \Lambda,$$

$$(f * g)(x) = \sum_{\sigma \in \Lambda} \sum_{\tilde{\sigma} \in \Lambda} S_{\sigma,\tilde{\sigma}}(x), \quad x \in \mathbf{R}^n. \tag{1.23}$$

Functions $S_{\sigma,\tilde{\sigma}}$ can be written in the form

$$S_{\sigma,\tilde{\sigma}}(z_\sigma) = \frac{1}{(2\pi)^n} \sum_{\sigma \in \Lambda} \int_{\mathbf{R}^n} e^{iz_\sigma \zeta_{\tilde{\sigma}}} \mathcal{F}(\chi_\sigma F_\sigma)(\zeta_{\tilde{\sigma}})/J^2(\zeta_{\tilde{\sigma}})\chi_s(\zeta_{\tilde{\sigma}})d\xi, \ z_\sigma \in \mathbf{R}^n + I_\sigma.$$

Let $s \in \Lambda$. Then functions

$$S_{\sigma,\tilde{\sigma},s}(z_\sigma) = \frac{1}{(2\pi)^n} \int_{\mathbf{R}^n} e^{iz_\sigma \zeta_{\tilde{\sigma}}} \mathcal{F}(\chi_{\tilde{\sigma}} F_\sigma)(\zeta_{\tilde{\sigma}})/J^2(\zeta_{\tilde{\sigma}})\chi_s(\zeta_{\tilde{\sigma}})d\xi,$$

$z_\sigma \in \mathbf{R}^n + iI_\sigma$, $\sigma, \tilde{\sigma}, s \in \Lambda$, are also continuable to the real axis. The corresponding functions $x \to S_{\sigma,\tilde{\sigma},s}(x)$ are continuous and of infra exponential type on $\mathbf{R}^n$. Moreover, on $\mathbf{R}^n$,

$$S_{\sigma,\tilde{\sigma},s}(x) \cong S_{\sigma,\tilde{\sigma},s}(x + i\Gamma_\sigma 0),$$

$$S_{\sigma,\tilde{\sigma}}(x) = \sum_{s \in \Lambda} S_{\sigma,\tilde{\sigma},s}(x). \tag{1.24}$$

Let us analyze functions

$$I_{s\varepsilon}(\zeta) = J^{-2}(\zeta)e^{-\varepsilon s\zeta}\chi_s(\zeta), \quad \zeta \in (\mathbf{R}^n + i\{|\eta| < 1\}),$$

where $s \in \Lambda$ and $\varepsilon > 0$. These functions are elements of $\mathcal{P}_*$ because of

$$|I_{s,\varepsilon}(\zeta)| = |J^{-2}(\zeta)|\exp\left(-\varepsilon \sum_{i=1}^n s_i\zeta_i\right)\prod_{i=1}^n|\chi_{s_i}(\zeta_i)|$$

$$\leq |J^{-2}(\zeta)|\prod_{i=1}^n|\chi_{s_i}(\zeta_i)\exp(-\varepsilon s_i\zeta_i)$$

$$\leq C\exp(-\varepsilon \sum_{i=1}^n|\xi_i|), \quad |\eta| < 1.$$

Therefore, $I_{s,\varepsilon} \in \tilde{\mathcal{O}}^{-\varepsilon}(\mathbf{D}^n + i\{|\eta| < 1\})$, $s \in \Lambda$. Since the Fourier transform maps $\mathcal{P}_*$ onto $\mathcal{P}_*$, there exists $\psi_{s,\varepsilon} \in \mathcal{P}_*$ such that $\mathcal{F}(\psi_{s,\varepsilon}) = I_{s,\varepsilon}$. By Proposition 8.2.2 in [75],

$$\psi_{s,\varepsilon} \in \tilde{\mathcal{O}}^{-1}(\mathbf{D}^n + i\{|y| < \varepsilon\}), \quad s \in \Lambda.$$

Denote by

$$q_s(x) = \sum_{\sigma \in \Lambda} \sum_{\tilde{\sigma} \in \Lambda} S_{\sigma,\tilde{\sigma},s}(x)$$

$$\cong \sum_{\sigma \in \Lambda} \sum_{\tilde{\sigma} \in \Lambda} \mathcal{F}^{-1}(\mathcal{F}(F_\sigma \chi_{\tilde{\sigma}}) J^{-2} \chi_{\tilde{\sigma}})(x + i(\Gamma_\sigma \cup \Gamma_s)0), x \in \mathbf{R}^n. \quad (1.25)$$

We prove that functions q_s, $s \in \Lambda$, have the properties cited in Theorem 1.11.

Property 1 follows from (1.24) and (1.25).

By (1.22) and (1.23) property 2 is satisfied.

It remains to prove property 3.

If $f \in \mathcal{Q}(\mathbf{D}^n)$ and $\varphi \in \mathcal{P}_*$, then $f * \varphi \in \tilde{\mathcal{O}}(\mathbf{D}^n + iI')$, where I' is an interval containing zero. We shall use this fact and the properties of functions $I_{s,\varepsilon}$, already analyzed.

For a fixed $s \in \Lambda$ there exists $\varepsilon_0 > 0$ such that εs belongs to all infinitesimal wedges of the form $\mathbf{D}^n + i(\Gamma_\sigma \cup \Gamma_s)0$ which appear in (1.25). For $\varepsilon \in (0, \varepsilon_0]$, we have

$$q_s(x + i\varepsilon s) = \sum_{\sigma \in \Lambda} \sum_{\tilde{\sigma} \in \Lambda} \frac{1}{(2\pi)^n} \int_{\mathbf{R}^n} e^{i(x+i\varepsilon s)\zeta_{\tilde{\sigma}}} \mathcal{F}(F_\sigma \chi_{\tilde{\sigma}})(\zeta_{\tilde{\sigma}}) J^{-2}(\zeta_{\tilde{\sigma}}) \chi_{si}(\zeta_{\tilde{\sigma}}) d\xi$$

$$= \sum_{\sigma \in \Lambda} \sum_{\tilde{\sigma} \in \Lambda} \frac{1}{(2\pi)^n} \int_{\mathbf{R}^n} e^{ix\zeta_{\tilde{\sigma}}} \mathcal{F}(F_\sigma \chi_{\tilde{\sigma}})(\zeta_{\tilde{\sigma}}) \mathcal{F}(\psi_{s,\varepsilon})(\zeta_{\tilde{\sigma}}) d\xi$$

$$= \sum_{\sigma \in \Lambda} \sum_{\tilde{\sigma} \in \Lambda} ((F_\sigma \chi_{\tilde{\sigma}}) * \psi_{s,\varepsilon})(x)$$

$$= \left(\left(\sum_{\sigma \in \Lambda} F_\sigma \right) * \psi_{s,\varepsilon} \right)(x) = (f * \psi_{s,\varepsilon})(x).$$

Now, for every fixed $\varepsilon \in (0, \varepsilon_0]$, and $s \in \Lambda$,

$$\lim_{x \in \Gamma, \|x\| \to \infty} q_s(x + i\varepsilon s)/c(x) = \lim_{x \in \Gamma, \|x\| \to \infty} (f * \psi_{s,\varepsilon})(x)/c(x)$$

$$= \lim_{x \in \Gamma, \|x\| \to \infty} \langle f(t+x)/c(x), \psi_{s,\varepsilon}(t) \rangle. \qquad \square$$

We cite some papers related to this problem: ([131], [132], [155],[151], [123]).

1.10 *S-asymptotic expansions in* $\mathcal{F}'_g$

A sequence $\langle \psi_n \rangle_n$ of positive real-valued functions $\psi_n(t)$, $t \in (t_0, \infty)$, $t_0 \geq 0$ (defined on $(0, t_0)$, $t_0 > 0$) is said to be *asymptotic* if and only if $\psi_{n+1}(t) = o(\psi_n(t))$, $t \to \infty$ ($t \to 0$). The formal series $\sum_{n \geq 1} u_n(t)$ is an asymptotic expansion of the function u related to the asymptotic sequence $\{\psi_p(t)\}$ if

$$u(t) - \sum_{n=1}^{k} u_n(t) = o(\psi_k(t)), \quad t \to \infty \ (t \to 0) \tag{1.26}$$

for every $k \in \mathbf{N}$. We write in this case:

$$u(t) \sim \sum_{n=1}^{\infty} u_n(t) | \{\psi_n(t)\}, \quad t \to \infty \ (t \to 0). \tag{1.27}$$

If $u_n(t) = c_n \psi_n(t)$, for every $n \in \mathbf{N}$, where c_n are complex numbers, then expansion (1.27) is unique. Indeed, the numbers c_n can be unambiguously computed from (1.26). In this case, we omit from the notation $\{\psi_n(t)\}$ in (1.27). A series which is an asymptotic expansion of a function f can be also convergent. However, the series is divergent in general; nevertheless, several terms of it can give valuable information, and very often its good approximation properties come actually from the fact that the series is divergent. Sometimes, if we take more terms from the asymptotic series, we obtain a worse approximation; consequently, the determination of the optimal number of terms for a good approximation depends on a careful analysis of the problem under consideration.

An asymptotic expansion does not determine only one function. The following example illustrates this fact:

$$\exp\left(\frac{1}{x}\right) \sim \sum_{k=0}^{\infty} \frac{x^{-k}}{k!} \quad \text{and} \quad \exp\left(\frac{1}{x}\right) + \exp(-x^2) \sim \sum_{k=0}^{\infty} \frac{x^{-k}}{k!}, \quad x \to \infty.$$

In many problems of applied mathematics one is led to the use of asymptotic series. (See [44], [10], [15], [14], [192], [69]). A clear exposition of the theory and the use of asymptotic series of functions and distributions can be found in ([50]–[56]).

We shall discuss in this section the S-asymptotic expansion of generalized functions belonging to $\mathcal{F}'_g$.

1.10.1 *General definitions and assertions*

In this section Γ will be a convex cone with the vertex at zero belonging to $\mathbf{R}^d$ and $\Sigma(\Gamma)$ the set of all real-valued and positive functions $c(h), h \in \Gamma$. We shall consider the asymptotic expansion when $\|h\| \to \infty, \ h \in \Gamma$.

Definition 1.2. A distribution $T \in \mathcal{F}'_g$ has a S-asymptotic expansion related to the asymptotic sequence $\langle c_n(h) \rangle_n \subset \Sigma(\Gamma)$, if for every $\varphi \in \mathcal{F}$

$$\langle T(t+h), \varphi(t) \rangle \sim \sum_{n=1}^{\infty} \langle U_n(t,h), \varphi(t) \rangle | \{c_n(h)\}, \|h\| \to \infty, \ h \in \Gamma, \quad (1.28)$$

where $U_n(t,h) \in \mathcal{F}'_g$ (with respect to t) for $n \in \mathbf{N}$ and $h \in \Gamma$. We write in short:

$$T(t+h) \overset{s}{\sim} \sum_{n=1}^{\infty} U_n(t,h) | \{c_n(h)\}, \|h\| \to \infty, \ h \in \Gamma. \quad (1.29)$$

Remarks. 1) In the special case $U_n(t,h) = u_n(t)c_n(h), \ u_n \in \mathcal{F}'_g, \ n \in \mathbf{N}$, we simply write

$$T(t+h) \overset{s}{\sim} \sum_{n=1}^{\infty} u_n(t)c_n(h), \ \|h\| \to \infty, \ h \in \Gamma. \quad (1.30)$$

In this case the given S-asymptotic expansion is unique.

2) Brychkov's general definition is in $\mathcal{S}'(\mathbf{R})$ and slightly different from ours ([15], [16] and [20]); his idea reformulated in $\mathcal{F}'_g(\mathbf{R})$ gives the following definition. Suppose that $f \in \mathcal{F}'_g(\mathbf{R})$ and that the function $\exp(ixh)$, where h is a real parameter, is a multiplier in $\mathcal{F}'_g(\mathbf{R})$.

Definition 1.3. Suppose that $f \in \mathcal{F}'_g(\mathbf{R})$. It is said that $f(x)e^{ixh}$ has an asymptotic expansion related to the asymptotic sequence $\langle \psi_n(h) \rangle_n$ if for every $\varphi \in \mathcal{F}_g(\mathbf{R})$

$$\langle f(x)e^{ixh}, \varphi(x) \rangle \sim \sum_{n=1}^{\infty} \langle C_n(x,h), \varphi(x) \rangle | \{\psi_n(h)\}, \ h \to \infty,$$

where $C_n(x,h) \in \mathcal{F}'_g(\mathbf{R})$ (with respect to x), $n \in \mathbf{N}, \ h \geq h_0$. We write in short:

$$f(x)e^{ixh} \sim \sum_{n=1}^{\infty} C_n(x,h) | \{\psi_n(h)\}, \ h \to \infty.$$

To obtain an equivalent definition of this asymptotic expansion, Brychkov has supposed that $\mathcal{F}'_g(\mathbf{R}) = \mathcal{S}'(\mathbf{R})$ and $\mathcal{F}_g(\mathbf{R}) = \mathcal{S}(\mathbf{R})$. Then, by putting $g = \hat{f} = \mathcal{F}(f)$, $c_n(\cdot, t) = \mathcal{F}(C_n(\cdot, t))$ and $\phi = \mathcal{F}(\varphi)$, Definition 1.3 reduces to:

Definition 1.4. A distribution $g \in \mathcal{S}'(\mathbf{R})$ has an asymptotic expansion related to the sequence $\langle \psi_n(h) \rangle_n$ if for every $\phi \in \mathcal{S}(\mathbf{R})$

$$\langle g(h-t), \phi(t) \rangle \sim \sum_{n=1}^{\infty} \langle c_n(t,h), \phi(t) \rangle |\{\psi_n(h)\}, h \to \infty,$$

where $c_n(\cdot, h) \in \mathcal{S}'(\mathbf{R})$, $n \in \mathbf{N}$ and $h \geq h_0$.

Definition 1.4 is, of course, a particular case of Definition 1.2 if we use $T(t) = g(-t)$ and the cone $\Gamma = \mathbf{R}_-$.

In [15] and [20] authors studied asymptotic expansions of tempered distributions given by Definition 1.3. In [16] Brychkov extended Definition 1.4 to the n-dimensional case, but only on a ray $\{\varepsilon y; \varepsilon > 0\}$ for a fixed $y \in \mathbf{R}^n$.

We study in this section the asymptotic expansion not only in $\mathcal{S}'(\mathbf{R})$ and not only on a ray, but on a cone in $\mathbf{R}^n$. The next remarks state some motivations for such investigations.

Remarks. 1) A distribution in $\mathcal{S}'(\mathbf{R})$ can have an S-asymptotic expansion in $\mathcal{D}'(\mathbf{R})$ without having the same S-asymptotic expansion in $\mathcal{S}'(\mathbf{R})$. Such an example is the regular distribution f defined by

$$f(t) = H(t) \exp(1/(1+t^2)) \exp(-t), \quad t \in \mathbf{R},$$

where H is the Heaviside function.

It is easy to prove that for $h \in \mathbf{R}_+$

$$f(t+h) \overset{s}{\sim} \sum_{n=1}^{\infty} \frac{1}{(n-1)!}(1+(t+h)^2)^{1-n} \exp(-t-h) |\{e^{-h}h^{2(1-n)}\}, h \to \infty,$$

while

$$t \mapsto U_n(t,h) = (1+(t+h)^2)^{1-n} \exp(-t-h), \quad n \in \mathbf{N}, h > 0$$

do not belong to $\mathcal{S}'(\mathbf{R})$.

2) The regular distribution g defined by

$$g(t) = \exp(1/(1+t^2)) \exp(t), \quad t \in \mathbf{R}$$

belongs to $\mathcal{D}'(\mathbf{R})$ but not to $\mathcal{S}'(\mathbf{R})$. It has the S-asymptotic expansion in $\mathcal{D}'(\mathbf{R})$:

$$g(t+h) \overset{s}{\sim} \sum_{n=1}^{\infty} \frac{1}{(n-1)!}(1+(t+h)^2)^{1-n}\exp(t+h)|\{e^h h^{2(1-n)}\}, \ h \to \infty,$$

where $\Gamma = \mathbf{R}_+$.

3) We can distinguish two cases of S-asymptotic expansions. If in Definition 1.2,

$$U_n(t,h) = u_n(t)c_n(h), \ \ n \in \mathbf{N},$$

then the *S-asymptotic expansion is* called *of second type*. (See Remark after Definition 1.2). If $U_n(t,h) = u_n(t+h)$, $n \in \mathbf{N}$, then the *S-asymptotic expansion is of first type*.

The following example illustrates the difference between these two types of S-asymptotic expansions.

Let $f(x) = \sqrt{x^2 + x}$, $x > 0$ and $f(x) = 0$, $x \leq 0$. A S-asymptotic expansion of the first type for this distribution is

$$f(x+h) \overset{s}{\sim} \sum_{n=1}^{\infty} \binom{1/2}{n-1}(x+h)^{2-n}|\{h^{2-n}\}, \ h \to \infty,$$

but the sequence $u_n(x) = \binom{1/2}{n-1}x^{2-n}$ cannot give a S-asymptotic expansion of the second type for f.

S-asymptotic expansions have similar properties as those of S-asymptotics.

Theorem 1.12. *Let $T \in \mathcal{F}'_g$ and*

$$T(t+h) \overset{s}{\sim} \sum_{n=1}^{\infty} U_n(t,h)|\{c_n(h)\}, \ \|h\| \to \infty, \ h \in \Gamma.$$

Then:

a) $\qquad T^{(k)}(t+h) \overset{s}{\sim} \sum_{n=1}^{\infty} U_n^{(k)}(t,h)|\{c_n(h)\}, \ \|h\| \to \infty, \ h \in \Gamma.$

b) Let the open set Ω have the property: for every $r > 0$ there exists a β_0 such that the closed ball $B(0,r) = \{x \in \mathbf{R}^n, \|x\| \le r\}$ is in $\{\Omega - h, h \in \Gamma, \|h\| \ge \beta_0\}$. If $T, T_1 \in \mathcal{F}'_0$ and $T_1 = T$ over Ω, then

$$T_1(t + h) \overset{s}{\sim} \sum_{n=1}^{\infty} U_n(t,h)|\{c_n(h)\}, \quad \|h\| \to \infty, \ h \in \Gamma,$$

as well.

*c) Assume additionally that $\mathcal{F}_g$ is a Montel space and that in $\mathcal{F}_g$ the convolution is well defined (see **0.6**) and hypocontinuous. Let $S \in \mathcal{F}'_g$, supp S being compact. Then*

$$(S * T)(t + h) \overset{s}{\sim} \sum_{n=1}^{\infty} (S * U_n)(t,h)|\{c_n(h)\}, \quad \|h\| \to \infty, \ h \in \Gamma.$$

Proof. We prove only a). The proofs of b) and c) are the same as in the proof of Theorem 1.2. We have

$$\lim_{h \in \Gamma, \|h\| \to \infty} \frac{\left\langle T^{(k)}(t+h) - \sum_{n=1}^{m} U_n^{(k)}(t,h), \varphi(t) \right\rangle}{c_m(h)}$$

$$= \lim_{h \in \Gamma, \|h\| \to \infty} \frac{\left\langle T(t+h) - \sum_{n=1}^{m} U_n(t,h), (-1)^{|k|}\varphi^{(k)}(t) \right\rangle}{c_m(h)} = 0. \qquad \square$$

A relation between the asymptotic expansion of a locally integrable function f and its S-asymptotic expansion when seen as a regular generalized function is provided in the following proposition (see [152]).

Proposition 1.8. *Let $f(t), U_n(t,h)$ and $V_n(t), t \in \mathbf{R}^n, n \in \mathbf{N}, h \in \Gamma$, be locally integrable functions such that for every compact set $K \subset \mathbf{R}^n$ the following ordinary asymptotic expansion holds,*

$$f(t+h) \sim \sum_{n=1}^{\infty} U_n(t,h)|\{c_n(h)\}, \quad \|h\| \to \infty, h \in \Gamma, t \in K$$

and for every $k \in \mathbf{N}$

$$\frac{1}{c_k(h)}\left| f(t+h) - \sum_{n=1}^{k} U_n(t,h) \right| \le V_k(t), \quad t \in K, h \in \Gamma, \|h\| \ge r(k,K).$$

Then for $f \in \mathcal{F}'_0$, we have

$$f(t+h) \overset{s}{\sim} \sum_{n=1}^{\infty} U_n(t,h)|\{c_n(h)\}, \quad \|h\| \to \infty, h \in \Gamma.$$

Proposition 1.9. *Suppose that Γ has the nonempty interior. Let $T \in \mathcal{F}'$ and*

$$T(t+h) \overset{s}{\sim} \sum_{n=1}^{\infty} u_n(t)c_n(h), \ \|h\| \to \infty, \ h \in \Gamma.$$

If $u_m \neq 0$, $m \in \mathbf{N}$, then u_m has the form

$$u_m(t) = \sum_{k=1}^{m} P_k^m(t) \exp(a^k \cdot t), \ t \in \mathbf{R}^n, \ m \in \mathbf{N},$$

where $a^k = (a_1^k, \ldots, a_n^k) \in \mathbf{R}^n$ and P_k^m are polynomials with degrees less than k at every t_i, $i = 1, \ldots, n$.

Proof. Definition 1.2 and the given asymptotics implies

$$\lim_{h \in \Gamma, \|h\| \to \infty} T(t+h)/c_1(h) = u_1(t) \neq 0 \quad \text{in } \mathcal{F}'.$$

Now Proposition 1.2 implies the explicit form of u_1.

The following limit gives u_2:

$$\lim_{h \in \Gamma, \|h\| \to \infty} \frac{\langle T(t+h), \varphi(t) \rangle - \langle u_1(t), \varphi(t) \rangle c_1(h)}{c_2(h)} = \langle u_2, \varphi \rangle, \ \varphi \in \mathcal{F}.$$

Note,

$$\lim_{h \in \Gamma, \|h\| \to \infty} \frac{\langle (D_{t_i} - a_i^1)T(t+h), \varphi(t) \rangle}{c_2(h)} = \langle (D_{t_i} - a_i^1)u_2(t), \varphi(t) \rangle, \ \varphi \in \mathcal{F}.$$

Two cases are possible.

a) If $(D_{t_i} - a_i^1)u_2 = 0$, $i = 1, \ldots, n$, then $u_2(t) = C_2 \exp(a^1 \cdot t)$.

b) If $(D_{t_i} - a_i^1)u_2 \neq 0$ for some i, then by Proposition 1.2, $(D_{t_i} - a_i^1)u_2(t) = c \exp(a^2 \cdot t)$ and u_2 has the form

$$C_2 \exp(a^1 \cdot t) + P_2^2(t_1, \ldots, t_n) \exp(a^2 \cdot t),$$

where P_2^2 is a polynomial of the degree less than 2 with respect to each t_i, $i = 1, \ldots, n$.

In the same way, we prove the assertion for every u_m. $\square$

We will give an example of a function which has S-asymptotic expansion of the first type but does not have the asymptotic expansion as a function:

Let $\psi(t) = 1, t \in (n - 2^{-n}, n + 2^{-n}), n \in \mathbf{N}$, and $\psi(t) = 0$ outside of these intervals. Let

$$\psi_\alpha(x) = e^{\alpha x} \int_0^x \psi(t)dt, \ x \in \mathbf{R}, \ \alpha \in \mathbf{R}.$$

Since $\int_0^x \psi(t)dt \to 2$ as $x \to \infty$, we have that $\psi_\alpha(x) \sim 2e^{\alpha x}, x \to \infty$ but $\psi'_\alpha(x)$ does not have an ordinary asymptotic behavior (see Example **5.** in 1.5.1).

Let $\langle \alpha_j \rangle_j$ be a strictly decreasing sequence of positive numbers. Let θ be a function, $\theta \in C^\infty, \theta \equiv 1$ for $x > 1, \theta \equiv 0$ for $x < 1/2$ and $f(x) = \sum_{i=1}^{\infty} \psi_{\alpha_i}(x)\theta(x-i), x \in \mathbf{R}$. We have

$$f(x) \sim \sum_{i=1}^{\infty} \psi_{\alpha_i}(x)|\{2\exp(\alpha_i x)\}, \ x \to \infty.$$

This implies that f has the S-asymptotic expansion of the first type:

$$f(x+h) \overset{s}{\sim} \sum_{i=1}^{\infty} \psi_{\alpha_i}(x+h)|\{2e^{\alpha_i h}\}, \ h \to \infty$$

and

$$F(x+h) = f'(x+h) \overset{s}{\sim} \sum_{i=1}^{\infty} \psi'_{\alpha_i}(x+h)|\{2e^{\alpha_i h}\}, \ h \to \infty$$

but F does not have the ordinary asymptotic expansion.

For S-asymptotic expansions see also ([108], [152]).

1.10.2   *S-asymptotic Taylor expansion*

Estrada and Kanwal have introduced in [53] and [54] the asymptotic Taylor expansion of distributions (see also [56]). We generalize it to $\mathcal{F}'_g$.

Definition 1.5. Let $f \in \mathcal{F}'_g$ and $D^k f$ be the k-th partial derivative of f. A formal series

$$\sum_{|k|=0}^{\infty} \frac{D^k f(x)y^k}{k!}\varepsilon^{|k|}, \ \text{ where } \ y \in \mathbf{R}^n \text{ is fixed,}$$

is the asymptotic Taylor expansion of f, as $\varepsilon \to 0$.

It means that for any test function $\phi \in \mathcal{F}_g$

$$\langle f(x+\varepsilon y), \phi(x) \rangle - \sum_{|k|=0}^{N} \frac{\langle D^k f(x), \phi(x) \rangle y^k}{k!}\varepsilon^{|k|} = O(\varepsilon^{N+1}), \ \text{ as } \ \varepsilon \to 0.$$

We write in short

$$f(x + \varepsilon y) \sim \sum_{|k|=0}^{\infty} (D^k f(x) y^k) \frac{\varepsilon^{|k|}}{k!}, \quad \varepsilon \to 0. \tag{1.31}$$

In fact, the asymptotic Taylor expansion is the S-asymptotic expansion related to the asymptotic sequence $\{c_n(h) = h^n; \ n \in \mathbf{N}\}$, where $h \to 0$, with $U_n(x, h) = u_n(x) c_n(h)$ and $\Gamma = \{\varepsilon y; \ \varepsilon > 0\}$, where $y \in \mathbf{R}^n$ is fixed (compare with Definition 1.2). We illustrate Definition 1.5 by two examples:

1. $\delta(x + \varepsilon y) \sim \sum\limits_{|\mathbf{N}|=0}^{\infty} \frac{1}{N!} (D^N \delta(x) y^N) \varepsilon^N, \ \text{ as } \ \varepsilon \to 0,$

where $\varepsilon^N = \varepsilon^{N_1 + \cdots + N_n}$.

2. If $\alpha \notin \mathbf{Z}$, then the distribution $x_+^\alpha \in \mathcal{D}'(\mathbf{R})$ has the asymptotic Taylor expansion

$$(x - \varepsilon)_+^\alpha \sim \sum_{k=0}^{\infty} \binom{\alpha}{k} (-1)^k x_+^{\alpha-k} \varepsilon^k, \ \varepsilon \to 0,$$

This means that if $\phi \in \mathcal{D}(\mathbf{R})$, then

$$\text{F.p.} \int_0^\infty (x - \varepsilon)^\alpha \phi(x) dx \sim \sum_{k=0}^{\infty} \left[\binom{\alpha}{k} (-1)^k \text{F.p.} \int_0^\infty x^{\alpha-k} \phi(x) dx \right] \varepsilon^k, \ \varepsilon \to 0,$$

where F.p. stands for the *finite part* (cf. [146] , [56]).

We now discuss the problem of the convergence of an asymptotic Taylor expansion.

If a regular distribution is defined by a real analytic function f, then the asymptotic Taylor expansion for f is a convergent power series in an appropriate domain. But if f is only smooth, then the asymptotic Taylor expansion for f does not necessarily have a domain of convergence. The asymptotic Taylor expansion of δ distribution, given in Example 1, is not convergent in $\mathcal{D}'(\mathbf{R})$; we know that the series $\Sigma a_k \delta^{(k)}$ diverges unless $a_k = 0$, $\|k\| \geq n_0 \in \mathbf{N}$.

One could attempt to solve this problem by trying to interpret the asymptotic Taylor expansion for a distribution in a wider space $\mathcal{F}'_g$ of generalized functions asking the question: *Find a necessary and sufficient condition for the asymptotic Taylor expansion of a generalized function to be convergent in $\mathcal{F}'_g$, i.e., so that it becomes its Taylor series.*

In the following theorems we give the answer to this question if $\mathcal{F}'_g$ is the space of distributions $(\mathcal{F}'_g = \mathcal{D}')$, the space of ultradistributions in which

(M.1), (M.2) and (M.3) are satisfied (see **0.5.2**) $(\mathcal{F}'_g = \mathcal{D}'^*)$ and the space of Fourier hyperfunctions $(\mathcal{F}'_g = \mathcal{Q})$.

Theorem 1.13. ([163]). *Let $f \in \mathcal{D}'$ $(f \in \mathcal{D}'^*)$ and $y = (y_1, \ldots, y_n)$ $y_i \neq 0$, $i = 1, \ldots, n$. The asymptotic Taylor expansion for f, on the straight line $\{\varepsilon y; \varepsilon \in \mathbf{R}\}$, is a convergent Taylor series in $\mathcal{D}'$ (in $\mathcal{D}'^*$) if and only if there exists an $r = (r_1, \ldots, r_n), r_i > 0, i = 1, \ldots, n$, such that f is a real analytic function on $\mathbf{R}^n$ which can be extended as a holomorphic function on $\{z \in \mathbf{C}^n; |\mathrm{Im}\, z_i| < r_i, i = 1, \ldots, n\}$.*

Proof. Suppose that $f \in \mathcal{D}'$ $(f \in \mathcal{D}'^*)$ and that $w \in \mathcal{D}$ $(w \in \mathcal{D}^*)$. Then

$$\langle f(x + \varepsilon y), w(x)\rangle = (f * \breve{w})(\varepsilon y), \quad y \in \mathbf{R}^n, \varepsilon \in \mathbf{R} \tag{1.32}$$

and

$$(f * \breve{w})(\varepsilon y + y_0) = (f(t) * \breve{w}(t - y_0))(\varepsilon y), \quad y, y_0 \in \mathbf{R}^n, \tag{1.33}$$

where $*$ is the sign of convolution and $\breve{w}(t - y_0) = w(-t + y_0)$. This shows that the expansion in a polydisc around y_0 can be transferred to the expansion around 0.

In the sequel we shall use the following results proved in ([146], Chap. VI, Theorem XXIV and [161], Theorem 1):

The generalized function $f \in \mathcal{D}'$ $(f \in \mathcal{D}'^*)$ is a real analytic function if and only if $f * w$ is real analytic for every $w \in \mathcal{D}$ $(w \in \mathcal{D}^*)$.

$\Leftarrow$ Suppose that there exists $r = (r_1, \ldots, r_n)$, $r_i > 0, i = 1, \ldots, n$ such that f is holomorphic in $\{x \in \mathbf{C}^n;\ |\mathrm{Im}\, z_i| < r_i, i = 1, \ldots, n\}$. Then its Taylor series converges in $D(a, r) = D(a_1, r_1) \times \cdots \times D(a_n, r_n)$ for every $a \in \mathbf{R}^n$. Therefore for every $r' = (r'_1, \ldots, r'_n)$, $0 < r'_i < r_i$, $i = 1, \ldots, n$ and every compact set $K \subset \mathbf{R}^n$, by the characterization of a real analytic function,

$$|f^{(k)}(x)| \leq \mathbf{C}(r')^{-k}k!, \quad k \in \mathbf{N}_0^n, \ x \in K.$$

Let us prove that

$$\sum_{|k|=0}^{\infty} \frac{f^{(k)}(x)}{k!}(r'')^k \tag{1.34}$$

converges in $\mathcal{D}'$ (in $\mathcal{D}^{*'}$) for any r'', $0 < r''_i < r'_i$, $i = 1, \ldots, n$. It is enough to prove that for every $w \in \mathcal{D}$ ($w \in \mathcal{D}^*$) and every $p \in \mathbf{N}$

$$\lim_{N \to \infty} \left\langle \sum_{|k|=N}^{N+p} \frac{f^{(k)}(x)}{k!}(r'')^k, w(x) \right\rangle$$

$$= \lim_{N \to \infty} \sum_{|k|=N}^{N+p} \int_{\mathbf{R}^n} (f^{(k)}(x)/(k!))w(x)dx(r'')^k = 0.$$

Since for every $w \in \mathcal{D}$ ($w \in \mathcal{D}^*$)

$$\left| \int_{\mathbf{R}^n} (f^{(k)}(x)/(k!))w(x)dx \right| \le C(r')^{-k} \int_{\mathbf{R}^n} |w(x)|dx, \; |k| \ge n_0(K),$$

the series (1.34) converges for every r'', $0 < r''_i < r'_i$. Consequently, there exists $\varepsilon \ne 0$ such that $0 < |\varepsilon y_i| < r_i$ and that series (1.31) converges in $\mathcal{D}'$ (in $\mathcal{D}'^*$).

$\Rightarrow$ Suppose now that the series (1.31) converges in $\mathcal{D}'$ (in $\mathcal{D}'^*$) for a fixed $y \in \mathbf{R}^n$ and a fixed $\varepsilon > 0$. Then for every $w \in \mathcal{D}$ ($w \in \mathcal{D}^*$)

$$\sum_{|k|=0}^{\infty} \frac{\langle D^k f(x), w(x) \rangle}{k!}(y\varepsilon)^k$$

is a convergent series in the polydisc $D(0, r)$, where $r_i = y_i \varepsilon, i = 1, \ldots, n$. By (1.33), $(f * w)(z)$ is a holomorphic function on $\{z \in \mathbf{C}^n; |\text{Im } z_i| < r_i, i = 1, \ldots, n\}$. By the cited theorem, f is a real analytic function. We now continue the proof dividing it into two cases.

Case $f \in \mathcal{D}'$.

By (VI, 6; 22) in [146]

$$f = \Delta^k(gE * f) - (v * f), \tag{1.35}$$

where Δ is the Laplacian, $v \in \mathcal{D}$, and k is large enough so that for a fixed $m, gE \in \mathcal{D}_V^m$, where V is a relatively compact neighborhood of zero.

By the property of $f * w$ we just proved, it follows that for every $w \in \mathcal{D}$ and for every $y_0 \in \mathbf{R}^n$ there exists $M > 0$ such that

$$(r'')^k|(f^{(k)} * \omega)(y_0)| \le M k!,$$

where $k \in \mathbf{N}_0^n$ and $0 < r'' < r$. By Theorem XXII Chap. VI in [146], there exists $m \ge 0$ such that this inequality is also true if $w = \tilde{w} \in \mathcal{D}_V^m$, where V

is a relatively compact neighborhood of zero. Consequently, it follows that $f * \tilde{w}$ is a holomorphic function in $B(y_0, r)$. By (1.35), f is also holomorphic in $B(y_0, r)$ for every $y_0 \in \mathbf{R}^n$.

Case $f \in \mathcal{D}'^*$. In the proof of this case we use the following Theorem ([81]):

Let the sequence M_p satisfy conditions (M1), (M2), and (M3). For a given H_p and a compact neighborhood Q of zero in $\mathbf{R}^n$ there exist an ultradifferential operator $P(D)$ of class $*$ and functions $\varphi \in C^\infty$, and $w \in \mathcal{D}_Q^*$ such that

$$P(D)\varphi = \delta + w$$

$$\mathrm{supp}\varphi \subset Q, \quad \sup_{x \in Q}|\varphi^{(k)}(x)|/H_{|k|}M_{|k|} \to 0, \quad |k| \to \infty.$$

Following the proof of Theorem 1 in [161], one can conclude that if

$$|(D^k f * w)(y_0)| = |(f * w)^{(k)}(y_0)| \leq Ck!\,(r')^{-k},$$

where $w \in \mathcal{D}^*$ and $k \in \mathbf{N}_0^n$, then there exists H_p such that the same inequality holds for $w = \tilde{w} \in \tilde{\mathcal{D}}_K^{H_p M_p}$, where $\tilde{\mathcal{D}}_K^{H_p M_p}$ is the completion of $\mathcal{D}_K^*$ under the norm $g_{H_p M_p}$. Therefore $f * \varphi$ is a holomorphic function in $B(y_0, r)$.

By the cited theorem

$$f = P(D)(\varphi * f) + f * w, \tag{1.36}$$

where $\varphi * f$ and $f * w$ are holomorphic in a ball around any point of $\mathbf{R}^n$. So the proof will be finished if we prove that for a real analytic function θ and an ultradifferential operator $P(D)$ of (*)-class, $P(D)\theta$ is real analytic.

Therefore, we will prove first the next assertion.

Lemma 1.3. *Let $P(D)$ be an ultradifferential operator of (*)-class and θ be real analytic in a neighborhood of $x_0 \in \mathbf{R}^n$. Then $P(D)\theta$ is real analytic in a neighborhood of x_0.*

Proof. We will prove the assertion in the case $n = 1$ and $P(D)$ being of $\{M_p\}$-class which is equivalent to $P(D) = \Sigma\, a_k D^k$, where $a_k \in \mathbf{C}$, $k \in \mathbf{N}_0$ and there exist $C > 0$ and $h > 0$ such that

$$|a_k| \leq \frac{Ch^k}{M_k}, \quad k \in \mathbf{N}_0.$$

We have to prove that there exists $H > 0$ such that in a ball $B_r = \{x; |x - x_0| \leq r\}$

$$\sup_{\alpha \in \mathbf{N}_0^n, x \in B_r} \frac{H^\alpha |(P(D)\theta)^{(\alpha)}(x)|}{M_\alpha} < \infty.$$

We have $(x \in B_r, \ \alpha \in \mathbf{N}_0)$

$$\frac{1}{M_\alpha} \left| H^\alpha D^\alpha \sum_{k=0}^\infty a_k D^k \theta(x) \right| = \left| \frac{H^\alpha}{M_\alpha} \sum_{k=0}^\infty a_k D^{\alpha+k} \theta(x) \right|$$

$$\leq \left(\frac{CC_1 H^\alpha}{M_\alpha} \sum_{k=0}^\infty \frac{h^k}{M_k} C_1^{k+\alpha} (k+\alpha)! \right),$$

where we have used $|D^k \theta(x)| \leq C_1^{k+1} k!$, $x \in B_r$ which holds for some $C_1 > 0$. Since $(k+\alpha)! \leq e^{k+\alpha} k! \, \alpha!$, we continue

$$\leq CC_1 \sum_{k=0}^\infty e^{\alpha+k} \frac{H^\alpha h^k C_1^{k+\alpha} k!}{M_\alpha M_k} \leq \tilde{C} \sup_\alpha \frac{(C_1 e)^\alpha \alpha! \, H^\alpha}{M_\alpha} \sum_{k=0}^\infty \frac{(C_1 h)^k k!}{M_k}.$$

The right side is bounded for every $H > 0$. This proves the lemma and consequently the Theorem. $\qquad\square$

Theorem 1.14. ([164]) *Let* $q \in \mathcal{Q}$, $\xi = (\xi_1, \ldots, \xi_n)$, $\xi_i > 0$, $i = 1, \ldots, n$ *and* $\varepsilon > 0$. *The asymptotic Taylor expansion (1.31) for* q *on the straight line* $\{h\xi; \ h \in \mathbf{R}\}$ *is a convergent Taylor series in the topology of* $\mathcal{Q}$ *if and only if there exists an* $r = (r_1, \ldots, r_n)$, $r_i > 0$, $i = 1, \ldots, n$, *such that* q *is given by a real analytic function which can be extended as a holomorphic function on* $\{z \in \mathbf{C}^n; \ |\mathrm{Im}\, z_i| < r_i, \ i = 1, \ldots, n\}$.

Proof. In the first part of the proof, we suppose that series (1.31) converges in $\mathcal{Q}$ for a fixed $y = \xi = (\xi_1, \ldots, \xi_n)$, $\xi_i > 0$, $i = 1, \ldots, n$, and a fixed $\varepsilon > 0$. Since $\mathcal{Q}$ is an FS-space it is equivalent to suppose that series (1.31) converges weakly in $\mathcal{Q}$.

By Theorem 1.3 and Remark 1.4 in [73], there exists an elliptic local operator $J_1(D)$ and an infinitely differentiable function g *rapidly decreasing* ($|g(x)| \leq C \exp(-\alpha\|x\|)$, $x \in \mathbf{R}^n$ for some $\alpha > 0$) such that

$$\delta = J_1(D)g \quad (\delta \text{ is the delta distribution}).$$

Also (cf. Theorem in [74]) for every Fourier hyperfunction $q \in \mathcal{Q}$, we can find an elliptic local operator $J_2(D)$ and an infinitely differentiable function f of infra-exponential growth such that $q = J_2(D)f$. Then, by the properties of the convolution:

$$q = (J_2(D)f) * (J_1(D)g) = J_2(D)J_1(D)((\ell f) * (\ell g)), \qquad (1.37)$$

where $J(D) = J_2(D)J_1(D)$ is also an elliptic operator. (ℓf denotes the hyperfunction defined by the function f).

From the proof of two mentioned theorems and by Theorem 8.2.6 in [75] it follows that there exist two sets of functions $\{F_\sigma; \ \sigma \in \Lambda\}$ and $\{G_{\tilde\sigma}; \ \tilde\sigma \in \Lambda\}$ such that:

a) $F_\sigma \in \tilde{\mathcal{O}}(\mathbf{D}^n + iI_\sigma)$ and $G_{\tilde\sigma} \in \tilde{\mathcal{O}}^{-\alpha}(\mathbf{D}^n + iI_{\tilde\sigma})$, $\alpha > 0$, $\sigma, \tilde\sigma \in \Lambda$;

b) the functions F_σ and $G_{\tilde\sigma}$ can be extended to the real axis $\mathbf{R}^n$ as infinitely differentiable functions $F_\sigma(x)$ and $G_{\tilde\sigma}(x)$, respectively. $F_\sigma(x)$ are infra-exponential and $G_{\tilde\sigma}(x)$ are rapidly decreasing, $\sigma, \tilde\sigma \in \Lambda$.

c) $f(x) = \sum\limits_{\sigma \in \Lambda} F_\sigma(x)$ and $g(x) = \sum\limits_{\tilde\sigma \in \Lambda} G_{\tilde\sigma}(x)$, $x \in \mathbf{R}^n$; g is analytic outside $\{0\}$.

By Carlemann's theorem, we have

$$f(x) := \sum_{\sigma \in \Lambda} F_\sigma(x + i\Gamma_\sigma 0), \ g(x) := \sum_{\tilde\sigma \in \Lambda} G_{\tilde\sigma}(x + i\Gamma_{\tilde\sigma}0), \ x \in \mathbf{R}^n$$

(cf. [75] Lemma 8.4.7, and [145], §7).

Our first step is to prove that the Fourier hyperfunction $(\ell f) * (\ell g)$ is defined by the infra-exponential function

$$\int\limits_{\mathbf{R}^n} f(x - u)g(u)du, \ x \in \mathbf{R}^n \,.$$

By the definition of the convolution, we have

$$(lf) * (lg)(u) := \sum_{\sigma \in \Lambda} \sum_{\tilde\sigma \in \Lambda} \int\limits_{Imw = v_{\tilde\sigma}} F_\sigma(z - w)G_{\tilde\sigma}(w)du, \ w = u + iv, \ v_{\tilde\sigma} \in I_{\tilde\sigma} \,,$$

$$(1.38)$$

where $z - w \in \mathbf{R}^n + i\Gamma_\sigma 0$. Hence z can move inside $\mathbf{R}^n + i(\Gamma_\sigma + \Gamma_{\tilde\sigma})0$. We can shift the integral path in the last integral to the real axis $\mathbf{R}^n$. This change of the path is justified by Cauchy's integral formula and the growth rate of the functions f and g at infinity. Then we have

$$(\ell f) * (\ell g)(x) := \sum_\sigma \sum_{\tilde\sigma} \int\limits_{\mathbf{R}^n} F_\sigma(z - u)G_{\tilde\sigma}(u)du \,.$$

Every integral in the last sum is a function into $\tilde{\mathcal{O}}(\mathbf{D}^n + i(\Gamma_\sigma + \Gamma_{\tilde{\sigma}})0)$, respectively, and can be extended to the real axis as a slowly increasing continuous function. Therefore by the same Carlemann's theorem

$$(\ell f) * (\ell g)(x) = \ell \left(\sum_\sigma \sum_{\tilde{\sigma}} \int_{\mathbf{R}^n} F_\sigma(x - u) G_{\tilde{\sigma}}(u) du \right)$$

$$= \ell \left(\int_{\mathbf{R}^n} f(x - u) g(u) du \right). \tag{1.39}$$

Hence, the first step is proved.

In the second step, we shall prove that $(lf) * (lg)(x)$ is a real analytic function in a neighborhood of zero, that is it can be extended on a complex neighborhood of zero as a holomorphic function.

By definition of an infinitesimal wedge U_{σ_1} of type $\mathbf{D}^n + \Gamma_{\sigma_1} 0$ for every proper subcone $\Gamma'_{\sigma_1} \subset\subset \Gamma_{\sigma_1}$ there exists $w > 0$ such that $U_{\sigma_1} \supset \mathbf{D}^n + i(\Gamma'_{\sigma_1} \cap \{|y| < w\})$. Put $\check{G}(u) = G(-u + iw/2)$. Then $\check{G} \in \tilde{\mathcal{O}}^{-\alpha}(\mathbf{D}^n + i\{|v| < w/2\})$ and $\check{G} \in \mathcal{P}_*$, as well.

Let us consider an addend in (1.38) and the corresponding integral in (1.39). We have

$$\ell(F_\sigma * G_{\tilde{\sigma}})(x) := \int_{\mathbf{R}^n} F_\sigma \left(z - \left(u + \frac{1}{2} iv_{\tilde{\sigma}} \right) \right) G_{\tilde{\sigma}}(u + \frac{1}{2} iv_{\tilde{\sigma}}) du, \ x \in \mathbf{R}^n,$$

where $v_{\tilde{\sigma}} \in \Gamma'_{\tilde{\sigma}} \cap \{|y| < \frac{1}{2} w\}$. Thus, for $x \in \mathbf{R}^n$,

$$(F_\sigma * G_{\tilde{\sigma}}) \left(x + \frac{1}{2} iv_{\tilde{\sigma}} \right) = \int_{\mathbf{R}^n} F_\sigma(x - u) \check{G}_{\tilde{\sigma}}(-u) du$$

$$= \int_{\mathbf{R}^n} F_\sigma(x + u) \check{G}_{\tilde{\sigma}}(u) du$$

$$= \int_{\mathbf{R}^n} F_\sigma \left(x + u + \frac{1}{2} iv_{\tilde{\sigma}} \right) \check{G}_{\tilde{\sigma}} \left(u + \frac{1}{2} iv_{\tilde{\sigma}} \right) du. \tag{1.40}$$

By (1.38)–(1.40), we have

$$((\ell f) * (\ell g)) \left(x + \frac{1}{2} iv_{\tilde{\sigma}} \right) = \sum_{\tilde{\sigma}} \langle \ell f(x + u), \ \check{G}_{\tilde{\sigma}}(u) \rangle, \ \check{G}_{\tilde{\sigma}} \in \mathcal{P}_*.$$

Let $\varepsilon > 0$ be fixed. The assumption of the theorem, implies that $\langle q(x + u), G_{\tilde{\sigma}}(u) \rangle, G_{\tilde{\sigma}} \in \mathcal{P}_*$, as a function in $x \in \mathbf{R}^n$, can be extended on $B(0, \varepsilon\xi) \subset \mathbf{C}^n$ as a holomorphic function. Then

$$\langle q(x+u), G_{\tilde{\sigma}}(u)\rangle = \langle J_2(D)f(x+u), G_{\tilde{\sigma}}(u)\rangle$$
$$= J_2(D)\langle f(x+u), G_{\tilde{\sigma}}(u)\rangle, \ x \in \mathbf{R}^n.$$

Since $J_2(D)$ is an elliptic local operator, it follows that for every $\tilde{\sigma} \in \Lambda$, $\langle f(\cdot + u), G_{\tilde{\sigma}}(u)\rangle$ is also real analytic. Consequently, $((\ell f) * (\ell g))$ is holomorphic in $B(0, \beta)$ for $\beta, \beta_i > 0$, $i = 1, \ldots, n$.

Now it is easy to prove that $((\ell f) * (\ell g))(z)$ is holomorphic on $\mathbf{R}^n + B(0, \beta)$.

In order to finish the first part of the proof, we have only to remark that from (1.37) and from the property of an elliptic local operator it follows that q is a real analytic function which can also be extended as a holomorphic function on $\mathbf{R}^n + B(0, \beta)$.

For the second part of the proof suppose that q is given by a real analytic function which can be extended on $\mathbf{R}^n + B(0, \beta)$, $\beta_i > 0$, $i = 1, \ldots, n$, as a holomorphic function. Then for $\varepsilon > 0$ and $\xi \in \mathbf{R}^n$,

$$\ell q(\varepsilon\xi + x) := q(\varepsilon\xi + x + iI_{\sigma_0}), \ x \in \mathbf{R}^n$$

and, for $y_0 \in I\sigma_0$,

$$\langle \ell q(\varepsilon\xi + x), \varphi(x)\rangle = \int_{Imz=y_0} q(\varepsilon\xi + x + iy_0)\varphi(x+iy_0)dx, \ \varphi \in \mathcal{P}_*.$$

By the same arguments which we have used to transform (1.38) into (1.39), we have

$$\langle \ell q(\varepsilon\xi + x), \varphi(x)\rangle = \int_{\mathbf{R}^n} q(\varepsilon\xi + x)\varphi(x)dx, \ \varphi \in \mathcal{P}_*.$$

Since $q(z)$ is holomorphic on $\mathbf{R}^n + B(0, \beta)$, the same holds for $\langle \ell q(\varepsilon\xi + x), \varphi(x)\rangle$. $\qquad\square$

For asymptotic Taylor expansions, see also [163], [1], [50], [56].

1.11   *S-asymptotics in subspaces of distributions*

We discuss in this section the following problem: *Let $\mathcal{A}'$ be a subspace of $\mathcal{D}'$. If $T \in \mathcal{A}'$ and if T has the S-asymptotics in $\mathcal{D}'$, is it true that T has the S-asymptotics in $\mathcal{A}'$, as well?* The answer is not simple. We shall analyze two cases, $\mathcal{A}' = \mathcal{S}'$ and $\mathcal{A}' = \mathcal{K}'_1$. First, we shall illustrate the problem by an example.

The distribution $H(t)\exp(-t)$ (H is Heaviside's function) belongs to $\mathcal{S}'(\mathbf{R})$. It has S-asymptotics in $\mathcal{D}'(\mathbf{R})$ related to $c(h) = \exp(-h)$ with limit $U = \exp(-t)$. We know that $\exp(-t)$ does not belong to $\mathcal{S}'(\mathbf{R})$ and $\mathcal{D}(\mathbf{R})$ is dense in $\mathcal{S}(\mathbf{R})$. Hence $H(t)\exp(-t)$ cannot have the S-asymptotics in $\mathcal{S}'(\mathbf{R})$.

To answer our question, we need to introduce some additional conditions over $c(h)$.

Although we defined the S-asymptotics of distributions in a general cone Γ, we shall restrict ourselves to the cone $\Gamma = \mathbf{R}^n_+$ for the sake of simplicity. Recall, for $a, b \in \mathbf{R}^n$, $a \geq b$ means $a_i \geq b_i, i = 1, \ldots, n$.

The set of real-valued functions $h \to c(h)$, $h \in \mathbf{R}^n$, defined on $\mathbf{R}^n$, different from zero for $h \in \mathbf{R}^n_+$, is denoted by $\Sigma(\mathbf{R}^n)$. We assume, without losing generality, that function c in $\Sigma(\mathbf{R}^n)$ are positive and equal to 1 in $\mathbf{R}^n_+ \setminus (\mathbf{R}^n_+ + a)$, where $a \in \mathbf{R}^n_+$ depends on c and $\mathbf{R}^n_+ + a = \{x + a; x \in \mathbf{R}^n_+\}$.

By $\Sigma_e(\mathbf{R}^n)$ is denoted the subset of $\Sigma(\mathbf{R}^n)$ such that $c \in \Sigma_e(\mathbf{R}^n)$ if and only if for some $C > 0, d > 0, k > 0$ and $h_0 = (h_{0,1}, \ldots, h_{0,n}) \in \mathbf{R}^n_+$

$(E.1)$ \qquad $c(h + r) \leq Cc(h)\exp(k\|r\|), \quad h > h_0, \ r \in \mathbf{R}^n.$

$(E.2)$ \qquad $c(h)\exp(k\|h\|) \geq d, \quad h > h_0.$

By $\Sigma_p(\mathbf{R}^n)$ is denoted another subset of $\Sigma(\mathbf{R}^n)$ defined as follows, $c \in \Sigma_p(\mathbf{R}^n)$ if and only if for some $C > 0, d > 0, k > 0$ and $h_0 = (h_{0,1}, \ldots, h_{0,n}) \in \mathbf{R}^n_+$:

$(P.1)$ \qquad $c(h + r) \leq Cc(h)(1 + \|r\|)^k, \quad h > h_0, \ r \in \mathbf{R}^n.$

$(P.2)$ \qquad $c(h)(1 + \|h\|)^k \geq d, \quad h > h_0.$

Obviously, $\Sigma_p(\mathbf{R}^n) \subset \Sigma_e(\mathbf{R}^n)$.

We begin with some properties of the sets $\Sigma_e(\mathbf{R}^n)$ and $\Sigma_p(\mathbf{R}^n)$, we restrict our attention to $n = 1$ in order to analyze explicit representations for their elements.

The set $\Sigma_p(\mathbf{R})$.

By Remark 3 after Proposition 1.2, a function c from Definition 1.1 has the form $c(h) = \exp(\alpha h)L(\exp(h))$, $h > h_0 > 0$, where $\alpha \in \mathbf{R}$ and L is a slowly varying function, and the limit distribution $U(x) = C\exp(\alpha x)$. For the S-asymptotics in $\mathcal{S}'$, α has to be zero: Then $c(h) = L(\exp h)$, $h > h_0$.

Proposition 1.10. *Suppose that $c \in \Sigma(\mathbf{R})$.*

a) *If there exists $T \in \mathcal{S}'(\mathbf{R})$ such that $T(x + h) \overset{s}{\sim} c(h)U(x), h \in \mathbf{R}_+$ with $U \neq 0$, then $c(h) = L(\exp h)$ where L is a slowly varying function, and so c satisfies (P.2).*

b) *If $c = h^{\nu}L(h), h > h_1$ and $c(h) = 1, h \leq h_1$, where $\nu \in \mathbf{R}, h_1 > 0$ and L is a slowly varying and monotonous function for $h > h_1$, then c satisfies (P.1).*

c) *If c is of the same form as in b), where L is only slowly varying, then c satisfies (P.1) but for $r \in \mathbf{R}_+$.*

For the proof cf. [107].

The set $\Sigma_e(\mathbf{R})$.

Proposition 1.11. *Suppose that $c \in \Sigma(\mathbf{R})$.*

a) *If there exists $T \in \mathcal{D}'(\mathbf{R})$ such that $T(x+h) \overset{s}{\sim} c(h)U(x), U \neq 0$, then (E.2) holds for c.*

b) *If $c(h) = \exp(\alpha|h|)h^{\nu}L(h), h > h_1, c(h) = 1, h \leq h_1$, where $\alpha, \nu \in \mathbf{R}, h_1 > 0$ and L is a monotonous slowly varying function for $h > h_1$, then (E.1) holds for c.*

c) *If c is of the same form as in b), where L is only a slowly varying function, then c satisfies (E.1) but for $r \in \mathbf{R}_+$.*

For the proof cf. [107].

We prove now that if $T \in \mathcal{K}'_1(\mathbf{R})$ $(T \in \mathcal{S}'(\mathbf{R}))$ has the S-asymptotics in the cone $\mathbf{R}_+$ related to some $c(h) \in \sum_e(\mathbf{R})$ $(c(h) \in \sum_p(\mathbf{R}))$ with limit U in the space $\mathcal{D}'(\mathbf{R})$, then the limit

$$\lim_{h \to \infty} T(x + h)/c(h) = U(x)$$

also exists in $\mathcal{K}'_1(\mathbf{R})$ (in $\mathcal{S}'(\mathbf{R})$). If $n > 1$, we have to assume some additional conditions which imply the same assertion for the multidimensional case.

First, we shall prove a theorem which extends Proposition 1.7 b) and which will be used later.

Theorem 1.15. *Let $T \in \mathcal{D}'$. If for every rapidly exponentially decreasing function r (for every $k > 0, r(x) \exp(k\|x\|) \to 0$ as $\|x\| \to \infty$) the set $\{r(h)T(x + h); h \in \mathbf{R}^n\}$ is bounded in $\mathcal{D}'$, then $T \in \mathcal{K}'_1$.*

Proof. Let K be an arbitrary compact set in $\mathbf{R}^n$. For every $\phi \in \mathcal{D}_K \subset \mathcal{D}$,

$$h \mapsto r(h)\langle T(x+h), \phi(x)\rangle, h \in \mathbf{R}^n,$$

is a bounded function. This implies that for some $k_1 = k_1(\phi) > 0$ and some $C = C(\phi) > 0$

$$|\langle T(x+h), \phi(x)\rangle| \leq C \exp(k_1\|h\|), \ h \in \mathbf{R}^n. \tag{1.41}$$

If $h \in \mathbf{R}^n$ is fixed, then

$$\log^+ |\langle T(x+h), \phi(x)\rangle| \, / (1 + \|h\|), \quad \phi \in \mathcal{D}_K,$$

a defines a continuous function on $\mathcal{D}_K$. It follows from (1.41) that $\{\phi \mapsto \log^+ |< T(x+h), \phi(x) >|/(1 + \|h\|); h \in \mathbf{R}^n\}$ is a bounded family of continuous functions on $\mathcal{D}_K$; from the classical theorem of Baire, it follows the existence of some $k > 0$, which does not depend on $\phi \in \mathcal{D}_K$, such that the set $\{\langle T(x+h)\exp(-k\|h\|), \phi(x)\rangle; h \in \mathbf{R}^n\}$ is bounded for each $\phi \in \mathcal{D}_K$, and hence, for each $\phi \in \mathcal{D}$. This implies (see Theorem XXII, Chapter VI in [146]), that for a given open bounded set $\Omega \subset \mathbf{R}^n, 0 \in \Omega$, there exists a compact neighborhood of zero K and $m \in \mathbf{N}_0$ such that for every $\phi \in \mathcal{D}_K^m$,

$$\{x \mapsto (T(t+h) * \phi(t))(x)/\exp(k\|h\|); h \in \mathbf{R}^n\}$$

is a bounded family of continuous bounded functions on Ω.

Since $(T(t+h) * \phi(t))(x) = (T * \phi)(x+h)$, setting $x = 0$, we obtain that $h \mapsto (T * \psi)(h)/\exp(k\|h\|), h \in \mathbf{R}^n$, is a bounded function on $\mathbf{R}^n$ for any $\psi \in \mathcal{D}_K^m$. Now, by (VI, 6; 22) in [146], we obtain

$$T = \Delta^N(\gamma E * T) - \psi * T, \tag{1.42}$$

where E is the fundamental solution of $\Delta^N E = \delta$ (Δ is the Laplacian), $\gamma \in \mathcal{D}_K, \gamma \equiv 1$ in a neighborhood of 0 and $\psi \in \mathcal{D}_K$. If N is sufficiently large, $\gamma E \in \mathcal{D}_K^m$. Thus, $\gamma E * T$ and $\psi * T$ are in $\mathcal{K}'_1$, and this completes the proof. $\qquad\square$

Theorem 1.16. *Let* $T \in \mathcal{K}'_1(\mathbf{R})$ *($T \in \mathcal{S}'(\mathbf{R})$) and* $c \in \sum_e(\mathbf{R})$ *($c \in \sum_p(\mathbf{R})$). a) If the set* $\{T(x+h)/c(h); h > a\}$ *is bounded in* $\mathcal{D}'(\mathbf{R})$, *then this set is bounded in* $\mathcal{K}'_1(\mathbf{R})$ *(in* $\mathcal{S}'(\mathbf{R})$*) as well.*
b) If there exists the limit

$$\lim_{h \to \infty} \langle T(x+h)/c(h), \varphi(x)\rangle = \langle S, \varphi\rangle, \varphi \in \mathcal{D}(\mathbf{R}),$$

then this limit exists for every $\varphi \in \mathcal{K}_1(\mathbf{R})$ *(for every* $\varphi \in \mathcal{S}(\mathbf{R})$*). In particular,* $S \in \mathcal{K}'_1(\mathbf{R})(S \in \mathcal{S}'(\mathbf{R}))$.

Proof. We prove the theorem for $T \in \mathcal{K}'_1(\mathbf{R})$, because for $T \in \mathcal{S}'(\mathbf{R})$ it can be done in a similar way. (We have to replace $\exp(k\|\cdot\|)$ by $(1 + \|\cdot\|)^k$).

a) Using the last part of the proof of Theorem 1.15, we obtain that for some $m_1 \in \mathbf{N}_0$ and some compact neighborhood of zero, $K_1, h \mapsto (T *\\ \psi)(h)/c(h), h > 0$ is a bounded function for every $\psi \in \mathcal{D}^{m_1}_{K_1}(\mathbf{R})$. Since $T \in \mathcal{K}'_1(\mathbf{R})$, it holds that for some $k > 0$, $m_2 \in \mathbf{N}_0$ and some compact neighborhood of zero $K_2, h \mapsto (T * \psi)(h)/\exp(k|h|), h \in \mathbf{R}$, is a bounded function for every $\psi \in \mathcal{D}^{m_2}_{K_2}(\mathbf{R})$. Thus, by taking N in (1.42) sufficiently large ($\Delta = d^2/dx^2$) and $K = K_1 \cap K_2$, we obtain that for some $m \in N_0$

$$T = \sum_{i=0}^{m} F_i^{(i)},$$

where $F_i, i = 0, \ldots, m$, are continuous functions on $\mathbf{R}$ such that for some $M_1 > 0$, $M_2 > 0$ and $k > 0$

$$\sup\{|F_i(x)/c(x)|; x > 0, i = 0, \ldots, m\} \leq M_1, \qquad (1.43)$$

$$\sup\{|F_i(x)|\exp(k|x|); x \in \mathbf{R}, i = 0, \ldots, m\} \leq M_2,$$

Let $\phi \in \mathcal{K}_1(\mathbf{R})$ and $h > h_0$ be fixed. We put

$$I_i(h, \phi) \equiv I_i = \int_{-\infty}^{\infty} (|\phi^{(i)}(x)\|F_i(x+h)|/c(h))dx$$

$$= \left(\int_{-\infty}^{-h} + \int_{-h}^{\infty} \right) (|\phi^{(i)}(x)\|F_i(x+h)|/c(h))dx$$

$$= I_i(-\infty, -h) + I_i(-h, \infty), i = 0, \ldots, m, h > h_0.$$

If $x \in (-\infty, -h)$, then $|x + h| = |x| - h$ and by (E.2), we obtain

$$I_i(-\infty, -h) \leq d^{-1} \int_{-\infty}^{-h} |\phi^{(i)}(x)|\exp(2k)|x|)|F_i(x+h)/\exp(k|x+h|)|dx$$

$$\leq Md^{-1} \int_{-\infty}^{\infty} |\phi^{(i)}(x)|\exp(k|x|)dx.$$

From the definition of the space $\mathcal{K}_1(\mathbf{R})$ it follows that the last integral is finite.

Because of (E.1) and (1.43), we obtain that for some $k_1 > 0$,

$$I_i(-h, \infty) \le \int_{-h}^{\infty} |\phi^{(i)}(x)| \|F_i(x+h)/c(x+h)|exp(k_1|x|)dx$$

$$\le M_1 \int_{-\infty}^{\infty} |\phi^{(i)}(x)|exp(k_1|x|)dx < \infty.$$

Since

$$|\langle T(x+h)/c(h), \phi(x)\rangle| \le \sum_{i=0}^{m} I_i(h, \phi),$$

from the preceding inequalities, we obtain that for some $A > 0$ which does not depend on $h > h_0$ and for $\phi \in \mathcal{K}_1$,

$$|\langle T(x+h)/c(h), \phi(x)\rangle| \le A, \quad \phi \in \mathcal{K}_1, \ h > h_0.$$

Thus the proof of a) is complete.

b) If the limit given in b) exists, then for an $a \in \mathbf{R}$, the set $\{T(x + h)/c(h); h \ge a\}$ is bounded in $\mathcal{K}'(\mathbf{R})$ (in $\mathcal{S}'(\mathbf{R})$). Since $\mathcal{D}(\mathbf{R})$ is dense in $\mathcal{K}_1(\mathbf{R})$ (in $\mathcal{S}(\mathbf{R})$), it is now enough to use the assertion of a) and the Banach–Steinhaus theorem. □

If we assume more on a distribution T, then we can assume less on the function c.

Theorem 1.17. *Let $T \in \mathcal{D}'(\mathbf{R})$ and $\text{supp}\, T \subset [0, \infty)$. Suppose that $c \in \Sigma(\mathbf{R})$ and satisfies (E.2) and (E.1) with $r \in \mathbf{R}_+$ (satisfies (P.2) and (P.1) with $r \in \mathbf{R}_+$).*

a) If the set $\{T(x+h)/c(h); h > a\}$ is bounded in $\mathcal{D}'(\mathbf{R})$, then this set is bounded in $\mathcal{K}'_1(\mathbf{R})$ (in $\mathcal{S}'(\mathbf{R})$), as well.

b) If there exists the limit

$$\lim_{h \to \infty} \langle T(x+h)/c(h), \varphi(x)\rangle = \langle S, \varphi\rangle, \quad \varphi \in \mathcal{D}(\mathbf{R}),$$

then this limit exists for every $\varphi \in \mathcal{K}_1(\mathbf{R})$ (for every $\varphi \in \mathcal{S}(\mathbf{R})$). In particular, $S \in \mathcal{K}'_1(\mathbf{R})$ ($S \in \mathcal{S}'(\mathbf{R})$).

For the proof see [106].

We give other versions of Theorem 1.17 because we want to emphasize that the space $\mathcal{S}'(\mathbf{R})$ is "natural" for the distributions having the S-asymptotics related to $c(h) = h^\nu L(h)$, $h > 0$, $\nu \in \mathbf{R}$, while the space $\mathcal{K}'_1(\mathbf{R})$ is "natural" for those distributions having the S-asymptotics related to $c(h) = \exp(ah)L(\exp(h))$, $h > 0$, $a \in \mathbf{R}$, where L is a slowly varying function.

Theorem 1.18. *Let* $T \in \mathcal{D}'(\mathbf{R})$, $\operatorname{supp} T \subset (-\omega, \infty)$, $\omega > 0$ *and* $T(x + h) \overset{s}{\sim}$ $e^{ah}L(e^h) \cdot Ce^{ax}$, $h \in \mathbf{R}_+$ *in* $\mathcal{D}'(\mathbf{R})$ *(and* $T(x + h) \overset{s}{\sim} h^\nu L(h) \cdot C$, $h \in \mathbf{R}_+$ *in* $\mathcal{D}'(\mathbf{R})$), *then* $T \in \mathcal{K}'_1(\mathbf{R})$ $(T \in \mathcal{S}'(\mathbf{R}))$ *and* T *has the S-asymptotics in* $\mathcal{K}'_1(\mathbf{R})$ *(in* $\mathcal{S}'(\mathbf{R})$) *related to the same* c *and with the same limit.*

Proof. By Theorem 1.9 and Lemma 1.2 there exist functions F_i, $i = 0, \ldots, m$, continuous on $(-\omega, \infty)$, such that

$$T = \sum_{i=0}^{m} D^i F_i \text{ on } (-\omega, \infty),$$

where

$$|F_i(x + h)/c(h)| \leq M_i, \ h \in \mathbf{R}_+, \ x \in (-\omega, \omega), \ i = 0, \ldots, m,$$

$$\operatorname{supp} F_i \subset (-\omega - \delta, \infty), \ i = 0, \ldots, m, \ \delta > 0.$$

For a slowly varying function L there exists a slowly varying function $L^* \in C^\infty_{(c, \infty)}(\mathbf{R})$, $\alpha > 0$, such that $L^*(x)/L(x) \to 1, x \to \infty$. We refer to the Remark after Proposition 1.2 for the construction of L^*. Therefore, we can suppose, without any restriction, that L, which appeared in c, is in $C^\infty_{(\alpha, \infty)}(\mathbf{R})$, $\alpha > 0$. Let us denote by $M_i(h) = F_i(h)/c(h), h \in \mathbf{R}; M_i$ is a continuous and bounded function on $\mathbf{R}$ and $F_i = cM_i$. This implies that $T \in \mathcal{K}'_1$. The next step is to prove that for $h_0 \in \mathbf{R}$ the set $\{T(x + h)/c(h); h \geq h_0\}$ is weakly bounded in $\mathcal{K}'_1$ which is equivalent to the strong boundedness in $\mathcal{K}'_1$.

We need the following inequality:

$$c(x + h)/c(h) = e^{\alpha x}L(e^x e^h)/L(e^h) \leq Ae^{ax + \delta|x|}, x, h \in \mathbf{R},$$

where $\delta > 0$ (see [9], p. 25). Let $\varphi \in \mathcal{K}_1$, then

$$|\langle T(x + h)/c(h), \varphi(x)\rangle| \leq \sum_{i=0}^{m} \int_{\mathbf{R}} |F_i(x + h)/c(h)||\varphi^{(i)}(x)|dx$$

$$\leq \sum_{i=0}^{m} \int_{\mathbf{R}} |M_i(x+h)c(x+h)/c(h)| |\varphi^{(i)}(x)| dx$$

$$\leq \sum_{i=0}^{m} \int_{\mathbf{R}} e^{ax+\delta|x|} |\varphi^{(i)}(x)| dx.$$

This proves that $\{T(x+h)/c(h); h \geq h_0\}$ is weakly bounded in $\mathcal{K}'_1$. Since $\mathcal{D}$ is dense in $\mathcal{K}'_1$, the Banach–Steinhaus theorem implies the S-asymptotics in $\mathcal{K}'_1$ related to c. $\qquad\square$

In order to extend Theorem 1.17 to the multidimensional case, we have to introduce the following notation.

We denote by Λ the set of all $n-th$ class variations of elements $\{-1, 1\}$. If $(a_1, \ldots, a_n) \in \Lambda$, then we put

$$\Gamma(a_1, \ldots, a_n) = \{h \in \mathbf{R}^n; \sum_{i=1}^{n} \text{sgn}(a_i \cdot h_i) = n\}.$$

(This means if $a_i = 1, (a_i = -1)$, then $h_i > 0 (h_i < 0)$). For example $\Gamma(1, \ldots, 1) = \mathbf{R}^n_+$ and $\Gamma(-1, \ldots, -1) = \mathbf{R}^n_-$. Let

$$c(h) = c_1(h_1) \ldots c_n(h_n), \ h_i \in \mathbf{R}, \ i = 1, \ldots, n, \qquad (1.44)$$

where

$$c_i \in \Sigma_e(\mathbf{R}) \ (c_i \in \Sigma_p(\mathbf{R})), i = 1, \ldots, n.$$

Obviously, $c \in \Sigma_e(\mathbf{R}^n) \ (c \in \Sigma_p(\mathbf{R}^n))$. Let $(a_1, \ldots, a_n) \in \Lambda$ be given. We denote by $j_i, i = 1, \ldots, r$ the components of $(a_1, \ldots, a_n)$ which are equal to 1, and those which are equal to -1 by $s_i, i = 1, \ldots, m$ $(r + m = n)$.

Let $h \in \Gamma(a_1, \ldots, a_n)$ and $k > 0$. We put

$$c^k_{(a_1, \ldots, a_n)}(h) = c_j(h_j) \ldots c_{j_r}(h_{j_r}) \cdot \exp(k(|h_{s_1}| + \cdots + |h_{s_m}|))$$

Theorem 1.19. *Let $T \in \mathcal{K}'_1$ $(T \in \mathcal{S}')$ and c be of the form (1.44).*

a) If there exists $k > 0$ such that for every $(a_1, \ldots, a_n) \in \Lambda$

$$\{T(x+h)/c^k_{(a_1, \ldots, a_n)}(h); \ h \in \Gamma(a_1, \ldots, a_n)\}$$

is bounded in $\mathcal{D}'$, then $\{T(x+h)/c(h); \ h > 0\}$ is bounded in $\mathcal{K}'_1$ (in $\mathcal{S}'$).

b) If

$$\lim_{h \to \infty} \langle T(x+h)/c(h), \phi(x) \rangle = \langle S(x), \phi(x) \rangle, \phi \in \mathcal{D},$$

and if for some $k > 0$ and every $(a_1, \ldots, a_n) \in \Lambda \backslash \{(1, \ldots, 1), (-1, \ldots, -1)\}$
the sets

$$\{T(x+h)/c^k(a_1, \ldots, a_n)(h); \ h \in \Gamma(a_1, \ldots, a_n)\},$$

are bounded in $\mathcal{D}'$, then $T(x+h)/c(h)$ converges to S in $\mathcal{K}'_1$ (in $\mathcal{S}'$), as $h \to \infty$. In particular, $S \in \mathcal{K}'_1$ ($S \in \mathcal{S}'$).

For the proof see [106].

Instead of analyzing the two special subspaces $\mathcal{S}$ and $\mathcal{K}_1$ of $\mathcal{D}$ we can consider a general subspace $\mathcal{A}$ of $\mathcal{D}$.

Let Γ be a convex cone. We denote $\Sigma_q(\Gamma)$ a subset of $\Sigma(\Gamma)$ such that $c \in \Sigma_q(\Gamma)$ if and only if there exist $C > 0$ and a positive locally integrable function p such that

$$c(h+x) \leq Cc(h)p(x), \quad h, x \in \Gamma \backslash B(0, r). \tag{1.45}$$

In the sequel, put $G = \{x \in \mathbf{R}^n \backslash (\Gamma \cup B(0, r))\}$. We denote by $\mathcal{A}$ a barrelled vector space of smooth functions such that $\mathcal{D}$ is dense in $\mathcal{A}$ with its topology finer than the topology induced by $\mathcal{A}$; $\mathcal{A}'$ is the dual space of $\mathcal{A}$, $\mathcal{A}' \subset \mathcal{D}'$. We will suppose that the elements ϕ of $\mathcal{A}$ satisfy the following condition: for every $y \in B(0, r)$, $p(x)\phi(x+y) \in L^1$.

Theorem 1.20. *Suppose that $T \in \mathcal{A}'$ and $c \in \Sigma_q(\Gamma)$.*

a) If the sets:

$$Q_1 = \{T(x+h)/c(h); \ h \in \Gamma\}$$

$$Q_2 = \{T(x+k+h)/(c(h)p(k)); \ h \in \Gamma, k \in G\}$$

are weakly bounded in $\mathcal{D}'$, then the set Q_1 is weakly bounded in $\mathcal{A}'$ as well.

b) If T has S-asymptotics in $\mathcal{D}'$ related to $c(h)$ with the limit U and if the set Q_2 is weakly bounded in $\mathcal{D}'$, then T has the S-asymptotics in $\mathcal{A}'$ related to $c(h)$ and with the limit U.

We remark that the well-known basic spaces as $\mathcal{K}_1, \mathcal{S}, \mathcal{D}_{L^p} (1 \leq p < \infty)$ and $\mathcal{B}$ satisfy our conditions assumed in Theorem 1.20 for the space $\mathcal{A}$. The space $\mathcal{B} = \mathcal{D}_{L^\infty}$ is an example of one which does not satisfy them, due to the fact that $\mathcal{D}$ is not dense in $\mathcal{B}$.

If a distribution f has the S-asymptotic expansion in $\mathcal{D}'$, with adequate additional conditions, it can have the S-asymptotic expansion in a subspace of $\mathcal{D}'$. The following theorem gives such conditions.

Theorem 1.21. *Suppose that $f \in \mathcal{K}'_1(\mathbf{R})$ ($f \in \mathcal{S}'(\mathbf{R})$) and $\{u_i\} \subset \mathcal{K}'_1(\mathbf{R})$ ($\{u_i\} \subset \mathcal{S}'(\mathbf{R})$). If*

$$f(x+h) \overset{s}{\sim} \sum_{i=1}^{\infty} u_i(x+h)|\{c_i(h)\}, \quad h \to \infty \quad in \quad \mathcal{D}'(\mathbf{R})$$

and $c_i(h) = \exp(\alpha_i h) L_i(\exp h)$ (resp. $c_i = h^{\nu_i} L_i(h)$), $h > h_0$), where L_i, $i \in \mathbf{N}$, are monotonous slowly varying functions, then

$$f(x+h) \overset{s}{\sim} \sum_{i=1}^{\infty} u_i(x+h)|\{c_i(h)\}, \quad h \to \infty, \quad in \ \mathcal{K}'_1(\mathbf{R}) \ (in \ \mathcal{S}'(\mathbf{R})),$$

as well.

Proof. The proof is based on Proposition 1.11 and Theorem 1.18, we leave the details to the reader. □

The next two propositions give the relation between the asymptotic behavior of functions and the S-asymptotics of distributions defined by these functions.

Proposition 1.12. *Let $f \in L^1_{loc}(\mathbf{R})$ be such that f defines a regular distribution in $\mathcal{K}'_1(\mathbf{R})$, i.e., $f\phi \in L^1(\mathbf{R})$ for every $\phi \in \mathcal{K}_1(\mathbf{R})$. Further, assume*

$$f(x) \sim L(\exp x) \exp(\alpha x) \quad as \quad x \to \infty,$$

where L is a slowly varying monotonous function defined on (a, ∞). If $c(h) = L(\exp h) \exp(\alpha h), h > A$, and $g(x) = \exp(\alpha x), x \in \mathbf{R}$, then $f(x+h) \overset{s}{\sim} c(h)g(x), h \in \mathbf{R}_+$ in $\mathcal{K}'_1$.

Proof. Let $\phi \in \mathcal{D}$ and supp $\phi \subset [a, b]$. Since $(L(\lambda h)/L(h)) \to 1$, $h \to \infty$, uniformly on any compact interval contained in $(0, \infty)$, we obtain

$$\int_{-\infty}^{\infty} \frac{f(x+h)}{\exp(\alpha h) L(\exp h)} \phi(x) dx = \int_{a}^{b} \frac{f(x+h)}{\exp(\alpha(x+h)) L(\exp(x+h))}$$

$$\cdot \exp(\alpha x) \frac{L(\exp x \cdot \exp h)}{L(\exp h)} \phi(x) dx \to \int_{a}^{b} \exp(\alpha x) \phi(x) dx, h \to \infty.$$

The assertion follows now from Proposition 1.11 and Theorem 1.16.   □

Proposition 1.13. *Suppose that the functions f and $u_i, i \in \mathbf{N}$, are locally integrable and define regular distributions belonging to $\mathcal{K}'_1(\mathbf{R})$, i.e., $f\varphi$ and $u_i\varphi$ are in $L^1(\mathbf{R})$ for every $\varphi \in \mathcal{K}_1(\mathbf{R})$. Let $c_i(h) = \exp(\alpha_i h)L_i(\exp h)$, where L_i are monotonous slowly varying functions and $\alpha_i \in \mathbf{R}, i \in \mathbf{N}$. If*

$$f(x) \sim \sum_{i=1}^{\infty} U_i(x)|\ \{c_i(x)\}, \ x \to \infty,$$

then

$$f(x+h) \overset{s}{\sim} \sum_{i=1}^{\infty} u_i(x+h)|\{c_i(h)\}, \ h \to \infty.$$

Proof. The proof is similar to that of Proposition 1.13.   □

We shall prove a structural theorem for a distribution $T \in \mathcal{B}'$ having the S-asymptotics in a cone Γ with the nonempty interior.

Theorem 1.22. *Suppose $T_0 \in \mathcal{B}'$ and $T_0(x+h) \overset{s}{\sim} 1 \cdot U(x)$, $h \in \Gamma$ in $\mathcal{D}'$, then*

a) $U = C$.

b) $T_0 = \sum\limits_{i=0}^{2} \Delta^{ik} F_i$, where F_i are continuous functions belonging to L^{∞};

c) For every $0 \le i \le 2$ functions $F_i(x+h)$ of part b) converge uniformly to a constant when x belongs to a compact set K and $h \in \Gamma$, $\|h\| \to \infty$.

d) T_0 has the S-asymptotics in $\mathcal{B}'$, related to $c = 1$ and with the limit $U = C$ in the cone Γ.

Proof. a) U has to be a constant because $c = 1$ (Proposition 1.2).

b) From the fact that $T_0 \in \mathcal{B}'$ it follows that $(T_0 * \check{\zeta}) = \langle T_0(x+\cdot), \zeta(x)\rangle \in L^{\infty}$ for every $\zeta \in \mathcal{D}$ (see [146], VI, §8) and the set of distributions $Q = \{T_h \equiv T_0(x+h); \ h \in \mathbf{R}^n\}$ is weakly bounded and bounded in $\mathcal{D}'$.

We will construct another bounded set of distributions. Denote $S = \{\psi \in \mathcal{D}; \ \|\psi\|_{L^1} \le 1\}$. We have seen that for a fixed $\zeta \in \mathcal{D}$, $(T_0 * \check{\zeta}) \in L^{\infty}$. Now, for every $\psi \in S$:

$$|\langle T_0 * \check{\psi}, \zeta\rangle| = |\langle T_0 * \check{\zeta}, \psi\rangle| = |\int_{\mathbf{R}} (T_0 * \check{\zeta})(t)\psi(t)dt| \le \|T_0 * \check{\zeta}\|_{L^{\infty}} \|\psi\|_{L^1}.$$

Hence, the set of regular distributions, defined by the set of continuous functions $\{U_\psi \equiv T_0 * \check{\psi}; \ \psi \in S\}$ is weakly bounded and therefore bounded in $\mathcal{D}'$.

A set $W' \in \mathcal{D}'$ is bounded if and only if for every $\alpha \in \mathcal{D}$ the set of functions $\{T * \alpha; \ T \in W'\}$ is bounded on every compact set M belonging to $\mathbf{R}^n$ (see [146], VI, §7). Hence $\{T * \alpha; \ T \in W'\}$ defines a bounded set of regular distributions. In such a way $\{T_h * \zeta; \ T_h \in Q\}$ and $\{U_\psi * \zeta; \ U_\psi \in H\}$ are bounded sets of regular distributions. Now, for these two sets we can repeat twice a part of the proof of Theorem XXII, Chapter VI in [146].

We denote by Ω an open neighborhood of zero in $\mathbf{R}^n$ which is relatively compact in $\mathbf{R}^n$, namely, $\bar{\Omega} = K$ is a compact set. Then, by the mentioned part of the proof, there exist $m_1 \geq 0$ and $m_2 \geq 0$, such that the mappings $(\alpha, \beta) \to U_\psi * (\alpha * \beta)$ or $(\alpha, \beta) \to T_h * (\alpha * \beta)$, $h \in \mathbf{R}^n$, are equicontinuous and map $\mathcal{D}_\Omega^{m_1} \times \mathcal{D}_\Omega^{m_1}$ and $\mathcal{D}_\Omega^{m_2} \times \mathcal{D}_\Omega^{m_2}$ into L_B^∞ respectively; B is the ball $B(0, r)$, where r is a positive number. Hence, for every $x \in B$ and $h \in \mathbf{R}^n$ the function $(T_h * \alpha * \beta)(x) = (T_0 * \alpha * \beta)(x + h)$ is continuous.

Let $Z(0, \rho)$ be a ball in L_B^∞. Then there exists a neighborhood $V_1(m_1, \varepsilon_1, K_1)$ in $\mathcal{D}_\Omega^{m_1}$, such that $U_\psi * (\alpha * \beta) \in Z(0, \rho)$ for $\alpha, \beta \in V_1(m_1, \varepsilon_1, K_1)$, $U_\psi \in H$ and a neighborhood $V_2(m_2, \varepsilon_2, K_2) \subset \mathcal{D}_\Omega^{m_2}$, such that $T_h * (\alpha * \beta) \in Z(0, \rho)$ for $\alpha, \beta \in V_2(m_2, \varepsilon_2, K_2)$, $T_h \in Q$. Let $K_0 = K_1 \cap K_2$, $\varepsilon_0 = \min(\varepsilon_1, \varepsilon_2)$ and $m = \max(m_1, m_2)$. We shall use (VI, 6; 23) in [146].

$$T_0 = \Delta^{2k} * (\gamma E * \gamma E * T_0) - 2\Delta^k * (\gamma E * \xi * T_0) + (\xi * \xi * T_0),$$

where E is a solution of the iterated Laplace equation; $\Delta^k E = \delta$; $\gamma, \xi \in \mathcal{D}_\Omega$, $\operatorname{supp} \gamma$ and $\operatorname{supp} \xi$ belonging to $K_0 = K_1 \cap K_2$. We have only to choose the number k large enough so that $\gamma E \in \mathcal{D}_\Omega^m$. Now, we can take: $F_2 = \gamma E * \gamma E * T_0$; $F_1 = \gamma E * \xi * T_0$ and $F_0 = \xi * \xi * T_0$. All of these functions are of the form: $F_i = T_0 * \alpha_i * \beta_i$; $\alpha_i, \beta_i \in V(m, \varepsilon'_0, K_0)$, $\varepsilon'_0 > 0$.

We have to prove that F_i, $i = 0, 1, 2$, have the properties given in Theorem 1.22. For $\alpha_i, \beta_i \subset V(m, \varepsilon'_0, K_0)$ and $\psi \in S$

$$|\langle (T_0 * \alpha_i * \beta_i), \psi \rangle| = |[T_0 * \check{\psi}) * (\check{\alpha}_i * \check{\beta}_i)](0)| \leq \rho(\varepsilon'_0/\varepsilon_0)^2 \equiv M.$$

Now let $\mu \neq 0$ be any element in L^1. Then $\mu/\|\mu\|_{L^1} \in S$ and $|\langle (T_0 * (\alpha_i * \beta_i)), \mu \rangle| \leq M \|\mu\|_{L^1}$ which proves that $T_0 * (\alpha_i * \beta_i)$, $i = 0, 1, 2$, belong to L^∞. Since $F_i = T_0 * (\alpha_i * \beta_i)$, $\alpha_i, \beta_i \in V(m, \varepsilon'_0, K_0)$, F_i, $i = 0, 1, 2$, are continuous and belong to L^∞.

c) We shall continue the investigations of the properties of F_i. By the properties of the convolution, we have

$$F_i(x+h) = F_i * \tau_{-h} = T_0 * (\alpha_i * \beta_i) * \tau_{-h} = T_h * (\alpha_i * \beta_i), \ \alpha_i, \beta_i \in \mathcal{D}_\Omega^m,$$

where τ_{-h} is the translation operator.

We have proved that the mappings $(\alpha, \beta) \to T_h * \alpha * \beta$, $T_h \in Q$, are equicontinuous and map $\mathcal{D}_\Omega^m \times \mathcal{D}_\Omega^m$ into L_B^∞. $\mathcal{D}$ is a dense subset of $\mathcal{D}^m$, $m \geq 0$. We can construct a subset of $\mathcal{D}_K$, $cl\Omega = K$, which is dense in $\mathcal{D}_\Omega^m$. Since $T_h * (\zeta * \psi) \to C * \zeta * \psi$ for $\zeta * \psi \in \mathcal{D}_\Omega \times \mathcal{D}_\Omega$, then $T_h * \alpha_i * \beta_i$ converges to $C * \alpha_i * \beta_i$, as well (see [146], VI, §7),when $\|h\| \to \infty$, $h \in \Gamma$, $i = 0, 1, 2$.

d) It remains to prove the last part of Theorem 1.22. For $\mu \in \mathcal{D}_{L^1}$ and $T \in \mathcal{B}$, we have:

$$|\langle T_0(x+h), \mu(x) \rangle| \leq \sum_{i=0}^2 \int_{\mathbf{R}^n} |F_i(x+h) \Delta^{ik} \mu(x)| dx \leq \sum_{i=0}^2 M_i \int_{\mathbf{R}^n} |\Delta^{ik} \mu(x)| dx,$$

where $M_i = \sup|F_i(x)|$, $x \in \mathbf{R}^n$. Hence the set $\{T_0(x+h), \ h \in \mathbf{R}^n\}$, is weakly bounded in $\mathcal{B}'$. Since $\mathcal{D}$ is dense in $\mathcal{D}_{L^1}$, by the Banach–Steinhaus theorem the limit:

$$\lim_{h \in \Gamma, h \to \infty} \langle T_0(x+h), \mu(x) \rangle, \ \mu \in \mathcal{D}_{L^1},$$

exists, as well, and it is equal to $\langle C, \mu \rangle$. $\square$

1.12   *Generalized S-asymptotics*

Definition 1.1 corresponds to some subsets of distributions of a limited growth (see Theorem 1.16). This is the reason for introducing a generalization of the S-asymptotics ([119]).

Definition 1.6. Suppose that e is a function on $\mathbf{R}^n$ such that $e^{-1} \in M_{(.)}$ (see **0.6**). Then, it is said that $T \in \mathcal{F}_g'$ has generalized S-asymptotics in the cone Γ, related to e, if there exists

$$w. \lim_{\|h\| \to \infty, h \in \Gamma} T(x+h)/e(x+h) = 1 \text{ in } \mathcal{F}_g'.$$

We write in short $T(x) \overset{gs}{\sim} e(x)$, $\|x\| \to \infty$, $x \in \Gamma$.

If we compare Definition 1.1 and Definition 1.6, we obtain.

Proposition 1.14. *Let e be a function on $\mathbf{R}^n$ such that $e^{-1} \in M_{(\cdot)}$. Then, $T \in \mathcal{F}'_g$ has generalized S-asymptotics in Γ, related to e if and only if T/e has the S-asymptotics in Γ related to $c = 1$ and with limit $U = 1$.*

In such a way, dividing $T \in \mathcal{F}'_g$ by e, we may apply our results on the S-asymptotics to this new context. So, all the assertions on the S-asymptotics can be transferred for the generalized S-asymptotics by simply using Proposition 1.14. For this reason, we underline only some of the key properties of generalized S-asymptotics.

The generalized S-asymptotics is a local property, as well (see Theorem 1.4). From Theorems 1.2,d) and 1.3,4) it follows:

Proposition 1.15. *a) Let $S \in \mathcal{E}'$ and $T \in \mathcal{D}'$. If $T(x) \overset{gs}{\sim} e(x)$, $\|x\| \to \infty$, $x \in \Gamma$, then $(S * (T/e))(c + h) \overset{s}{\sim} 1 \cdot (S * 1)(x)$, $h \in \Gamma$.*

As consequence of $T(x) \overset{gs}{\sim} e(x), \|x\| \to \infty, x \in \Gamma$, it follows that for any partial derivative D_{x_i}

$$(D_{x_i}(T/e))(x + h) \overset{s}{\sim} 1 \cdot 0, \ \|h\| \to \infty, \ h \in \Gamma.$$

In order to compare the existence of the generalized S-asymptotics with that of the S-asymptotics of a $T \in \mathcal{F}'$, we need the following notation and lemma.

Let Γ be a cone with the vertex at zero. We employ the notation $\mathrm{pr}\,\Gamma$ for the intersection of Γ and the unit sphere in $\mathbf{R}^n$.

Lemma 1.4. *Let $T \in \mathcal{F}'$ and let $c \in \Sigma(\Gamma)$ where Γ is a convex cone with the nonempty interior, ($\int \Gamma \neq \theta$). Let Γ' be a closed cone, $\Gamma' \subset \mathrm{int}\,\Gamma$. Suppose that $T(x + h) \overset{s}{\sim} c(h) \cdot U(x), h \in \Gamma$ with $U \neq 0$. Then, there exist $e \in C^\infty$, positive on $\mathbf{R}^n$, and $a \in \mathbf{R}^n$ such that*

$$\lim_{h \in \Gamma', \|h\| \to \infty} c(h)/e(x + h) = \exp(-(a \cdot x)) \ \text{in} \ \mathcal{E}. \tag{1.46}$$

Proof. We know (see Proposition 1.2) that $U(x) = C \exp(a \cdot x), a \in \mathbf{R}^n$, $C \neq 0$. Let $\phi_0 \in \mathcal{F}$ such that $\langle U, \phi_0 \rangle \neq 0$. We introduce functions e_1, e_2, e_3, e_4 and e in the following way: e_1 is a smooth function

$$e_1(y) = \langle T(x + y), \phi_0(x) \rangle / \langle U, \phi_0 \rangle, \ y \in \mathbf{R}^n.$$

Since

$$\lim_{h \in \Gamma, \|h\| \to \infty} \langle T(x + h)/c(h), \phi_0(x) \rangle = \langle U, \phi_0 \rangle,$$

there exists $\beta_0 > 0$ such that $e_1(y) > 0, y \in \{z \in \Gamma', \|z\| \geq \beta_0\}$.

Functions e_2, e_3 and e_4 are defined as follows: $e_2(x) = \max\{e_1(x), 0\}$, $x \in \mathbf{R}^n$, e_3 is the characteristic function of the set $C_{\mathbf{R}^n}\{x \in \Gamma', \|x\| \geq \beta_0\}$, $e_4 = e_2 + e_3$. Thus e_4 is positive and locally integrable.

Let w be a non-negative function belonging to $\mathcal{C}^\infty$ such that $w(x) = 1$ for $x \in B(0,1)$, $w(x) = 0$ for $\|x\| > 2$, and

$$\int_{\mathbf{R}^n} \exp(-(a \cdot x))w(x)dx = 1.$$

Now, we can construct the sought function $e \in \mathcal{C}^\infty : e = (e_4 * w)(x)$, $x \in \mathbf{R}^n$. The function $e(x)$ is positive because of

$$e(x) = \int_{\mathbf{R}^n} e_4(x - t)w(t)dt \geq \int_{B(0,1)} e_4(x - t)dt > 0, \ x \in \mathbf{R}^n.$$

It only remains to prove (1.45). The number $m := d(\mathrm{pr}\,\Gamma', C_{\mathbf{R}^n} \mathrm{int}\,\Gamma)$ is positive because $\mathrm{pr}\,\Gamma'$ is a compact set in $\mathbf{R}^n$ and $C_{\mathbf{R}^n} \mathrm{int}\,\Gamma$ is a closed one. Every ball $B(b, m')$, $0 < m' < \min(m, 1)$, $b \in \mathrm{pr}\,\Gamma'$ is contained in $\mathrm{int}\,\Gamma$ and for every $\lambda > 0$ the ball $B(\lambda b, \lambda m'), 0 < m' < \min(m, 1)$, is contained in $\mathrm{int}\,\Gamma$, as well.

For every compact set $K \subset \mathbf{R}^n, K \subset B(0, \rho)$ and every $\beta_0 \in \mathbf{R}_+$ there exists $\beta > 0$ such that $x + h \in \{y \in \Gamma'; \|y\| \geq \beta_0\}$ for $x \in K$ and $h \in \{y \in \Gamma', \{y\| > \beta\}$. Since $h \in \Gamma'$, h can be written as $h = \lambda b, b \in \mathrm{pr}\,\Gamma', \lambda > 0$. Thus $B(\lambda b, \lambda m') \subset \Gamma$ for every $\lambda > 0$. If $\lambda > \rho/m'$, then $x + h \in B(\lambda b, \lambda m')$ for $x \in K \subset B(0, \rho)$ and $h = \lambda b$ because $\|x + h - \lambda b\| = \|x\| \leq \rho < \lambda m'$. Now,

$$\|x + h\| \geq \|\|x\| - \|h\|\| \geq \lambda(1 - m').$$

Then, we can take $\beta = \max\{\rho/m', \beta_0/(1 - m')\}$.

For a compact set $K_0 \subset \mathbf{R}^n$ the set $K = K_0 \backslash B(0, 2)$ is also a compact set and belongs to a ball $B(0, \rho)$. Let us suppose that we have found β which corresponds to K and β_0 as above. Then, by the definition of e_4 we have $e_4(x - t + h) = e_1(x - t + h)$ for $x - t \in K; h \in \Gamma', \|h\| \geq \beta$. Hence

$$\lim_{h \in \Gamma', \|h\| \to \infty} e_4(x - t + h)/(c(h)) = \lim_{h \in \Gamma', \|h\| \to \infty} \left\langle \frac{T(y + x - t + h)}{c(h)\langle U, \phi_0 \rangle}, \phi_0(y) \right\rangle$$

$$= \exp(a \cdot (x - t))$$

and this limit is uniform for $x - t \in K$. Now, for $x \in K_0$ the following limit

$$\lim_{h \in \Gamma', \|h\| \to \infty} \frac{e(x + h)}{c(h)} = \lim_{h \in \Gamma', \|h\| \to \infty} \left(\frac{e_4(t + h)}{c(h)} * w(t) \right)(x)$$

$$= (\exp(a \cdot t) * w(t))(x) = \exp(a \cdot x)$$

is uniform, as well. □

Proposition 1.16. *Let Γ be a convex cone with nonempty interior, $\int \Gamma \neq \emptyset$, and let Γ' be a closed cone such that $\Gamma' \cap \partial \bar{B}(0, 1) \subset \operatorname{int} \Gamma$. Let $c \in \Sigma(\Gamma)$ and $T \in \mathcal{F}'$. If for $U(x) = C \exp(\alpha \cdot x)$, $x \in \mathbf{R}$, $C \in \mathbf{R}$, $C \neq 0$,*

$$T(x + h) \overset{s}{\sim} c(h) \cdot U(x), \quad h \in \Gamma, \ U \neq 0,$$

then there exists $e \in \mathcal{C}^\infty$ such that $e(x) \neq 0$, $x \in \mathbf{R}^n$, and $T(x) \overset{gs}{\sim} Ce(x)$, $\|x\| \to \infty$, $x \in \Gamma'$.

Proof. Let e be as in Lemma 1.4. It is enough to apply Proposition 1.2 and Theorem 1.2,b) to

$$\langle T(x + h)/e(x + h), \varphi(x) \rangle = \left\langle \frac{c(h)}{e(x + h)} \frac{T(x + h)}{c(h)}, \varphi(x) \right\rangle, \varphi \in \mathcal{F}. \quad \square$$

The next proposition also gives a relation between the S-asymptotics and the generalized S-asymptotics.

Proposition 1.17. *Suppose that $T \in \mathcal{F}', \Gamma = \{x \in \mathbf{R}^n, x = (0, \ldots, x_k, 0, \ldots, 0)\}, T/c = D_{x_k} S$ and $T(x) \overset{gs}{\sim} ac(x), \|x\| \to \infty, x \in \Gamma$. Then $S(x + h) \overset{s}{\sim} h_k \cdot 1, h \in \Gamma$.*

Proof. By L'Hospital's rule with Stolz's improvement, we have (with $h = (0, \ldots, h_k, 0, \ldots, 0)$ and $S * \phi(h) = \langle S(x + h), \phi(x) \rangle$)

$$\lim_{h_k \to \infty} \langle S(x + h)/h_k, \phi(x) \rangle = \lim_{h_k \to \infty} \frac{(S * \check{\phi}(h))}{h_k} = \lim_{h_k \to \infty} \left(\frac{T}{c} * \check{\phi} \right)(h)$$

$$= \lim_{h_k \to \infty} \left\langle \frac{T(x + h)}{c(x + h)}, \phi(x) \right\rangle = \langle a, \phi(x) \rangle. \quad \square$$

Finally, let us point out that if we take the limit

$$\lim_{h \in \Gamma, \|h\| \to \infty} T(x + h)/(c(h)e(x + h)),$$

in Definition 1.6., then nothing new is obtained (cf. Definition 1.6) for the following three cases: the one-dimensional case, when $\Gamma \subset \mathbf{R}^n$ is a ray, and when Γ has nonempty interior.

2   Quasi-asymptotics in $\mathcal{F}'$

2.1   *Definition of quasi-asymptotics at infinity over a cone*

Quasi-asymptotics has been originally defined and studied for tempered distributions. The motivation for such a choice can be found in Theorem 2.3 below. The first paper dealing with the analysis of the quasi-asymptotics was written by Zavyalov [200]. Thereafter, many results concerning the theory and applications of this notion have appeared. The main features of the theory (published until the year 1986) have been collected in the monograph [192]. We start with the most general definition of quasi-asymptotics of tempered distributions ([192]).

Let Γ be a closed convex acute solid cone (cf. **0.2**) in $\mathbf{R}^n$ and let $\{U_k; k \in I \subset \mathbf{R}\}$ be a family of linear nonsingular transforms of $\mathbf{R}^n$ which leave the cone Γ invariant (automorphisms of Γ). We assume that $J_k = \det U_k > 0$ and that I has ∞ as a limit point. Furthermore, let $\rho(k)$ be a positive function defined on I. Denote by $\mathcal{S}'_\Gamma$ the space of tempered distributions with supports in the cone Γ. It is important to mention that $\mathcal{S}'_\Gamma$ is isomorphic to the space $\mathcal{S}'(\Gamma)$ (see **0.5**).

Definition 2.1. Let $T \in \mathcal{S}'_\Gamma$. It is said that T has the quasi-asymptotics in the cone Γ over the family $\{U_k; k \in I\}$ related to the positive measurable function $\rho(k)$ if there exists a tempered distribution $g \neq 0$ such that

$$\lim_{k \in I, k \to \infty} \frac{1}{\rho(k)} T(U_k x) = g(x) \text{ in } \mathcal{S}'.$$

We will discuss the quasi-asymptotics in a particular case, namely, if $T \in \mathcal{F}'$, $U_k \Gamma = k\Gamma$ and $I = (0, \infty)$, where Γ is a closed, convex and acute cone in $\mathbf{R}^n$ (see [33]). For results related to Definition 2.1, see [192].

Definition 2.2. Let $T \in \mathcal{F}'_\Gamma$. It is said that T has the quasi-asymptotics in the cone Γ related to a positive measurable function ρ if there exists $g \neq 0$ such that

$$w\text{-}\lim_{k \to \infty} \frac{1}{\rho(k)} T(kx) = g(x) \text{ in } \mathcal{F}'.$$

We write in short $T(kx) \stackrel{q}{\sim} \rho(k)g(x)$, $k \to \infty$ in $\mathcal{F}'$; we will omit the space from the notation whenever it is clear from the context.

The definition itself forces ρ and g to have very specific forms. The following proposition states precisely to which classes of functions and distributions ρ and g must belong.

Proposition 2.1. *Let* $T \in \mathcal{F}'_\Gamma$. *If* $T(kx) \overset{q}{\sim} \rho(k)g(x)$, $k \to \infty$, *then:*

a) ρ *is a regularly varying function of the form* $\rho(k) = k^\alpha L(k)$, $k \geq k_0$, *(see* **0.3***) where* $\alpha \in \mathbf{R}$ *and* L *is a slowly varying function.*

b) g *is a homogeneous element of* $\mathcal{F}'_\Gamma$ *with degree* α, *that is,* $g(kx) = k^\alpha g(x)$, $k > 0$.

Proof. We take $\phi \in \mathcal{F}_\Gamma$ such that $\langle T, \phi \rangle \neq 0$. Let $K \subset (0, \infty)$ be any compact set. Then, by the assumption, we have

$$\lim_{k \to \infty} \left\langle \frac{T(ktx)}{\rho(kt)}, \phi(x) \right\rangle = \langle g(x), \phi(x) \rangle \tag{2.1}$$

uniformly in t, provided that t belongs to the compact set K. On the other hand, we have uniformly in $t \in K$

$$\lim_{k \to \infty} \left\langle \frac{T(kx)}{\rho(k)}, \frac{\phi(x/t)}{t} \right\rangle = \left\langle g(x), \frac{\phi(x/t)}{t} \right\rangle. \tag{2.2}$$

Combining relations (2.1) and (2.2), we obtain for $t > 0$,

$$\lim_{k \to \infty} \frac{\rho(kt)}{\rho(k)} = \lim_{k \to \infty} \frac{\langle T(kx), \frac{\phi(x/t)}{t} \rangle}{\langle T(ktx), \phi(x) \rangle} = \frac{\langle g(x), \frac{\phi(x/t)}{t} \rangle}{\langle g(x), \phi(x) \rangle} = C(t)$$

uniformly in $t \in K$. This is just the definition of a regularly varying function (see **0.3**). We thus have $C(t) = t^\alpha$, $t > 0$, for some $\alpha \in \mathbf{R}$; therefore, ρ has the desired form.

In order to prove that g is homogeneous of degree α, we have to use (2.1):

$$\langle g(tx), \varphi(x) \rangle = \lim_{k \to \infty} \left\langle \frac{T(ktx)}{\rho(k)}, \varphi(x) \right\rangle$$

$$= \lim_{k \to \infty} \frac{\rho(kt)}{\rho(k)} \left\langle \frac{T(ktx)}{\rho(kt)}, \varphi(x) \right\rangle$$

$$= \langle t^\alpha g(x), \varphi(x) \rangle, \quad \varphi \in \mathcal{F}_\Gamma, \ t > 0. \qquad \square$$

Remark. It is easy to prove that in case $\mathcal{F}'_\Gamma = \mathcal{S}'_{\mathbf{R}_+} \equiv \mathcal{S}'_+$, then $g = C f_{\alpha+1}$, where f_α is the homogeneous tempered distribution of degree $\alpha - 1$ given in **0.4**.

We emphasize that that all *homogeneous distributions* on the real line are explicitly known. Recall ([56]), they are of the form

$$g(x) = C_- x_-^\alpha + C_+ x_+^\alpha \quad \text{if } \alpha \notin \mathbf{N},$$

$$g(x) = \gamma \delta^{(k-1)}(x) + \beta x^{-k} \quad \text{if } \alpha = -k \in \mathbf{N}$$

(2.3)

where $C_-, C_+ \gamma, \beta \in \mathbf{C}$ and the distribution x^{-k} stands for the standard regularization of the corresponding function [56, 63], i.e., $x^{-1} = (\log|x|)'$, $-k\, x^{-k-1} = (x^{-k})'$. We refer to ([40, 56, 63]) for the explicit form of multidimensional homogeneous distributions.

An more general analysis of homogeneous generalized functions can be found in [85], but only in one dimension. We comment only part of it.

Let $W_\beta(I)$ be an abstract locally convex function space whose elements are defined on $I = (-\infty, \infty)$ or on $I = (0, \infty)$ and for which

$$\phi(\cdot) \mapsto \phi\left(\frac{\cdot}{y}\right), \quad y > 0,$$

is a continuous mapping from $W_\beta(I)$ to $W_\beta(I)$. If an element $g \in W_\beta'(I)$, where $W_\beta'(I)$ is the dual space of $W_\beta(I)$, satisfies the equation

$$g(y\cdot) = y^\alpha g(\cdot), \quad y > 0, \quad \alpha \in \mathbf{R},$$

in the sense that $\langle g(yx), \phi(x) \rangle = y^\alpha \langle g(x), \phi(x) \rangle$ for all $\phi \in W_\beta(I)$, then g is called a *homogeneous generalized function* from $W_\beta'(I)$ of degree α.

Let us suppose that $W_\beta(I)$ is given by the subspace of $C^\infty(I \setminus \{0\})$ for which all the seminorms

$$p_{n,\beta}(\phi) = \int_I |x|^\beta \left| x^n \phi^{(n)}(x) \right| dx, \quad n \in \mathbf{N}_0$$

are finite. In particular $p_{0,\beta}$ is a norm.

One can easily prove that for a given $\phi \in W_\beta(I)$ the sequence $\langle \eta_n \cdot \phi \rangle_{n \in \mathbf{N}}$, where η_n is an even positive function in $\mathcal{D}(I)$ such that $\eta_n(x) = 0$ for $x \in [0, 1/2n] \cup [n+1, \infty)$ and $\eta_n(x) = 1$ for $x \in [1/n, n]$, $n \in \mathbf{N}$, converges to ϕ in $W_\beta(I)$ as $n \to \infty$. This implies that the space $\mathcal{D}(I \setminus \{0\})$ is dense in $W_\beta(I)$. Thus, all homogeneous generalized functions of order $\alpha \notin -\mathbf{N}$ in $W_\beta(I)$ are of the form

$$A_1 x_+^\alpha + A_2 x_-^\alpha \text{ resp. } A_1 x_+^\alpha \text{ if } I = \mathbf{R} \text{ resp. } I = (0, \infty). \quad (2.4)$$

Let $\mathcal{U}_\beta(I)$, $\beta \in \mathbf{R}$, be a subspace of $W_\beta(I)$ with a locally convex topology, such that the inclusion mapping $i : \mathcal{U}_\beta(I) \to W_\beta(I)$ is continuous, as well as the following one

$$\mathcal{U}_\beta(I) \to \mathcal{U}_\beta(I) : \phi(\cdot) \mapsto \phi(\cdot/y), \quad y \in (0, \infty).$$

If $f \in W_\beta'(I)$ and if it is homogeneous of order α, then $f|_{\mathcal{U}_\beta(I)}$, the restriction of f on $\mathcal{U}_\beta(I)$, is of the form (2.4). A natural question is then to characterize spaces for which all homogeneous generalized functions are of the form (2.4). One can show that this is the case for $\mathcal{U}_\beta'(I)$. At a first sight, the introduction of these spaces may seem to be artificial, but it is naturally connected with the problem we are discussing.

Note that $\mathcal{D}(I \setminus \{0\})$ need not be dense in $\mathcal{U}_\beta(I)$. This is in fact the only interesting case. $\mathcal{U}_\beta(I)$ can contain $\mathcal{D}(I \setminus \{0\})$ but not as a dense subspace. Also, $\mathcal{U}_\beta(I)$ may not contain $\mathcal{D}(I)$ at all. If $\mathcal{U}_\beta(I)$ contains $\mathcal{D}(I \setminus \{0\})$ then clearly $\mathcal{U}_\beta(I)$ is dense in $W_\beta(I)$. Even in that case, we could not use this fact for the proof that any homogeneous element g on $\mathcal{U}_\beta(I)$ can be extended to $W_\beta(I)$, i.e., $g \in W_\beta'(I)$, and thus it is of the form (2.4).

For $\mathcal{U}_\beta(I)$, we assume: if $\phi \in \mathcal{U}_\beta(I)$ and $y_0 \in (0, \infty)$, then

$$\lim_{y \to y_0} \frac{\phi(\cdot/y) - \phi(\cdot/y_0)}{y - y_0} = \frac{d}{dy} \phi(\cdot/y) \Big|_{y=y_0}$$

in the sense of convergence in $\mathcal{U}_\beta(I)$. We have the following proposition (for the proof see [85]).

Proposition 2.2. *A generalized function* $f \in \mathcal{U}_\beta'(I)$ *is homogeneous of order* $\alpha \in \mathbf{R}$ *if and only if for each* $\phi \in \mathcal{U}_\beta(I)$

$$\langle f(x), (x\phi(x))' + \alpha\phi(x) \rangle = 0,$$

i.e., if and only if f fulfills the equation $xf' = \alpha f$ in $\mathcal{U}_\beta'(I)$.

A further analysis shows that $\mathcal{U}_\beta'(I)$ has the desired properties, for details see [85].

2.2 *Basic properties of quasi-asymptotics over a cone*

Recall, $\Gamma^* = \{y; y \cdot x \geq 0, \text{ for every } x \in \Gamma\}$ is the *conjugate cone* to the cone Γ (cf. **0.2**). Denote by $C = \text{int}\,\Gamma^*$. The *characteristic function of a* closed convex solid acute *cone* Γ is denoted by $\theta_{1,\Gamma}$. The function

$$K_C(z) = \int\limits_{\Gamma} e^{iz \cdot t} dt, \quad z \in \mathbf{R}^n + iC$$

is called *the Cauchy-Szegö kernel* of the tube domain $\mathbf{R}^n + iC$. *The closed convex acute solid cone* Γ *is regular* if $K_C(z)$ is a divisor of unity in the Vladimirov algebra $\mathcal{H}(C) = \mathcal{L}(\mathcal{S}'(\Gamma))$, where $\mathcal{L}$ denotes the Laplace transform (cf. [189], §12]). In case $n = 1, 2, 3$ it is well-known (see Chapter 1, §2.7 in [192]) that all (closed convex acute solid) cones are regular. For regular cones the distribution $\theta_{\alpha,\Gamma}$ is given by

$$\int\limits_{\Gamma} e^{iz \cdot t} \theta_{\alpha,\Gamma}(t) dt = K_C^\alpha(z), \quad z \in \mathbf{R}^n + iC, \quad \alpha \in \mathbf{R}.$$

This distribution has many interesting properties (see [33]):

1) $\operatorname{supp} \theta_{\alpha,\Gamma} \subset \Gamma$ and $\theta_{\alpha,\Gamma} \in \mathcal{S}'_\Gamma$;

2) $\theta_{\alpha,\Gamma}(kt) = k^{n(\alpha-1)} \theta_{\alpha,\Gamma}(t)$, $t \in \Gamma$, $k > 0$;

3) $\theta_{\alpha,\Gamma} * \theta_{\beta,\Gamma} = \theta_{\alpha+\beta,\Gamma}$;

4) For every $m \geq 0$ there exists n_0 such that $\theta_{\alpha,\Gamma} \in C^m(\mathbf{R}^n)$, $\alpha > n_0$;

5) $\{\operatorname{supp} \theta_{\alpha,\Gamma}(e - t)\} \cap \Gamma \subset \{\|t\| < R\}$, where $R > 0$ do not depend on α and $e \in \operatorname{pr} \Gamma = \Gamma \cap \{\|x\| = 1\}$.

6) For any $p \in \mathbb{N}_0$, there exists $N_0 > 0$ such that for some A and $q > 0$ and for every $n \geq N_0$, $\|\theta_{n,\Gamma}(x - t)\|_{p,\Gamma} \leq A(1 + \|x\|)^q$, $x \in \Gamma$ (cf. **0.5** for $\| \ \|_{p,\Gamma}$).

Operations of fractional derivatives (for $\alpha \leq -1$) and fractional integrals (for $\alpha > -1$) can be defined on $\mathcal{S}'_\Gamma$ via

$$\theta_{\alpha,\Gamma} : T \to \theta_{\alpha,\Gamma} * T,$$

the latter defines a continuous linear operator on $\mathcal{S}'_\Gamma$. We will use the notation $T^{(\alpha)}$ for $\theta_{-\alpha,\Gamma} * T$. By the properties of the convolution, we have

$$T^{(-\alpha)}(t) = \langle T(\tau), \theta_{\alpha,\Gamma}(t - \tau) \rangle,$$

if α is sufficiently large.

Definition 2.3. Suppose that $f \in L_{loc}(\Gamma)$. It is said that f has an asymptotic behavior in the cone Γ related to the positive function $\rho(k)$, $k \in (0, \infty)$ if there exists a function $g \neq 0$ such that

$$\lim_{k \to \infty} f(kx)/\rho(k) = g(x), \quad x \in \Gamma, \tag{2.5}$$

$$|f(kx)/\rho(k)|\le h(x), \quad x \in \Gamma, \ k \ge k_0 \ge 0, \qquad (2.6)$$

and $\varphi(x)h(x) \in L^1(\Gamma)$, for every $\varphi \in \mathcal{F}$.

Proposition 2.3. *If a locally integrable function f has the asymptotic behavior in the cone Γ related to ρ (Definition 2.3), and if f defines a regular element of $\mathcal{F}'_\Gamma$, then $f(kx) \overset{q}{\sim} \rho(k)g(x), \ k \to \infty$.*

Proof. We have

$$\lim_{k\to\infty} \langle f(kx)/\rho(k), \varphi(x)\rangle = \lim_{k\to\infty} \int_\Gamma f(kx)\varphi(x)/\rho(k)dx, \quad \varphi \in \mathcal{F}.$$

The assumption over f allows one to exchange the limit with the integral, by the Lebesgue theorem. □

Proposition 2.4. *Let $\mathcal{F}$ have the property that $\{\varphi\left(\dfrac{t}{k}\right); \ k \ge k_0 > 0\}$ is bounded in $\mathcal{F}$ for each $\varphi \in \mathcal{F}$. Suppose that $S, T \in \mathcal{F}'$, and $\rho(k) = k^\alpha L(k), \ \alpha > -n$, or $\alpha = -n$ and $L(k) \to \infty$ as $k \to \infty$.*

a) If S has compact support, then

$$w.\lim_{k\to\infty} S(kx)/\rho(k) = 0, \quad in \ \mathcal{F}'.$$

b) If $T = S$ on $\Gamma \cap \{\|x\| > R\}$ for some $R > 0$, and $S(kx) \overset{q}{\sim} \rho(k)g(x), \ k \to \infty$, then $T(kx) \overset{q}{\sim} \rho(k)g(x), \ k \to \infty$, as well.

Proof. a) For every $\varphi \in \mathcal{F}$

$$\langle S(kx)/\rho(k), \varphi(x)\rangle = \frac{1}{k^n\rho(k)} \left\langle S(x), \varphi\left(\frac{x}{k}\right)\right\rangle.$$

The set $\{\varphi(x/k); k \ge k_0 > 0\}$ is a bounded set in $\mathcal{F}$. Consequently, the set $\{\langle f(x), \varphi(x/k)\rangle; k \ge k_0 > 0\}$ is bounded in $\mathbf{R}$, too. Since $k^n\rho(k) \to \infty, k \to \infty$, the assertion in a) follows.

b) By assumption, $T - S$ has a compact support. Then, by a), for each $\varphi \in \mathcal{F}$

$$\lim_{k\to\infty} \langle T(kx)/\rho(k), \varphi(x)\rangle = \lim_{k\to\infty} \langle (T - S)(kx)/\rho(k), \varphi(x)\rangle$$

$$+ \lim_{k\to\infty} \langle S(kx)/\rho(k), \varphi(x)\rangle$$

$$= \lim_{k\to\infty} \langle S(kx)/\rho(k), \varphi(x)\rangle$$

which proves b). □

Remark. Proposition 2.4 b) asserts that the quasi-asymptotics related to $\rho(k) = k^\alpha L(k)$, $\alpha > -n$, is a local property. If $\alpha < -n$, then the quasi-asymptotics has no longer this property. The next example, in $\mathcal{S}'(\mathbf{R})$, illustrates this fact: If $f = \delta^{(m)} + (x_+^\beta)$, $m \in \mathbf{N}_0$, $\beta \notin N$, then f has quasi-asymptotics related to $\rho(k) = k^q$, where $q = \max(-m-1, -\beta)$ (see Examples **4** and **7** in **2.3**). On the other hand, if $\alpha = -1$, then this property depends on L. See also Example **5** of **2.3** in relation to Proposition 2.4 b).

Proposition 2.5. *Let* $m = (m_1, \ldots, m_n) \in (\mathbf{N}_0)^n$ *and* $T \in \mathcal{F}'_\Gamma$. *If* $T(kx) \overset{q}{\sim} \rho(k)g(x)$, $k \to \infty$ *and* $x^m \in M_{(\cdot)}$ *then*

$$(kx)^m T(kx) \overset{q}{\sim} k^{|m|}\rho(k)x^m g(x), \ k \to \infty.$$

Proof. We have

$$\lim_{k \to \infty} \left\langle \frac{(kx)^m T(kx)}{k^{|m|}\rho(k)}, \varphi(x) \right\rangle = \lim_{k \to \infty} \left\langle \frac{T(kx)}{\rho(k)}, x^m\varphi(x) \right\rangle$$

$$= \langle g(x), x^m\varphi(x) \rangle = \langle x^m g(x), \varphi(x) \rangle, \ \varphi \in \mathcal{F}. \ \square$$

Remark. If g is a distribution with support $\{0\}$, then $t^m g(t)$ may be equal zero.

One of the most useful and frequently used theorems that characterize the quasi-asymptotic behavior is the following one.

Theorem 2.1. *Let* $T \in \mathcal{S}'_\Gamma$ *and* Γ *be a regular cone. Then, T has quasi-asymptotics in* Γ *related to* ρ *if and only if there exists* $\alpha \in \mathbf{R}_+$ *such that* $T^{(-\alpha)}$ *has the quasi-asymptotics in* Γ *related to* $k^{n\alpha}\rho(k)$.

Proof. By the mentioned properties of $\theta_{\alpha,\Gamma}$,

$$\frac{1}{k^{n\alpha}\rho(k)}\langle T^{(-\alpha)}(kx), \varphi(x) \rangle = \frac{k^{-n(\alpha+1)}}{\rho(k)}\langle T^{(-\alpha)}(x), \varphi(x/k) \rangle$$

$$= \frac{k^{-n(\alpha+1)}}{\rho(k)}\langle T(x), \langle \theta_{\alpha,\Gamma}(\tau), \varphi((x+\tau)/k) \rangle \rangle$$

$$= \frac{1}{\rho(k)}\langle T(kx), \langle \theta_{\alpha,\Gamma}(\tau), \varphi(x+\tau) \rangle \rangle,$$

where $\varphi \in \mathcal{S}(\Gamma)$. Since the cone Γ is regular, the mapping: $\varphi(x) \to \langle \theta_{\alpha,\Gamma}(t), \varphi(x+t) \rangle$ is an automorphism of the space $\mathcal{S}(\Gamma)$. In order to complete the proof, it is enough to take limit in the last equality. $\qquad \square$

The same theorem can be proved in the same way for $\mathcal{S}'^*$ if (M.1), (M.2)' and (M.3)' are satisfied.

We now give a structural theorem for quasi-asymptotics on cones.

Theorem 2.2. *Suppose that Γ is a regular cone. Then $T \in \mathcal{S}'_\Gamma$ has quasi-asymptotics in Γ related to ρ if and only if there exists an integer N such that $T^{(-N)}$ is a continuous function and has the asymptotic behavior in Γ related to $k^{nN}\rho(k)$ (see Definition 2.3).*

Proof. The sufficiency follows from Theorem 2.1. Let us prove the necessity.

Since $T(kx)/\rho(k)$ converges to g in $\mathcal{S}'$ (in $\mathcal{S}'(\Gamma)$), there exists $p \in \mathbf{N}_0$ such that it converges in $\mathcal{S}'_p(\Gamma)$, as well (see **0.5.1**). Also, by properties 4) and 5) of $\theta_{\alpha,\Gamma}$ there exists $N_0 \in \mathbf{N}$ such that for every $t \in \mathbf{R}^n$ and $N \geq N_0$ the function $\theta_{N,\Gamma}(t-x) \in \mathcal{S}'_p(\Gamma)$. Then $T^{(-N)}$ is continuous and

$$\lim_{k \to \infty} \frac{1}{k^{nN}\rho(k)} T^{(-N)}(kx) = \lim_{k \to \infty} \frac{1}{k^{nN}\rho(k)} \langle T(t), \theta_{N,\Gamma}(kx - t) \rangle$$

$$= \lim_{k \to \infty} \langle T(kt)/\rho(k), \theta_{N,\Gamma}(x - t) \rangle$$

$$= \langle g(t), \theta_{N,\Gamma}(x - t) \rangle.$$

By property 6) of $\theta_{\alpha,\Gamma}$, there exist A and $q > 0$ such that

$$\left| \frac{1}{k^{nN}\rho(k)} T^{(-N)}(kx) \right| \leq C \|\theta_{N,\Gamma}(x - t)\|_{p,\Gamma} \leq A(1 + \|x\|)^q, \quad x \in \Gamma. \quad (2.7)$$

In this way we proved that $T^{(-N)}$ has the asymptotic behavior in Γ related to $k^{nN}\rho(k)$ (cf. Definition 2.3) . $\qquad \square$

We shall quote a remark given in [192]:

"In many cases it is important to know exactly which primitive of a distribution f, having quasi-asymptotics, already has an asymptotics. There is no simple or universal criterion. For instance, it would be natural to suppose that the condition $f \in \mathcal{S}'_p(\Gamma)$ guarantees the existence of such an N (depending perhaps on p, on a family $\{U_k, k \in I\}$ and on a function $\rho(k)$, $k \in I$) for which the primitive $f^{(-N)}$ has the asymptotics."

The following example shows that it is not the case, in general:

Let $\Gamma = \overline{\mathbf{R}}_+$ and let $\{f_n(\xi) = H(\xi)\exp(i\xi^{1/n}); \; n = 1, 2, \dots\}$ be a family of functions in $\mathcal{S}'_+$. Each function f_n has the quasi-asymptotics $-i^n n!\, \delta(\xi)$ over the family of transforms $\{U_k\xi = k\xi; \; k > 0\}$ related to the function $\rho(k) = 1/k$, $k > 0$ (cf. Definition 2.1). For every $n \in \mathbf{N}$ the function $f_n^{(-N)}$ has the asymptotics over the family of transforms $\{U_k\xi = k\xi, \; k > 0\}$ with respect to the function k^{N-1}, $k > 0$ if and only if $N - 1 > ((n-1)/n)N$; that is, when $N > n$ and, hence, $N \to \infty$, as $n \to \infty$. On the other hand, all the functions f_n, $n = 1, 2, \dots$ are infinitely differentiable for $\xi > 0$ and uniformly bounded (cf. [192]).

We prove now that the space of tempered distributions is naturally related to the concept of quasi-asymptotic behavior at infinity.

Theorem 2.3. *Let $T \in \mathcal{D}'$, $\operatorname{supp} T \subset \Gamma$, where Γ is regular. If*

$$\lim_{k \to \infty} \left\langle \frac{T(kx)}{\rho(k)}, \phi(x) \right\rangle = \langle g, \phi \rangle, \quad \phi \in \mathcal{D} \tag{2.8}$$

for some regularly varying function ρ and $g \neq 0$, then $T \in \mathcal{S}'_\Gamma$, and it has quasi-asymptotic behavior at infinity related to ρ.

Proof. By property 5) of $\theta_{\alpha,\Gamma}$ there exists a sufficiently large $R > 0$ such that $\theta_{m,\Gamma}(e - x) = 0$ for $e \in \operatorname{pr}\Gamma$, $x \in \Gamma$ and $\|x\| > R$. Let $\eta \in \mathcal{D}$ be such that $\eta(x) = 1$ for $\|x\| \leq R$. Then, $\left(\eta(x)\dfrac{T(kx)}{\rho(k)}\right), k \in \mathbf{N}$, is convergent in $\mathcal{S}'(\Gamma)$. Hence there exists a $p \in \mathbf{N}$ such that it converges in $\mathcal{S}'_p(\Gamma)$ (see **0.5.1**). By properties 4) and 5) of $\theta_{\alpha,\Gamma}$, we can find a sufficiently large $m \in \mathbf{N}$ such that

$$\lim_{k \to \infty} \frac{1}{\rho(k)}\langle \eta(x)T(kx), \theta_{m,\Gamma}(e - x)\rangle = g_m(e),$$

where $\theta_{m,\Gamma}(e-x)$ is in $\mathcal{S}'_p(\Gamma)$. Now the left-hand side can be written without η since $\theta_{m,\Gamma}(e - x) = 0$ for $|x| \geq R$, and this implies that

$$\lim_{k \to \infty} \frac{1}{\rho(k)}\frac{1}{k^{mn}}T^{(-m)}(ke) = \lim_{k \to \infty} \frac{1}{\rho(k)}\langle T(kx), \theta_{m,\Gamma}(e - x)\rangle = g_m(e) \tag{2.9}$$

and the last limit exists.

As in Theorem 2.2, it follows that $T^{(-N)} \in L^1_{loc}$ and that the limit in (2.5) for $f = T^{(-N)}$ and with $k^{Nm}\rho(k)$ instead $\rho(k)$ exists. Also (2.6) is satisfied.

Consequently, $T^{(-N)}$ is a locally integrable function having the asymptotic behavior related to $k^{nN}\rho(k)$ in the cone Γ. Now, we can use Theorem 2.2 to complete the proof. □

We now discuss some results in the context of ultradistributions.

Theorem 2.4. *Suppose that M_p satisfies (M.1), (M.2) and (M.3). If $f \in \mathcal{E}'^*(\mathbf{R})$ and supp $f \subset [0, \infty)$, then there exists a $p \in \mathbf{N}_0$ such that*

$$\lim_{k \to \infty} \langle k^{p+1} f(kx), \varphi(x) \rangle = C, \quad \varphi \in \mathcal{S}^*_{[0,\infty)},$$

where C can be zero, too.

Proof. By Theorem 10.3 in [79], there exist an ultradifferential operator $P(D) = \sum\limits_{n=0}^{\infty} a_n D^n$ of $*$ class and a compactly supported continuous function G, supp $G = K \supset$ supp$f$, such that

$$f = \sum_{n=0}^{\infty} a_n D^n G.$$

Suppose that $a_p \neq 0$ and $a_n = 0$, $n < p$. We have to analyze

$$\langle k^{p+1} f(kx), \varphi(x) \rangle = \left\langle k^{p+1} \left(\sum_{n=0}^{\infty} a_n D^n G \right)(kx), \varphi(x) \right\rangle$$

$$= k^p \sum_{n=p}^{\infty} (-1)^n a_n k^{-n} \left\langle G(x), \varphi^{(n)}\left(\frac{x}{k}\right) \right\rangle = (-1)^p a_p \int_K G(x)\varphi^{(p)}\left(\frac{x}{k}\right) dx$$

$$+ \sum_{n=p+1}^{\infty} (-1)^n a_n k^{-n+p} \int_K G(x)\varphi^{(n)}\left(\frac{x}{k}\right) dx, \quad \varphi \in \mathcal{S}^*_{[0,\infty)}.$$

Since

$$\int_K G(x)\varphi^{(p)}\left(\frac{x}{k}\right) dx \to \int_K G(x)dx \varphi^{(p)}(0), \quad k \to \infty,$$

it remains to prove that

$$\left| \sum_{n=p+1}^{\infty} (-1)^n a_n k^{-n+p} \int_K G(x)\varphi^{(n)}\left(\frac{x}{k}\right) dx \right| \leq C \sum_{n=p+1}^{\infty} k^{-n+p} \int_K |G(x)|\frac{L^n}{M_n} dx$$

$$\leq C_1 \sum_{n=p+1}^{\infty} k^{-n+p} \int_K |G(x)| dx \to 0, \quad k \to \infty,$$

in both cases, (M_p) and $\{M_p\}$.   □

Theorem 2.5. *Suppose that (M.1), (M.2) and (M.3) are satisfied. Let $T \in \mathcal{D}'^*(\mathbf{R}), \operatorname{supp} T \subset [0, \infty)$ and $c(k) = k^\alpha L(k)$, $k > 0, \alpha > -1$. If $T(x + h) \overset{s}{\sim} c(h)U$, then $T \in \mathcal{S}'^*_+$ and T has the quasi-asymptotics related to c, as well.*

Proof. Without loss of generality, we can suppose that c is continuous in $(0, \infty)$ (cf. Remarks after Proposition 1.2).

Denote by ω a function belonging to $\mathcal{D}^*$ such that $\omega(x) = 1$, $x \in [0, h_0]$, $h_0 > 0$. Then, $T = \omega T + (1 - \omega)T$. The support of ωT is compact, therefore $\omega T \in \mathcal{E}'^*$. By Theorem 2.4

$$\lim_{k \to \infty} \left\langle \frac{\omega(kt)T(kt)}{c(k)}, \varphi(t) \right\rangle = 0, \ \varphi \in \mathcal{S}^*_{[0,\infty)} \,.$$

The S-asymptotics is a local property if $\alpha > -1$ (Theorem 1.2, c)). Therefore $(1 - \omega)T$ has the same S-asymptotics as T and $\operatorname{supp}(1 - \omega)T \subset [h_0, \infty)$.

We shall use Theorem 1.10 with the assumptions that the set A of that theorem satisfies $A + \mathbf{R}_+ \subset [h_0, \infty)$, $h_0 > 0$, $(\Gamma = \mathbf{R}_+)$ and that $\operatorname{supp} f_i \subset (h_0 - \varepsilon, \infty), h_0 - \varepsilon > 0$, $i = 1, 2$.

Denote by $g_i(h) = f_i(h)/c(h)$, $h \in (h_0 - \epsilon, \infty)$. Then $f_i(h) = c(h)g_i(h), h \geq h_0 - \epsilon > 0$, $i = 1, 2$, and $\lim_{h \to \infty} g_i(h) = C_i$, $i = 1, 2$. Therefore,

$$((1 - \omega)T)(x) = P(D)c(x)g_1(x) + c(x)g_2(x), \ x > h_0 > 0 \,.$$

It follows that $(1 - \omega)T$ belongs to $\mathcal{S}'^*$. Consequently, $T \in \mathcal{S}'^*$. Suppose now that $\phi \in \mathcal{S}^*_{[0,\infty)}$. Then

$$\lim_{k \to \infty} \left\langle \frac{(1 - \omega(kt))T(kt)}{c(k)}, \phi(t) \right\rangle$$

$$= \lim_{k \to \infty} \left\{ \left\langle \frac{c(kt)g_1(kt)}{c(k)}, \sum_{i=1}^{\infty} \frac{(-1)^i}{k^i} a_i \left(\frac{d}{dt} \right)^i \phi(t) \right\rangle \right.$$

$$\left. + \left\langle \frac{c(kt)}{c(k)}(a_0 g_1(kt) + g_2(kt)), \phi(t) \right\rangle \right\}. \tag{2.10}$$

By the properties of g_1 and g_2, the last summand in (2.10) has a limit for every $\phi \in \mathcal{S}^*_{[0,\infty)}$. More precisely,

$$\lim_{k\to\infty} \left\langle \frac{c(kt)}{c(k)}(a_0 g_1(kt) + g_2(kt)), \phi(t) \right\rangle = \left\langle (a_0 C_1 + C_2)x^\alpha, \phi(x) \right\rangle.$$

Since $\left\{ \frac{c(k\cdot)g_1(k\cdot)}{c(k)}; k \geq 1 \right\}$ is a bounded set in $\mathcal{S}'^*_{[0,\infty)}$ and since

$$\sum_{i=1}^{\infty} \left(\frac{-1}{k}\right)^i a_i \left(\frac{d}{dx}\right)^i \phi(x)$$

tends to zero in $\mathcal{S}^*_{[0,\infty)}$ as $k \to \infty$, the first summand in (2.10) tends to zero as $k \to \infty$.

From the existence of the S-asymptotics of T related to c, it follows that $a_0 C_1 + C_2 \neq 0$.

Thus, we have proved that $T \in \mathcal{S}'^*_+$ and that T has the quasi-asymptotics related to c. $\qquad\square$

Let $\varphi \in \mathcal{S}^*_{[0,\infty)}$ and let f_γ, $\gamma \in \mathbf{R}$, be the function defined in **0.4.** Then $(f_\gamma * \varphi)(\xi) = \psi(\xi)$, $\xi \geq 0$ belongs to $\mathcal{S}^*_{[0,\infty)}$. One can prove that the mapping $\varphi \mapsto f_\gamma * \varphi$ is an automorphism of $\mathcal{S}^*_{[0,\infty)}$. The method for proving this property of f_γ is the same as for the space $\mathcal{S}[0,\infty)$.

Let $f \in \mathcal{S}'^*_{[0,\infty)}$. Recall, we denote by $f^{(-m)}$ an element belonging to $\mathcal{S}'^*_{[0,\infty)}$ defined by

$$\langle f^{(-\gamma)}(x), \varphi(x) \rangle = \langle f(x), (\check{f}_\gamma * \varphi)(x) \rangle, \quad \varphi \in \mathcal{S}^*_{[0,\infty)}.$$

Proposition 2.6. *Suppose that $f \in \mathcal{S}'^*_{[0,\infty)}$ and that γ is a real number. Then, f has quasi-asymptotics related to ρ if and only if $f^{(-\gamma)}$ has the quasi-asymptotics related to $k^\gamma \rho(k)$.*

Proof. By the definition of $f^{(-\gamma)}$,

$$\langle f^{(-\gamma)}, \varphi \rangle = \langle f(\xi), (\check{f}_\gamma * \varphi)(\xi) \rangle$$

and this exists for every $\varphi \in \mathcal{S}^*_{[0,\infty)}$. Therefore, for every $\varphi \in \mathcal{S}^*_{[0,\infty)}$ and $k > 0$, we have

$$\left\langle f^{(-\gamma)}(k\xi), \varphi(\xi) \right\rangle = \frac{1}{k}\left\langle f^{(-\gamma)}(\xi), \varphi\left(\frac{\xi}{k}\right) \right\rangle$$

$$= \left\langle \frac{k^\gamma}{k} f(\xi), (\check{f}_\gamma * \varphi)\left(\frac{\xi}{k}\right) \right\rangle$$

$$= k^\gamma \langle f(k\xi), (\check{f}_\gamma * \varphi)(\xi) \rangle.$$

Since the mapping: $\varphi \to \check{f}_\gamma * \varphi$ is an automorphisms of $\mathcal{S}^*_{[0,\infty)}$, this completes the proof. □

Theorem 2.6. *Suppose that (M.1), (M.2) and (M.3) are satisfied and that* $c(h) = h^\alpha L(h)$, $h > h_0$, $\alpha \le -1$. *If* $T \in \mathcal{D}'^*(\mathbf{R})$, $T(x+h) \overset{s}{\sim} c(h)U$, $h \in \mathbf{R}_+$ *and* $\operatorname{supp} T \subset [0,\infty)$, *then* $T \in \mathcal{S}'^*_+$.

If $\alpha = -1$ *and* $\widehat{L}(x) = \int\limits_{h_1}^{x} t^{-1}L(t)dt \to \infty$, *as* $x \to \infty$, *then* T *has quasi-asymptotics related to* $t^{-1}\widehat{L}(t)$. *In the other cases,* T *has the quasi-asymptotics related to* t^{-p}, $p \in \mathbf{N}$, *but the limit may be zero.*

Proof. Let w be the function used in Theorem 2.5. Then $T = wT + (1-w)T$. Since wT has a compact support, by Theorem 2.4, wT has quasi-asymptotics related to $c(k) = k^{-n}$ for an $n \in \mathbf{N}$.

The support of $(1-w)T$ belongs to $[h_0, \infty)$, $h_0 > 0$. Since the S-asymptotics is a local property, $(1-w)T$ has the S-asymptotics related to c, as well, and

$$((1-w)T)(t) = \sum_{i=0}^{\infty} a_i D^i(t^\alpha L(t)E_1(t)) + (t^\alpha L(t)E_2(t)), \ t > h_0, \quad (2.11)$$

where $\operatorname{supp} E_i \subset (h_1, \infty)$, $0 < h_1 \le h_0$ and $\lim\limits_{t\to\infty} E_i(t) = C_i$, $i = 1,2$. We know that $a_0 C_1 + C_2 \ne 0$ because $(1-w)T$ has the S-asymptotics related to $h^\alpha L(h)$.

We can choose h_0 and h_1 in such a way that $a_0 E_1(t) + E_2(t)$ does not change the sign when $t \in (h_1, \infty)$. Using (2.11) and the convolution with f_1 (cf. **0.4**) it follows

$$((1-w)T)^{(-1)} = f_1 * (t^\alpha L(t)(a_0 E_1(t) + E_2(t)))$$

$$+ \sum_{i=1}^{\infty} a_i D^i(f_1 * (t^\alpha L(t)E_1(t))).$$

We have to analyze the function $F = f_1 * (t^\alpha L(t)(a_0 E_1(t) + E_2(t)))$. This function is equal to zero in $(0, h_1)$ and

$$F(x) = \int\limits_{h_1}^{x} t^\alpha L(t)(a_0 E_1(t) + E_2(t))dt, \ x \ge h_1.$$

Case 1. Let $\alpha < -1$ or $\alpha = -1$ and $\int\limits_{h_1}^{\infty} x^{-1}L(x)dx < \infty$. Then,

$$F(x) \to \int\limits_{h_1}^{\infty} t^{\alpha}L(t)(a_0E_1(t) + E_2(t))dt, \quad \text{as } x \to \infty.$$

Consequently, F has S-asymptotics related to 1. Since $\int\limits_{h_1}^{\infty} t^{-1}L(t)E_1(t)dt$ $< \infty$, $((1-w)T)^{(-1)}$ has S-asymptotics related to 1, too. By Theorem 2.5, $((1-w)T)^{(-1)}$ has quasi-asymptotics related to 1, and by Proposition 2.6, $(1-wT)$ has quasi-asymptotics related to h^{-1}.

Case 2. Let $\alpha = -1$ and $\int\limits_{h_1}^{\infty} t^{-1}L(t) \to \infty$, as $x \to \infty$. Then, $F(x) = \widehat{L}(x)$, where $\widehat{L}$ is also a slowly varying function (see Proposition 1.5.9 in [9]); f has S-asymptotics related to $\widehat{L}(x)$. We have the same situation with $\int\limits_{h_1}^{\infty} t^{-1}L(t)E_1(t)dt = \widehat{L}_1(x)$. By Theorem 2.5, $((1-w)T)^{(-1)}$ has quasi-asymptotics related to $L_0 = \max(\widehat{L}, \widehat{L}_1)$ and by Proposition 2.6, $(1-w)T$ has quasi-asymptotics related to $k^{-1}L_0(k)$.

Taking care of the quasi-asymptotics of wT and $(1-w)T$, we have established the assertion of Theorem 2.6. □

The following example illustrates different possibilities in case $\alpha < -1$.

We use three functions

$$f(x) = \begin{cases} 0, & 0 \le x \le 1 \\ x^{-1-\varepsilon}, & 1 < x, \ 1 < \varepsilon < 2, \end{cases}$$

and $F(x) = H(x)H(1-x)$, $x \in \mathbf{R}$; H is the Heaviside function.

The quasi-asymptotics are given by:

$$\langle f(kx), \varphi(x) \rangle = k^{-1-\varepsilon} \int\limits_{1/k}^{a} x^{-1-\varepsilon}\varphi(x)dx$$

$$= \frac{1}{\varepsilon}\varphi\left(\frac{1}{k}\right)\frac{1}{k} - \frac{1}{\varepsilon}\frac{1}{1-\varepsilon}\varphi'\left(\frac{1}{k}\right)\frac{1}{k^2} - \frac{1}{\varepsilon}\frac{1}{1-\varepsilon}\frac{1}{k^{1+\varepsilon}}\int\limits_{1/k}^{a} x^{1-\varepsilon}\varphi''(x)dx, \ k > 0;$$

$$k\langle F(kx), \varphi(x) \rangle = k\int\limits_{0}^{1/k} \varphi(x)dx = \int\limits_{0}^{1} \varphi\left(\frac{t}{k}\right)dt \to \varphi(0), \ k \to \infty;$$

$$k^2 \langle (DF)(kx), \varphi(x) \rangle = k \langle F(kx), \varphi'(x) \rangle \to \varphi'(0), \; k \to \infty, \; \varphi \in \mathcal{S}'^{*}_{+}.$$

Take now the distribution $T = f + C_1 F + C_2 DF$. For appropriated constants C_1 and C_2, T can have the quasi-asymptotics related to $k^{-1}, k^{-2}, k^{-1-\varepsilon}$.

The following problem is discussed in [38]: Let $f \in \mathcal{S}'_+$ and let $\{\varphi_k\}_{k \in \mathbb{N}}$ be a sequence in $\mathcal{S}$. Assume that the following limit exists

$$\lim_{k \to \infty} \frac{1}{\rho(k)} \langle f(kt), \varphi_k(t) \rangle = c,$$

where ρ is a regularly varying function.

The question is to find conditions under which the limit

$$\lim_{k \to \infty} \frac{1}{\rho(k)} \langle f(kt), \varphi(t) \rangle = c_\varphi$$

exists for all $\varphi \in \mathcal{S}$, i.e., that $f \in \mathcal{S}'_+$ has the quasi-asymptotics related to ρ.

The authors proved that $\{\varphi_k(t)\}_{k \in \mathbb{N}}$ cannot belong to $\mathcal{S}$. But they constructed a new space $\mathcal{S}^{a,M}_{b,N,\delta}$ of test functions and the space $(\mathcal{S}^{a,M}_{b,N,\delta})'$ of distributions in which the asked question has a positive answer. (cf. Theorem 7 in [38]).

The relation between the spaces $\mathcal{S}_+$ and $\mathcal{S}^{a,M}_{b,N,\delta}$ is given by the following projective limits:

$$\bigcap_{M, M \in Z_+} \mathcal{S}^{a,M}_{b,N,\delta} = \mathcal{S}^{a,\infty}_{b,\infty} \equiv \mathcal{S}^a_b, \quad \bigcap_{a,b \in \mathbf{R}} \mathcal{S}^a_b = \mathcal{S}_+.$$

The results of Theorem 7 in [38] can be applied in the analysis of quasi-asymptotics of solutions to differential equations, and to Abelian and Tauberian type theorems for integral transforms of distributions.

2.3   Quasi-asymptotic behavior at infinity of some generalized functions

We examine first the case when a regular distribution T is defined by a locally integrable function F on $\mathbf{R}$ and has support in an interval $[a, \infty), a > 0$. We write $T = H(x - a)F$, where H is the Heaviside function.

  1. *Let $T = H(x - a)F(x)$, $x \in \mathbf{R}$, for $a > 0$, where F is a locally integrable function satisfying $\int_a^\infty |F(x)| dx < \infty$. Then T has the quasi-asymptotics related to k^{-1}.*

It follows at once from

$$\lim_{k\to\infty} \langle kT(kx), \phi(x)\rangle = \lim_{k\to\infty} \int_a^\infty F(x)\phi\left(\frac{x}{k}\right) dx = \langle C\delta, \phi\rangle,$$

where $C = \int_a^\infty F(x)dx$.

We obtain special cases if either $F(x) \sim x^\alpha L(x)$ as $x \to \infty$ and $\alpha < -1$ or $F(x) \sim L(x)/x$, $x \to \infty$, provided that $\int_a^\infty |L(x)/x|dx < \infty$.

2. Let $a > 0$ and $T(x) = H(x - a)F(x), x \in \mathbf{R}$, where F is a locally integrable function such that $F(x) \sim L(x)/x$ as $x \to \infty$. If

$$L^*(x) := \int_a^x \frac{L(t)}{t}dt, \quad x > a$$

diverges to infinity as $x \to \infty$, then T has the quasi-asymptotics related to $L^*(k)/k$, $k \to \infty$.

L^* is also slowly varying at infinity (see **0.3** and [9]). Let

$$G(x) := (H * T)(x) = \int_a^x F(t)dt, \quad x \in \mathbf{R}.$$

Since

$$\lim_{x\to\infty} \frac{G(x)}{L^*(x)} = \lim_{x\to\infty} \frac{F(x)}{L(x)/x} = 1$$

and $\dfrac{d}{dx} G = T$, the Structural Theorem 2.2 implies the claim.

3. Let $a > 0$ and $T(x) = H(x - a)F(x)$, $x \in \mathbf{R}$, where F is a locally integrable function such that $F(x) \sim x^\alpha L(x)$ as $x \to \infty$ for $\alpha > -1$. Then T has the quasi-asymptotics related to $\rho(k) = k^\alpha L(k)$, $k \to \infty$.

It is obvious, since $G = (H * T)$ is a continuous function on $\mathbf{R}$ such that $G(x) \sim (x^{\alpha+1}/(\alpha + 1)) L(x)$, as $x \to \infty$.

Now we give the quasi-asymptotics of some distributions.

4. Denote by $\delta(x - a)$, $a \geq 0$, the delta distribution with support in a. Then $\delta^{(m)}(kx - a) \overset{q}{\sim} k^{-m-1}\delta^{(m)}(x)$, $k \to \infty$.

Indeed, observe that

$$\langle \delta(kx - a), \phi(x)\rangle = \frac{1}{k}\phi\left(\frac{a}{k}\right) = \frac{\phi(0) + O(1/k)}{k} \sim \frac{\phi(0)}{k}, \quad k \to \infty, \ \phi \in \mathcal{S};$$

therefore, $\delta(kx - a) \overset{q}{\sim} k^{-1}\delta(x)$, $k \to \infty$. The result now follows by differentiating m-times the last quasi-asymptotics (cf. Theorem 2.1).

5. *For every* $S \in \mathcal{E}' \cap \mathcal{S}'_+$ *there exists a natural number p such that S has the quasi-asymptotics related to k^{-p}, $k \to \infty$.*

For a given $S \in \mathcal{E}' \cap \mathcal{S}'_+$ there exists $m \in \mathbf{N}_0$ and a continuous function G on $\mathbf{R}$ with $\mathrm{supp}\, G \subset [0, \infty)$ such that $S = D^m G$ $(D = \dfrac{d}{dx})$. If $\mathrm{supp}\, S \subset [0, a]$, $a \geq 0$, then we have that G is equal to some polynomial of the order $\leq m - 1$ on the interval (a, ∞). Thus for some $0 \leq q \leq m - 1$ and some $C \neq 0$

$$G(x) \sim Cx^q \quad \text{as} \quad x \to \infty.$$

This implies that G has the quasi-asymptotics related to k^q, $k \to \infty$. The Structural Theorem 2.2 implies that S has the quasi-asymptotics related to k^{q-m}, $k \to \infty$; in fact $S(kx) \overset{q}{\sim} Ck^{q-m}\delta^{(m-q-1)}(x)$.

6. *Let F be a locally integrable function on $\mathbf{R} \setminus \{0\}$ equal to zero outside of some interval $[0, a]$, $a > 0$, such that*

$$F(x) \sim x^\alpha L(x) \quad \text{as} \quad x \to 0^+,$$

where $\alpha \leq -1$ and L is a slowly varying function at zero.

This function can be identified with a distribution S defined by

$$\langle S, \phi \rangle := \int_0^a F(x)(\phi(x) - \phi(0) - \cdots - \frac{x^{m-1}}{(m-1)!}\phi^{(m-1)}(0))dx \qquad (2.12)$$

if $-(m+1) < \alpha \leq -m$, $m \in \mathbf{N}$ and $\phi \in \mathcal{S}(\mathbf{R})$ (see [[138], p. 13]).

The distribution S defined by (2.12) has quasi-asymptotics related to $\dfrac{1}{k^{m+1}}$, $k \to \infty$. Let us prove this. For $\phi \in \mathcal{S}(\mathbf{R})$, we have

$$\langle k^{m+1}S(kx), \phi(x) \rangle = k^m \int_0^a F(x)\left(\phi\left(\frac{x}{k}\right) - \phi(0) - \cdots - \frac{x^{m-1}}{k^{m-1}(m-1)!}\phi^{(m-1)}(0)\right)dx$$

$$= k^m \int_0^a F(x)\frac{1}{m!}\left(\frac{x}{k}\right)^m \phi^{(m)}\left(\frac{\xi_x}{k}\right)dx, \quad 0 < \xi_x < x,$$

hence

$$\lim_{k \to \infty} \langle k^{m+1}S(kx), \phi(x) \rangle = \frac{(-1)^m}{m!}\langle \delta^{(m)}, \phi \rangle \int_0^a x^m F(x)dx, \quad \text{as} \quad x \to \infty.$$

Let us remark that if $\alpha > -1$, then the distribution S given by F is regular and $S(kx) \overset{q}{\sim} Ck^{-1}\delta$, $k \to \infty$, with $C = \int_0^a F(x)dx$.

7. *Let L be slowly varying at zero and at infinity.*

The distribution $R(x) = (x^\alpha L(x))_+$ (see [138]) has the quasi-asymptotics related to $\rho(x) = x^\alpha L(x)$, $x \to \infty$, if $\alpha \notin \mathbf{Z}_- = \{-1, -2, \dots\}$ and related to $\rho_1(x) = x^\alpha L^*(x)$ if $\alpha \in \mathbf{Z}_-$, where $L^*(x) = \int_a^x L(t)t^{-1}dt$, $x \geq a$.

8. Let f denote a measurable function on $\mathbf{R}$ with support in $[0, \infty)$ satisfying

$$f(x) \sim x^\beta L_1(x) \quad \text{as} \quad x \to 0^+, \tag{2.13}$$

where L_1 is a slowly varying function at zero and $-m - 1 < \beta \leq -m$. We suppose that f satisfies the following additional condition

$$x^m f(x) \quad \text{is integrable on} \quad (a, \infty), \, a > 0. \tag{2.14}$$

We denote by $\tilde{f}$ the following distribution in $\mathcal{S}'(\mathbf{R})$ defined by f:

$$\langle \tilde{f}, \phi \rangle = \int_0^\infty f(x)(\phi(x) - \phi(0) - \cdots - \frac{x^{m-1}}{(m-1)!}\phi^{(m-1)}(0))dx, \; \phi \in \mathcal{S}(\mathbf{R}), \tag{2.15}$$

if $-m - 1 < \beta < -m$, and for some $a > 0$, and if $\beta = -m$,

$$\langle \tilde{f}, \phi \rangle = \int_0^\infty f(x)(\phi(x) - \phi(0) - \cdots - \frac{x^{m-2}}{(m-2)!}\phi^{(m-2)}(0)$$

$$- \frac{x^{m-1}}{(m-1)!}\phi^{(m-1)}(0)H(a - x))dx, \quad \phi \in \mathcal{S}(\mathbf{R}). \tag{2.16}$$

The distribution $\tilde{f}$ from (2.15), respectively (2.16), has quasi-asymptotics related to k^{-m-1}, resp. k^{-m}, provided that both (2.13) and (2.14) hold.

9. *Let f satisfy (2.13) for $-m - 1 < \beta < -m$ and*

$$f(x) \sim x^\gamma L_2(x) \quad \text{as} \quad x \to \infty \; (\gamma < -m),$$

where L_2 is slowly varying at infinity. Then $\tilde{f}$ defined by (2.15) $(-m - 1 < \beta < -m)$, respectively by (2.16) $(\beta = -m)$, has quasi-asymptotics related to k^{-m-1} and k^{-m} respectively.

Observe that this quasi-asymptotics does not depend on the functions L_1 and L_2.

10. *Let f satisfy (2.13) for $-m - 1 < \beta < -m$ and*

$$f(x) \sim x^\nu L_2(x) \quad as \quad x \to \infty \quad (\nu > 0), \tag{2.17}$$

where L_2 is slowly varying at infinity.

Then $\tilde{f}$ defined by (2.15) has the quasi-asymptotics related to $x^\nu L_2(x)$.

We suppose in the next example, as usual, that M_p satisfies (M.1), (M.2)' and (M.3)'.

The quasi-asymptotics of ultradistributions is a natural extension of the same notion for distributions. Suppose that $f \in \mathcal{S}'_{[0,\infty)}$; it defines an ultradistribution $f \in \mathcal{S}'^*_{[0,\infty)}$. If it has the quasi-asymptotics as a distribution related to ρ, then it has also the quasi-asymptotics as an ultradistribution related to ρ, and with the same limit.

On the other hand, the next examples show that there exist elements of $\mathcal{S}'^*_{[0,\infty)}$ which are not in $\mathcal{S}'_{[0,\infty)}$ and have the quasi-asymptotics as ultradistributions.

We deal with the Beurling ultradistributions. In the case of Roumieu's ultradistributions the treatment is the same.

11. *Let $P(D) = \sum\limits_{n=0}^{\infty} a_n D^n$ be an ultradifferential operator of class (M_p) and let $a_0 \neq 0$ and $a_i \neq 0$ for infinitely many i. Then $P(D)\delta$ is an element of $\mathcal{S}'^{(M_p)}_{[0,\infty)}$ which is not a distribution. We shall show that $(P(D)\delta)(kx) \overset{q}{\sim} a_0 k^{-1} \delta(x)$.*

By Definition 2.2, we have to consider the following limit:

$$\lim_{k \to \infty} k\langle (P(D)\delta)(kx), \varphi(x) \rangle$$

$$= \lim_{k \to \infty} \langle \delta(y), \sum_{n=0}^{\infty} (-1)^n a_n k^{-n} \varphi^{(n)}(y/k) \rangle$$

$$= a_0 \varphi(0) + \lim_{k \to \infty} \sum_{n=1}^{\infty} (-1)^n a_n k^{-n} \varphi^{(n)}(0), \quad \varphi \in \mathcal{S}^{(M_p)}_{[0,\infty)}.$$

It remains to show that the last limit equals zero. Since for every $L > 0$

$$\sup_n L^n / M_n \sup_{x \in [0,\infty)} |\varphi^{(n)}(x)| < C_1,$$

it follows

$$\left| \sum_{n=1}^{\infty} (-1)^n a_n k^{-n} \varphi^{(n)}(0) \right| \leq C_1 \sum_{n=1}^{\infty} k^{-n} L^n / M_n |\varphi^{(n)}(0)|$$

$$\leq C_2 k^{-1} \to 0, \ k \to \infty.$$

Consequently, $(P(D)\delta)(kx) \overset{q}{\sim} (a_0/k)\delta(x)$ as $k \to \infty$ in $\mathcal{S}'^{(M_p)}_{[0,\infty)}$.

12. *There are also distributions which have no quasi-asymptotics as distributions, but they have the quasi-asymptotics as ultradistributions. One of such distributions is the following one*

$$f = \sum_{n=0}^{\infty} \delta^{(n)}(x - e^n)/M_n \,.$$

Suppose that there exist ρ and $g \in \mathcal{S}'_{[0,\infty)}$ such that $f(kx) \overset{q}{\sim} \rho(k)g(x)$ in $\mathcal{S}'_{[0,\infty)}$. Then, we would have

$$\lim_{k \to \infty} \frac{1}{\rho(k)} \left\langle \sum_{n=0}^{\infty} \delta^{(n)}(kx - e^n)/M_n, \ \varphi(x) \right\rangle = \langle g, \varphi \rangle$$

for every $\varphi \in \mathcal{S}_{[0,\infty)}$.

By Borel's theorem, there exists $\varphi_0 \in \mathcal{S}_{[0,\infty)}$ such that $\text{supp} \, \varphi_0 \subset [e^{-1}, e]$, and the set $\{\exp(-p^2 - p)/(\rho(e^p)M_p)|\varphi_0^{(p)}(1)|; \ p \in \mathbf{N}\}$ is not bounded in $\mathbf{R}$. For this φ_0 and for the subset $\{e^p; \ p \in \mathbf{N}\}$, we have

$$\left| \frac{1}{\rho(e^p)} \langle f(e^p x), \varphi_0(x) \rangle \right| = \left| \frac{1}{\rho(e^p)} \sum_{n=0}^{\infty} \frac{(-1)^n}{M_n \exp(np + p)} \varphi_0^{(n)}(e^{n-p}) \right|$$

$$= \frac{\exp(-p^2 - p)}{\rho(e^p) M_p} |\varphi_0^{(p)}(1)| \to \infty, \ p \to \infty.$$

This is in contradiction to our assumption $f(kx) \overset{q}{\sim} \rho(k)g(x)$ in $\mathcal{S}'_{[0,\infty)}$.

This distribution has the quasi-asymptotics as ultradistribution; the proof is the same as for $P(D)\delta$.

2.4 *Equivalent definitions of quasi-asymptotics at infinity*

Let $T \in \mathcal{S}'_+$. We denote by $\mathcal{L}T$ the *Laplace transform* of T: $\mathcal{L}(T)(z) = \langle T(t), e^{izt} \rangle$, $z \in \mathbf{R} + i\mathbf{R}_+$ (cf. [190], Ch. II, Part 9, [189] and [192], Ch. I, Part 2). We collect in the following theorem some equivalent ways to define quasi-asymptotics over the cone $\mathbf{R}_+$. The part c) is a prototype of a

Tauberian characterization for quasi-asymptotics; Tauberian theorems for various integral transforms will be the main subject of **4.2**.

Theorem 2.7. *Let $T \in \mathcal{S}'_+$ and $\rho(k) = k^\alpha L(k)$, $k \geq k_0$. The following statements are equivalent:*

a) $\displaystyle \lim_{k \to \infty} \frac{T(k \cdot)}{\rho(k)} = C f_{\alpha+1}$, *in* $\mathcal{S}'(\mathbf{R})$, $C \neq 0$.

b) $\displaystyle \lim_{k \to \infty} \frac{T(k \cdot + b)}{\rho(k)} = C f_{\alpha+1}$, *in* $\mathcal{D}'(\mathbf{R})$, $C \neq 0$, $b \in \mathbf{R}$.

c)   *A)* $\displaystyle \lim_{y \to 0^+} \frac{y}{\rho(1/y)} \mathcal{L}T(iy) = M \neq 0$;

   B)   *there exist* $D_1 > 0$, $m \in \mathbf{N}_0$, *and* $r_0 > 0$ *such that*

$$\left| \frac{r}{\rho(1/r)} \mathcal{L}T(re^{i\varphi}) \right| \leq \frac{D_1}{\sin^m \varphi}, \quad 0 < r < r_0, \ 0 < \varphi < \pi.$$

d) For every $\phi \in \mathcal{D}(\mathbf{R})$,

$$\lim_{k \to \infty} \frac{(T * \phi)}{\rho(k)}(k \cdot) = M_\phi f_{\alpha+1}, \quad \text{in } \mathcal{D}'(\mathbf{R}), \quad \text{where } M_\phi \neq 0.$$

e) There exists $\phi_0 \in \mathcal{D}(\mathbf{R})$ with the property $\mathcal{L}\phi_0(0) \neq 0$ such that

$$\lim_{k \to \infty} \frac{(T * \phi_0)(k \cdot)}{\rho(k)} = M_{\phi_0} f_{\alpha+1}, \quad \text{in } \mathcal{D}'(\mathbf{R}), \ M_{\phi_0} \neq 0.$$

*f) For a δ-sequence $\langle \delta_n \rangle_n$ (cf. **0.4**) there is a $C \neq 0$ such that*

$$\lim_{k \to \infty} \frac{(T * \delta_n)(k \cdot)}{\rho(k)} = C f_{\alpha+1}, \quad \text{in } \mathcal{D}'(\mathbf{R}), \quad \text{and uniformly for } n \in \mathbf{N}.$$

Proof. a) $\Rightarrow$ *b).* We start with the relation

$$\left\langle \frac{T(kx + b)}{\rho(k)}, \phi(x) \right\rangle = \left\langle \frac{T(kx)}{\rho(k)}, \phi\left(x - \frac{b}{k}\right) \right\rangle, \quad \phi \in \mathcal{D}(\mathbf{R}).$$

The set $\{\phi(\cdot - \frac{b}{k}); \ k \geq 1\}$ is bounded in $\mathcal{D}'(\mathbf{R})$. By using the equivalence of the weak and strong sequential convergence in $\mathcal{D}'(\mathbf{R})$ and the fact that $\phi(\cdot - b/k) \to \phi$, $k \to \infty$, in $\mathcal{D}(\mathbf{R})$, we obtain

$$\lim_{k \to \infty} \left\langle \frac{T(kx + b)}{\rho(k)}, \phi(x) \right\rangle = \lim_{k \to \infty} \left\langle \frac{T(kx)}{\rho(k)}, \phi\left(x - \frac{b}{k}\right) \right\rangle$$

$$= \langle C f_{\alpha+1}, \phi \rangle, \quad \phi \in \mathcal{D}(\mathbf{R}).$$

By Theorem 2.3 the last limit holds in $\mathcal{S}'(\mathbf{R})$, as well.

$b) \Rightarrow a)$ Take $b = 0$ and apply Theorem 2.3.

$a) \Rightarrow c)$ and $c) \Rightarrow a)$ is proved in [192].

$c) \Rightarrow d)$ For a $T \in \mathcal{S}'_+$, we have $T * \phi \in \mathcal{S}'_{[a,\infty)}$, $a \in \mathbf{R}$. We shall show that $T * \phi$ satisfies $c)$, it would imply that $T * \phi$ satisfies $a)$ and then T would satisfy $d)$. We have

$$\lim_{y\to0^+} \frac{y}{\rho(1/y)} \mathcal{L}T(yi)\mathcal{L}\phi(iy) = \lim_{y\to0^+} \frac{y}{\rho(1/y)}\mathcal{L}T(yi)\mathcal{L}\phi(0) = M_{\phi_0}.$$

Moreover, there exist $D_2 > 0$, $m \in \mathbf{N}_0$ and $r_0 > 0$ such that

$$\left| \frac{r}{\rho(1/r)} \mathcal{L}T(re^{i\varphi})\mathcal{L}\phi(re^{\varphi i}) \right|$$

$$\leq \left| \frac{r}{\rho(1/r)} \mathcal{L}T(re^{i\varphi}) \right| \max_{\substack{0<r\leq r_0 \\ 0\leq\varphi<\pi}} |\mathcal{L}\phi(re^{\varphi i})| \leq \frac{D_2}{\sin^m \varphi}.$$

$d) \Rightarrow e)$ It is obvious.

$e) \Rightarrow a)$ The assumptions in $e)$ imply $a)$ and consequently $c)$ for $T * \phi_0$. Now, we have

$$\lim_{y\to0^+} \frac{y}{\rho(1/y)} \mathcal{L}T(iy)\mathcal{L}\phi_0(iy) = \lim_{y\to0^+} \frac{y}{\rho(1/y)}\mathcal{L}T(iy)\mathcal{L}\phi_0(i0) = M \neq 0.$$

Taking care of the property that $\mathcal{L}\phi_0(0) \neq 0$, we have

$$\left| \frac{r}{\rho(1/r)} \mathcal{L}T(re^{i\varphi}) \right| \leq \frac{D'}{\sin^{m'} \varphi},$$

where $0 < r \leq r'_0$, and $0 \leq \varphi < \pi$ for appropriate r'_0, D' and m'. This gives $c)$ for T and consequently $a)$.

$a) \Rightarrow f)$ Let $\phi \in \mathcal{D}(\mathbf{R})$ and $n \in \mathbf{N}$. Then

$$\left\langle \frac{(T * \delta_n)(kx)}{\rho(k)}, \phi(x) \right\rangle = \frac{1}{k\rho(k)} \left\langle (T * \delta_n)(x), \phi\left(\frac{x}{k}\right) \right\rangle$$

$$= \frac{1}{k\rho(k)} \left\langle T(x), \left(\check{\delta}_n * \phi\left(\frac{\cdot}{k}\right) \right)(x) \right\rangle$$

$$= \frac{1}{\rho(k)} \left\langle T(kx), \int_{\mathbf{R}} \delta_n(-t)\phi\left(x - \frac{t}{k}\right) dt \right\rangle.$$

The set

$$\left\{ \int\limits_{\mathbf{R}} \delta_n(-t)\phi\left(x - \frac{t}{k}\right) dt; \ n \in \mathbf{N}, \ k \in (0, \infty) \right\}$$

is bounded in $\mathcal{D}(\mathbf{R})$ because of the properties of $\langle \delta_n \rangle_n$ (cf. **0.4**). Therefore, we have

$$\left\langle \frac{T(kx)}{\rho(k)}, \int\limits_{\mathbf{R}} \delta_n(-t)\phi\left(x - \frac{t}{k}\right) dt \right\rangle \to \left\langle Cf_{\alpha+1}, \left(\int\limits_{\mathbf{R}} \delta_n(-y)dy\right)\phi \right\rangle, \quad k \to \infty,$$

uniformly for $n \in \mathbf{N}$. Thus, for any $\varepsilon > 0$ there is a $k_0(\varepsilon)$ such that

$$\left| \left\langle \frac{(T * \delta_n)(kx)}{\rho(k)}, \phi(x) \right\rangle - \left\langle Cf_{\alpha+1}, \phi \right\rangle \right|$$

$$\leq \left| \left\langle \frac{T(kx)}{\rho(k)}, \int\limits_{\mathbf{R}} \delta_n(-t)\phi\left(x - \frac{t}{k}\right) dt \right\rangle \right.$$

$$\left. - \left\langle Cf_{\alpha+1}(x), \int\limits_{\mathbf{R}} \delta_n(-t)\phi\left(x - \frac{t}{k}\right) dt \right\rangle \right|$$

$$+ \left| \left\langle Cf_{\alpha+1}(x), \int\limits_{\mathbf{R}} \delta_n(-t)\phi\left(x - \frac{t}{k}\right) dt \right\rangle - \left\langle Cf_{\alpha+1}, \phi \right\rangle \right| < \varepsilon, \quad k > k_0(\varepsilon);$$

which proves a) $\Rightarrow$ f).

f) $\Rightarrow$ e) Since $\int\limits_{-\infty}^{\infty} \delta_n(x) = 1$, $n \in \mathbf{N}$, we have $\mathcal{L}\delta_n(0) \neq 0$. This implies that e) holds. $\qquad \square$

2.5 Quasi-asymptotics as an extension of the classical asymptotics

We have seen in Proposition 2.3 that if a locally integrable function f has asymptotic behavior related to a function ρ, then $\tilde{f}$ has the quasi-asymptotics related to ρ. The following theorem also goes in this direction.

Theorem 2.8. Let Γ be a closed convex acute solid cone in $\mathbf{R}^n$, $f \in \mathcal{S}'(\Gamma) \cap L^1_{loc}(\Gamma \cap \{\xi \in \mathbf{R}^n; \ |\xi| > R\})$ for some $R > 0$ and let $\rho(k) = k^\alpha L(k)$, $\alpha > -n$. If, for any $e \in \text{pr}\,\Gamma$ $(\text{pr}\,\Gamma = \{x \in \mathbf{R}^n; \ x \in \Gamma, \|x\| = 1\})$, the limits

$$\lim_{|\xi| \to \infty} \frac{1}{\rho(|\xi|)} f(|\xi|e) = g(e) \neq 0$$

exist, and for some $g_1 \in L^1(\mathrm{pr}\Gamma)$ the estimate

$$\frac{|f(\xi)|}{\rho(|\xi|)} \le g_1\left(\frac{\xi}{|\xi|}\right), \quad |\xi| > R, \; \xi \in \Gamma,$$

is satisfied, then f has quasi-asymptotics related to ρ.

The proof is similar to that of Proposition 2.3 (see Theorem 2, Chapter I, §3.3. in [192]).

Remark. The condition $\alpha > -n$ in Theorem 2.8 is essential. This fact is illustrated by the following two examples.

Let $f(x) = H(x-1)x^{-1-\varepsilon}$, $x \in \mathbf{R}, \varepsilon > 0$. Then $f(x) \sim x^{-1-\varepsilon}$, $x \to \infty$, but by Example 1 in **2.3** the regular distribution f has the quasi-asymptotics related to $\rho(k) = k^{-1}$ for any $\varepsilon > 0$.

Let $T(x) = H(x-2)(x\log^2 x)^{-1}, x \in \mathbf{R}$. Then $T(x) \sim (x\log^2 x)^{-1}$, $x \to \infty$. But by the same Example 1 in **2.3**, T has the quasi-asymptotics related to $\rho(k) = k^{-1}$.

A more difficult question is to find conditions under which the quasi-asymptotics implies the ordinary asymptotics of f. A partial answer is given in the ensuing theorem.

Theorem 2.9. *Let $T \in \mathcal{S}'_+$ be equal to a locally integrable function f in some interval $[b, +\infty)$, $b > 0$, with quasi-asymptotic behavior related to $\rho(k) = k^\alpha L(k)$, $\alpha > -1$. If for some $m \in \mathbf{N}$ the function $x^m f(x)$, $x \ge b$, is monotonous, then f has the asymptotic behavior at infinity related to the same regularly varying function ρ.*

Proof. By Proposition 2.5, the function $g(x) = x^m f(x)$, $x \ge b$, has the quasi-asymptotic behavior at infinity related to $\rho_1(k) = k^{\alpha+m}L(k)$. The monotonicity of g implies that its distributional derivative Dg can be written as $Dg = B + \mu$, where $B \in \mathcal{E}'_+$ and μ is a positive measure with support in $[b, \infty)$. By Theorem 2.1, Dg has the quasi-asymptotics related to $k^{\alpha+m-1}L(k)$. From Proposition 2.4 a), we see that Dg and μ have the same quasi-asymptotic behavior at infinity, related to $k^{\alpha+m-1}L(k)$, i.e.,

$$\lim_{k \to \infty} \frac{1}{k^{\alpha+m-1}L(k)} \langle \mu(kx), \phi(x) \rangle = C\langle f_{\alpha+m}, \phi \rangle, \quad \phi \in \mathcal{S}(\mathbf{R}). \qquad (2.18)$$

Choosing ϕ_ε and ψ_ε in $\mathcal{D}(\mathbf{R})$ with the properties

$$\phi_\varepsilon(x) = 1, \text{ for } |x| \le 1 - \frac{\varepsilon}{2} \text{ and } \phi_\varepsilon(x) = 0 \text{ for } |x| \ge 1,$$

$$\psi_\varepsilon(x) = 1, \text{ for } |x| \le 1 \text{ and } \psi_\varepsilon(x) = 0 \text{ for } |x| \ge 1 + \frac{\varepsilon}{2}$$

$(0 < \varepsilon < 1)$, we obtain

$$\frac{1}{k^{\alpha+m-1}L(k)}\langle\mu(kx), \phi_\varepsilon(x)\rangle \le \frac{1}{k^{\alpha+m-1}L(x)}\langle\mu(kx), H(1-x)\rangle$$

$$\le \frac{1}{k^{\alpha+m-1}L(k)}\langle\mu(kx)\psi_\varepsilon(x)\rangle, \qquad (2.19)$$

since

$$\phi_\varepsilon(x) \le H(1-x) \le \psi_\epsilon(x) \quad \text{for} \quad x > 0.$$

Using (2.18), we see that, both, the left and the right hand side of (2.19) tend to numbers which do not differ more than ε. Thus the expression in the middle of (2.18) tends to some limit independent of ε. Since $(H * \mu)$ is equal to $g(x)$ on (b_1, ∞), $b_1 \ge b$ and

$$\frac{1}{k^{\alpha+m-1}L(k)}\langle\mu(kx), H(1-x)\rangle = \frac{1}{k^{\alpha+m}L(k)}(H * \mu)(k),$$

we conclude that g has ordinary asymptotic behavior at infinity related to ρ_1, and this implies the statement. $\qquad\square$

Remark. It is easy to find a function which has quasi-asymptotics but not an asymptotic behavior. An example is given by $f(x) = H(x)\sin x, x \in \mathbf{R}$; $f(kx) \overset{q}{\sim} k^{-1}\delta(x), \ k \to \infty$.

In such a way, the quasi-asymptotics extends the notion of the asymptotic behavior of a locally integrable function.

2.6 Relations between quasi-asymptotics in $\mathcal{D}'(\mathbf{R})$ and $\mathcal{S}'(\mathbf{R})$

In this section we make a preliminary investigation of some questions raised by Theorem 2.3:

As in Theorem 2.3, suppose that $T \in \mathcal{D}'$ satisfies (2.8), with $g \in \mathcal{D}'$, but *we now remove the assumptions over the support of T* (we allow the support to be any subset of $\mathbf{R}^n$). It then is natural to ask: *Does $T \in \mathcal{S}'$? Furthermore, does the limit exist in $\mathcal{S}'$?* Let us observe that Theorem 2.3 does not give an answer to such a question because the cone Γ is acute and it cannot be the whole $\mathbf{R}^n$.

The following theorem and its corollary give a partial answer to this question. We will postpone the complete answer (in one dimension) for **2.10**.

Theorem 2.10. *Let $f \in \mathcal{D}'(\mathbf{R})$ and suppose that there exists, in the sense of convergence in $\mathcal{D}'(\mathbf{R})$, the limit*

$$\lim_{k \to \infty} f(kx)/\rho(k) = g(x) \neq 0, \tag{2.20}$$

where $\rho(k)$, $k \in (0, \infty)$, is a positive continuous function on $(0, \infty)$. Then,

(i) $\rho(x) = x^v L(x), x \in (0, \infty)$, for some $v \in \mathbf{R}$ and some slowly varying function L, and g is a homogeneous distribution with degree of homogeneity v.

(ii) $f \in \mathcal{S}'(\mathbf{R})$.

(iii) If $v > -1$, then the limit (2.20) exists in the sense of convergence in $\mathcal{S}'(\mathbf{R})$.

(iv) If $v = -1$ and $1/L(x), x \in (a, \infty)$, is bounded, then the limit (2.20) exists in the sense of convergence in $\mathcal{S}'(\mathbf{R})$, as well.

It should be noticed that the proof is quite different from that of Theorem 2.3 where we used the fact that $\mathcal{D}'_\Gamma$ and $\mathcal{S}'_\Gamma$ are convolution algebras. The absence of the convolution algebra structure makes the argument more complex.

Proof. (i) Let $\varphi \in \mathcal{D}(\mathbf{R})$ such that $\langle g, \varphi \rangle \neq 0$. We have

$$\lim_{k \to \infty} \langle f(kmx)/\rho(k), \varphi(x) \rangle = \langle g(mx), \varphi(x) \rangle;$$

$$\lim_{k \to \infty} (\rho(mk)/\rho(k)) \langle f(kmx)/\rho(km), \varphi(x) \rangle$$
$$= \langle g(x), \varphi(x) \rangle \lim_{k \to \infty} (\rho(mk)/\rho(k)), \quad m > 0.$$

This implies that, for every $m > 0$, we have the existence of the limit

$$\lim_{k \to \infty} (\rho(mk)/\rho(k)) = d(m).$$

By [[148], p. 17], we obtain that for some $v \in \mathbf{R}$ and some slowly varying function L, $\rho(k)$ and $d(k)$ are of the form

$$\rho(k) = k^v L(k), \quad d(k) = k^v, \quad k \in (a, \infty).$$

Since $\langle g(mx), \varphi(x)\rangle = m^v \langle g(x), \varphi(x)\rangle$, $m > 0$, we obtain that g is homogeneous of degree v.

(ii) The set $\{f(kx)/(k^v L(k)); \ k > 0\}$ is a bounded subset of $\mathcal{D}'(\mathbf{R})$. From Theorem XXII, Chapter VI in [146], it follows that for a given open bounded neighborhood of zero Ω there exists a compact neighborhood of zero K and a non-negative integer m such that for any $\varphi \in \mathcal{D}_K^m(\mathbf{R})$

$$\Omega \ni x \mapsto ((f(kt)/(k^v L(k))) * \varphi(t))(x), \quad k \in (0, \infty), \qquad (2.21)$$

is a family of functions which are continuous and uniformly bounded on Ω. Let $\Omega = (-2, 2)$ and $K = [-\varepsilon, \varepsilon]$. Since the weakly bounded family (2.21) is strongly bounded in $\mathcal{D}_K^m(\mathbf{R})$, we obtain that for every bounded set $A \subset \mathcal{D}_{[-\varepsilon,\varepsilon]}^m$, the set of functions

$$\{\Omega \ni x \mapsto ((f(kt)/(k^v L(k))) * \varphi(t))(x); \ k > 0, \ \varphi \in A\}$$

is a bounded family of continuous functions on Ω. Let $\psi \in \mathcal{D}_{[-\varepsilon,\varepsilon]}^m$ and let

$$\varphi_k(x) = \psi(kx)/k^m, \quad x \in \mathbf{R}, \ k \geq 1.$$

Since $\operatorname{supp} \varphi_k(x) \subset \{x; |x| \leq \varepsilon/k\} \subset [-\varepsilon, \varepsilon]$, we have that $A = \{\varphi_k(x); \ k \geq 1\}$ is a bounded family in $\mathcal{D}_{[-\varepsilon,\varepsilon]}^m$ and that

$$\{((f(kt)/(k^v L(k))) * \varphi_r(t))(x); \ k > 0, \ r \geq 1\}$$

is a bounded family of continuous functions on Ω. Taking $r = k$, we obtain that for some $M > 0$

$$| ((f(kt)/(k^v L(k))) * (\psi(kt)/k^m))(x) | \leq M, \quad x \in (-2, 2), \ k \geq 1.$$

From

$$(f(kt) * \psi(kt))(x) = \langle f(kt), \psi(k(x - t))\rangle = k^{-1}\langle f(t), \psi(kx - t)\rangle$$
$$= k^{-1}(f * \psi)(kx),$$

we obtain that

$$| (f * \psi)(kx)/(k^{v+m+1} L(k)) | < M \quad \text{for} \ x \in (-2, 2), \ k \geq 1.$$

Taking $x = 1$ and $x = -1$ it follows that for any $\psi \in \mathcal{D}_{[-\varepsilon,\varepsilon]}^m$ there exists $M_\psi > 0$ such that

$$| (f * \psi)(x) | \leq M_\psi(1 + |x|^{v+m+1} L(|x|)), \quad x \in \mathbf{R}.$$

By (VI, 6; 22) in [146], we obtain

$$f = \frac{d^{2s}}{dx^{2s}}(\gamma E * f) - \psi * f, \qquad (2.22)$$

where E is the fundamental solution of $d^{2s}E/dx^{2s} = \delta$, $\gamma \in \mathcal{D}_{[-\varepsilon,\varepsilon]}$, $\gamma \equiv 1$ in a neighborhood of zero and $\psi \in \mathcal{D}_{[-\varepsilon,\varepsilon]}$. If s is large enough, then $\gamma E \in \mathcal{D}^m_{[-\varepsilon,\varepsilon]}$. Thus (2.22) implies that $f \in \mathcal{S}'(\mathbf{R})$.

(iii) We can rewrite (2.22) in the form $f = d^{2s}f_1/dx^{2s} + f_2$, where f_1 is a continuous and f_2 a smooth function such that

$$\sup\{|f_1(x)|, \, |f_2(x)|\} \le M(1 + |x|^{v+m+1}L(|x|)), \; x \in \mathbf{R},$$

for some $M > 0$.

Since

$$f(kx)/(k^v L(k)) = \left(\left[\frac{d^{2s}}{dx^{2s}}f_1\right](kx) + f_2(kx)\right)/(k^v L(k))$$

$$= \frac{d^{2s}}{dx^{2s}}(f_1(kx))/(k^{v+2s}L(k)) + f_2(kx)/(k^v L(k)),$$

we obtain, for s large enough that for any $\varphi \in \mathcal{D}(\mathbf{R})$:

$$\lim_{k\to\infty} \langle f(kx)/(k^v L(k)), \varphi(x)\rangle$$

$$= \lim_{k\to\infty} \langle f_2(kx)/(k^v L(k)), \varphi(x)\rangle = \langle g(x), \varphi(x)\rangle. \qquad (2.23)$$

Let us set

$$f_{2+}(x) = \begin{cases} f_2(x), & x > 0 \\ 0, & x \le 0 \end{cases}, \qquad f_{2-}(x) = \begin{cases} f_2(x), & x < 0 \\ 0, & x \ge 0. \end{cases}$$

Clearly, for any $\varphi \in \mathcal{D}(0,\infty)$ ($\operatorname{supp}\varphi \subset (0,\infty)$),

$$\lim_{k\to\infty} \langle f_2(kx)/(k^v L(k)), \varphi(x)\rangle = \lim_{k\to\infty} \langle f_{2+}(kx)/(k^v L(k)), \varphi(x)\rangle.$$

If $\varphi \in \mathcal{D}(0,\infty)$, then $\psi(t) = \varphi(e^t)e^t$, $t \in \mathbf{R}$, is an element in $\mathcal{D}(\mathbf{R})$. Moreover, the mapping $\varphi \mapsto \psi$ defined above is a bijection. Since for $\varphi \in \mathcal{D}(0,\infty)$,

$$\int\limits_0^\infty f_{2+}(kx)\varphi(x)dx = \int\limits_{-\infty}^\infty f_{2+}(ke^t)\varphi(e^t)e^t dt,$$

by putting $k = e^r$, $r \in \mathbf{R}$, we obtain that for a function $F_{1,2}(t) = f_{2+}(e^t)$, $t \in \mathbf{R}$, there exists the limit

$$\lim_{r\to\infty} \langle F_{2,1}(t+r)/(e^{vr}L(e^r)), \; \psi(t)\rangle$$

for any $\psi \in \mathcal{D}(\mathbf{R})$. Using again Theorem XXII, Chapter VI in [146], we obtain that for any open set $\Omega \ni 0$ there exists a compact neighborhood of 0 and a non-negative integer m such that for any $\varphi \in \mathcal{D}_K^m(\mathbf{R})$,

$$x \mapsto (F_{2,1}(t+r) * \varphi(t))(x)/(e^{vr}L(e^r)), \quad r \geq 0,$$

is a bounded family of continuous functions on Ω. Since $(F_{2,1}(t+r) * \varphi(t))(x) = (F_{2,1}(t) * \varphi(t))(x+r), x \in \mathbf{R}$ taking $x = 0$ and using (VI; 6; 22) in [146], we obtain

$$F_{2,1} = \frac{d^{2l}}{dx^{2l}}H_{2,1} + G_{2,1},$$

where $H_{2,1}$ is a continuous function and $G_{2,1}$ is a smooth function on $\mathbf{R}$ such that

$$\sup\{|H_{2,1}(t)|, |G_{2,1}(t)|\} < Me^{vt}L(e^t), \quad t > 0. \tag{2.24}$$

If $t \in (-\infty, 0)$, then $e^t \in (0,1)$ and since $F_{2,1}(t)$ is bounded on $(-\infty, 0)$, we obtain that $H_{2,1}$ and $G_{2,1}$ are bounded on $(-\infty, 0)$. Namely, both functions are equal to the convolution of $F_{2,1}$ with suitable functions with compact supports.

From

$$f_{2+}(e^t) = \frac{d^{2l}}{dx^{2l}}H_{2,1}(t) + G_{2,1}(t), \quad t \in (-\infty, \infty),$$

we obtain

$$f_{2+}(x) = \sum_{p=1} a_p x^p \frac{d^p}{dx^p}(H_{2,1}(\log x)) + G_{2,1}(\log x), \quad x > 0,$$

where a_p are suitable constants.

Set now

$$\bar{G}_{2,1}(x) = \begin{cases} G_{2,1}(\log x), & x > 0 \\ 0, & x \leq 0, \end{cases} \qquad \bar{H}_{2,1}(x) = \begin{cases} H_{2,1}(\log x), & x > 0 \\ 0, & x \leq 0. \end{cases}$$

The distributions $f_{2+} \in \mathcal{D}'(-\infty, \infty)$ and $\sum\limits_{p=1}^{2l} a_p x^p \frac{d^p}{dx^p}\bar{H}_{2,1} + \bar{G}_{2,1}$ are equal to each other. For every $\varphi \in \mathcal{D}(\mathbf{R})$, we have

$$\int_0^\infty f_{2+}(kx)\varphi(x)dx = \langle f_{2+}(kx), \varphi(x) \rangle$$

$$= \sum_{p=1}^{2l} a_p \langle (kx)^p \frac{1}{k^p}(\bar{H}_{2,1}(kx))^{(p)}, \varphi(x) \rangle + \langle \bar{G}_{2,1}(kx), \varphi(x) \rangle$$

$$= \sum_{p=1}^{2l} a_p (-1)^p \langle \bar{H}_{2,1}(kx), (x^p \varphi(x))^{(p)} \rangle + \langle \bar{G}_{2,1}(kx), \varphi(x) \rangle .$$

The functions $H_{2,1}$ and $G_{2,1}$ are bounded on $(-\infty, 0)$. Thus, by (2.24), we have that for some $M > 0$

$$\sup\{|\bar{G}_{2,1}(kx)|, |\bar{H}_{2,1}(kx)|\} \leq M(kx)^v L(kx), \quad x > \frac{1}{k}, \qquad (2.25)$$

$$\sup\{|\bar{G}_{2,1}(kx)|, |\bar{H}_{2,1}(kx)|\} \leq M, \quad 0 < x \leq \frac{1}{k}, \qquad (2.26)$$

Since $v > -1$, these inequalities imply that for any $\varphi \in \mathcal{D}(\mathbf{R})$,

$$\int_0^\infty f_{2+}(kx)\varphi(x)dx/(k^v L(k)) < \infty . \qquad (2.27)$$

In a similar way as in the proof of Theorem 2.3, one can prove that (2.25) holds for every $\varphi \in \mathcal{S}(\mathbf{R})$. If we put $F_{2,2}(t) = f_{2-}(-e^t)$, $t > 0$, then, by the same arguments as above, one can prove that for every $\varphi \in \mathcal{S}(\mathbf{R})$,

$$\int_{-\infty}^0 f_{2-}(kx)\varphi(x)dx/(k^v L(k)) < \infty . \qquad (2.28)$$

By the Banach–Steinhaus Theorem, it follows from (2.23), (2.27) and (2.28) that

$$\lim_{k\to\infty} f(kx)/k^v L(k) = g(x)$$

in the sense of convergence in $\mathcal{S}'(\mathbf{R})$.

(iv) The proof is the same as (iii). Namely, in this case the estimates (2.25) and (2.26) imply the claim as well.   □

We can extend (ii) in Theorem 2.10 to the case $v \notin -\mathbf{N}$. First, we shall recall two assertions (see Lemma 1 and Lemma 2, §7.4 in [192]).

Lemma 2.1. *Let* $f \in \mathcal{S}'(\mathbf{R})$ *and let* ρ *be a regularly varying function of the degree* $\beta \neq -1, -2, \ldots$

a) *If there exists the limit*

$$f(kt)/\rho(k) \to g \neq 0, \quad k \to \infty, \quad \text{in } \mathcal{S}'(\mathbf{R}) ,$$

where $f = f_+ + f_-$ ($f_+ \in \mathcal{S}'_+$ *and* $f_- \in \mathcal{S}'_-$), *then there exists* $N \in \mathbf{N}_0$ *such that for* $f_{+,N} = t^N f_+$ *and* $f_{-,N} = t^N f_-$,

$$\lim_{k\to\infty} f_{\pm,N}(kx)/(k^N \rho(k)) = C_\pm f_{\beta+N+1}(\pm t) ,$$

where $(C_+, C_-) \neq (0,0)$.

b) If $f \in \mathcal{S}'_+$ *and if, for* $N \in \mathbf{N}_0$, $f_N = t^N f$ *has quasi-asymptotics at* ∞ *related to* $k^N \rho(k)$, *then:*

1^0 *If* $\beta > -1$, *then* f *has quasi-asymptotics at* ∞ *related to* $\rho(k), k > 0$.

2^0 *If* $\beta < -1$, *then there exist* $a_j \in \mathbf{R}$, $j = 0, 1, \ldots, p$, *such that*

$$g(t) = f(t) + \sum_{j=0}^{p} a_j \delta^{(j)}(t)$$

has the quasi-asymptotics related to $\rho(k)$.

Corollary of Theorem 2.10. *Let* $f \in \mathcal{D}'(\mathbf{R})$. *Suppose the following limit exists, in the sense of convergence in* $\mathcal{D}'(\mathbf{R})$,

$$\lim_{k \to \infty} f(kx)/(k^v L(k)) = g(x) \neq 0,$$

where $v \in \mathbf{R} \setminus \{-1, -2, -3, \ldots\}$. *Then this limit exists in the sense of convergence in* $\mathcal{S}'(\mathbf{R})$ *as well.*

Proof. Theorem 2.10. implies that $f \in \mathcal{S}'(\mathbf{R})$.

Let $n \in \mathbf{N}$ be such that $v + n > -1$. Clearly, for $f_1(x) = x^n f(x), x \in \mathbf{R}$, there holds

$$\lim_{k \to \infty} \langle f_1(kx)/(k^v L(k)), \varphi(x) \rangle = \langle x^n g(x), \varphi(x) \rangle, \quad \varphi \in \mathcal{D}(\mathbf{R}).$$

There exist distributions $f_+(x) \in \mathcal{S}'_+$ and $f_- \in \mathcal{S}'_-(\operatorname{supp} f_- \subset (-\infty, 0])$ such that $f = f_+ + f_-$. This decomposition of f implies the decomposition of f_1: $f_1(x) = x^n f_+(x) + x^n f_-(x)$, $x \in \mathbf{R}$, where $x^n f_+(x) \in \mathcal{S}'_+$ and $x^n f_-(x) \in \mathcal{S}'_-$.

From Lemma 2.1 a), it follows that for some $m \in \mathbf{N}$,

$$(t^{m+n} f_\pm(t))(kx)/(k^{m+n+v} L(k)) \to C_\pm f_{v+m+n+1}(x) \text{ in } \mathcal{S}'(\mathbf{R}) \text{ as } k \to \infty,$$

where $(C_+, C_-) \neq (0,0)$.

Now, Lemma 2.1 b) implies that for some constants a_α, $\alpha = 0, 1, \ldots, p$, and b_β, $\beta = 0, 1, \ldots, r$,

$$(f_+(kt) + \sum_{\alpha=0}^{p} a_\alpha \delta^{(\alpha)}(kt))/(k^v L(k)) \to C_1 f_{v+1}(t) \text{ in } \mathcal{S}'(\mathbf{R}) \text{ as } k \to \infty,$$

$$(f_-(kt) + \sum_{\beta=0}^{r} b_\beta \delta^{(\beta)}(kt))/(k^v L(k)) \to C_2 f_{v+1}(-t) \text{ in } \mathcal{S}'(\mathbf{R}) \text{ as } k \to \infty,$$

where $(C_1, C_2) \neq (0, 0)$.

Let us notice that in Lemma 2.1 b) it is assumed that the limit distribution g is different from 0. But this assertion also holds for $g = 0$.

Thus for suitable constants $\tilde{C}_\alpha$, $\alpha = 0, 1, \ldots, s$, $s = \max\{p, r\}$

$$(f(kt) + \sum_{\alpha=0}^{s} \tilde{C}_\alpha \delta^{(\alpha)}(kt))/(k^v L(k)) \to C_1 f_{v+1}(t) + C_2 f_{v+1}(-t),$$

as $k \to \infty$ in $\mathcal{S}'(\mathbf{R})$. The fact that $v \in \mathbf{R} \setminus \mathbf{N}$ implies that

$$\frac{1}{k^v L(k)} \sum_{\alpha=0}^{s} \tilde{C}_\alpha \delta^{(\alpha)}(kt) \to 0, \quad k \to \infty,$$

and this completes the proof. $\square$

2.7   Quasi-asymptotics at $\pm\infty$

In **2.1–2.5** the essential assumption was that the generalized functions were supported by an acute cone Γ. In this section, we relax this restriction over the support in the one-dimensional case. In fact, we already faced this situation in **2.6** and studied some problems occurring when the distributions are not supported by an acute cone. We will focus in the case of distribution spaces. Let us observe that some of the results below can be also generalized to $\mathcal{F}'(\mathbf{R}^n)$.

Definition 2.4. It is said that $f \in \mathcal{F}'(\mathbf{R})$ has quasi-asymptotics at $\pm\infty$ related to some positive measurable function $c(k)$, $k \in (a, \infty)$, $a > 0$, if for some $g \in \mathcal{F}'(\mathbf{R})$

$$\lim_{k \to \infty} \langle f(kx)/c(k), \phi(x) \rangle = \langle g(x), \phi(x) \rangle, \quad \phi \in \mathcal{F}(\mathbf{R}).$$

We write in short: $f(kx) \overset{q}{\sim} c(k)g(x)$, $k \to \infty$ in $\mathcal{F}'(\mathbf{R})$, or simply $f \overset{q}{\sim} g$ at $\pm\infty$ related to $c(k)$.

The results of **2.7** give us already some important properties of quasi-asymptotics at $\pm\infty$, they are stated in the following remark.

Remark. If $g \neq 0$ and $\mathcal{F}'(\mathbf{R}) = \mathcal{D}'(\mathbf{R})$ in Definition 2.4, then, by Theorem 2.10, $f \in \mathcal{S}'(\mathbf{R})$, $c(x) = x^v L(x)$, $x \in (0, \infty)$, for some $v \in \mathbf{R}$ and some slowly varying function L, and g is a homogeneous distribution with degree of homogeneity v.

Several properties of the quasi-asymptotics at $\pm\infty$ are listed in the following theorems.

Theorem 2.11. *Let $f \in \mathcal{D}'(\mathbf{R})$ and $f \overset{q}{\sim} g$ at $\pm\infty$ related to $k^\nu L(k)$. Then:*

(i) $f^{(m)} \overset{q}{\sim} g^{(m)}$ at $\pm\infty$ related to $k^{\nu-m}L(k)$, $k > a$, $m \in \mathbf{N}$;

(ii) if $m \in \mathbf{N}$ then $x^m f \overset{q}{\sim} x^m g$ at $\pm\infty$ related to $k^{\nu+m}L(k)$;

(iii) if $\phi \in \mathcal{E}(\mathbf{R})$ and c_1 is a measurable positive function on some interval (a, ∞), $a > 0$, such that

$$\phi(kx)/c_1(k) \to \phi_0(x) \quad \text{in } \mathcal{E}(\mathbf{R}), \quad k \to \infty, \; x \in \mathbf{R},$$

then $f\phi \overset{q}{\sim} g\phi_0$ at $\pm\infty$ related to $c_1(k)k^\nu L(k)$.

Proof. Properties (i) and (ii) follow easily from the definition. For (iii), observe that if $\varphi \in \mathcal{D}(\mathbf{R})$, then $\varphi(x)\phi(kx)/c_1(k) \to \varphi(x)\phi_0(x)$ in $\mathcal{D}(\mathbf{R})$. By the equivalence between weak and strong sequential convergence in $\mathcal{D}'$, we have

$$\lim_{k\to\infty} \left\langle \frac{f(kx)\phi(kx)}{c_1(k)k^\nu L(k)}, \varphi(x) \right\rangle = \lim_{k\to\infty} \left\langle \frac{f(kx)}{k^\nu L(k)}, \varphi(x)\frac{\phi(kx)}{c_1(k)} \right\rangle = \langle g\phi_0, \phi \rangle. \quad \square$$

Theorem 2.12. *Let $f \in \mathcal{E}'(\mathbf{R})$ and $f \overset{q}{\sim} g$ at $\pm\infty$ related to $k^\nu L(k)$, $g \neq 0$. Then $L(k) = 1$, $k > a$, $\nu \in -\mathbf{N}$, and $g(x) = C\delta^{(-\nu-1)}(x)$, for some constant C. Moreover, the limit in Definition 2.4 can be extended on $\mathcal{S}(\mathbf{R})$.*

Proof. It is the same as that of Example 5 in **2.3**. $\square$

Theorem 2.13. *Let F be a locally integrable function on $\mathbf{R}$ and $\nu \in \mathbf{R}$, $\nu > -1$, such that*

$$\lim_{x\to\pm\infty} \frac{F(x)}{|x|^\nu L(|x|)} = C_\pm,$$

where L is slowly varying at ∞. Then $F \overset{q}{\sim} g$ at $\pm\infty$ related to $k^\nu L(k)$, where

$$g(x) = C_+ x_+^\nu + C_- x_-^\nu.$$

Proof. Let us put $F_+(x) = H(x)F(x)$ and $F_-(x) = H(-x)F(x)$, $x \in \mathbf{R}$, we can now apply Example 3 in **2.3** to each $F_\pm$ (see also Theorem 2.8). $\square$

We have the following structural theorem, it will be extended in **2.10**.

Theorem 2.14. *Let $f \in \mathcal{D}'(\mathbf{R})$ and $f \overset{q}{\sim} g$ at $\pm\infty$ related to $k^\nu L(k)$, where $g \neq 0$ and $\nu \in \mathbf{R}\backslash(-\mathbf{N})$. There are $m \in \mathbf{N}_0$ and a locally integrable function F such that*

$$f = F^{(m)} \quad and \quad \lim_{x \to \pm\infty} \frac{F(x)}{|x|^{\nu+m}L(|x|)} = C_\pm \,,$$

where $(C_+, C_-) \neq (0, 0)$.

Proof. Since $f \in \mathcal{S}'(\mathbf{R})$ (Theorem 2.10), let $f = f_+ + f_-$, where $f_+ \in \mathcal{S}'_+$ and $f_- \in \mathcal{S}'_-$ (supp $f_- \subset (-\infty, 0]$). The Corollary of Theorem 2.10 implies that for every $\phi \in \mathcal{S}(\mathbf{R})$,

$$\left\langle \frac{f(kx)}{k^\nu L(k)}, \, \phi(x) \right\rangle \to \langle g(x), \phi(x) \rangle \quad \text{as } k \to \infty \,.$$

As in the proof of the Corollary of Theorem 2.10, one may choose $f_\pm$ satisfying the additional requirement

$$\left\langle \frac{f_\pm(kx)}{k^\nu L(k)}, \, \phi(x) \right\rangle \to \langle \tilde{C}_\pm f_{\nu+1}(\pm x), \phi(x) \rangle \quad \text{as } k \to \infty \,.$$

By Theorem 2.2, there exist locally integrable functions F_1 and F_2 with supp $F_1 \subset [0, \infty)$, supp $F_2 \subset (-\infty, 0]$, and $m \in \mathbf{N}_0$ such that

$$f_+(x) = F_1^{(m)}(x), \quad f_-(x) = F_2^{(m)}(x), \quad x \in \mathbf{R} \,,$$

and

$$\lim_{x \to \infty} \frac{F_1(x)}{x^{\nu+m}L(x)} = C_+, \quad \lim_{x \to -\infty} \frac{F_2(x)}{|x|^{\nu+m}L(|x|)} = C_- .$$

This completes the proof. $\square$

The proof of the next result can be found in [111].

Theorem 2.15. *Let $f \in \mathcal{S}'(\mathbf{R})$ and $\phi_0 \in \mathcal{D}(\mathbf{R})$ such that $\int \phi_0(t)dt = 1$. Let $f' \overset{q}{\sim} g$ at $\pm\infty$ related to $k^\nu L(k)$, $\nu \in \mathbf{R}$, $g \neq 0$, and*

$$\left\langle \frac{f(kx)}{k^{\nu+1}L(k)}, \, \phi_0(x) \right\rangle \to \langle g_0(x), \, \phi_0(x) \rangle \,,$$

where $g_0 \in \mathcal{S}'(\mathbf{R})$ and $g_0' = g$. Then $f \overset{q}{\sim} g_0$ at $\pm\infty$ related to $k^{\nu+1}L(k)$.

Recall that the Fourier transform of a tempered distribution f is denoted by $\mathcal{F}(f)$ or $\hat{f}$. We shall analyze in 2.8 the quasi-asymptotics of a distribution at zero (cf. Definition 2.5). In the following theorem, which is very useful, we already compare these two notions via the Fourier transform.

Theorem 2.16. *Let* $f \in \mathcal{D}'(\mathbf{R})$ *and* $v \in \mathbf{R} \setminus (-\mathbf{N})$. *If*

$$f \overset{q}{\sim} g \quad at \quad \pm\infty \quad related\ to \quad k^v L(k), \tag{2.29}$$

with $g \neq 0$, *then*

$$\lim_{k\to\infty} \frac{\hat{f}(x/k)}{(1/k)^{-v-1}L_1(1/k)} = \hat{g}(x) \quad in \quad \mathcal{S}'(\mathbf{R}), \tag{2.30}$$

where $L_1(\cdot) = L(1/\cdot)$ *is slowly varying at the origin.*

Conversely, if $f \in \mathcal{S}'(\mathbf{R})$ *and (2.30) holds with* $v \in \mathbf{R}$, *then (2.29) holds, as well.*

Proof. Let $\phi \in \mathcal{S}(\mathbf{R})$. By the Corollary of Theorem 2.10, $f \in \mathcal{S}'(\mathbf{R})$ and

$$\left\langle \frac{f(kx)}{k^v L(k)}, \widehat{\phi}(x) \right\rangle = \left\langle \frac{\hat{f}(x/k)}{(1/k)^{-v-1}L_1(1/k)}, \phi(x) \right\rangle, \quad k > 0.$$

This implies the assertion. $\qquad\qquad\qquad\qquad\qquad\qquad\qquad\qquad\qquad\qquad\square$

Theorem 2.17. *Let* $T \in \mathcal{E}'(\mathbf{R})$ *and* $T \overset{q}{\sim} g_1$ *at* $\pm\infty$ *related to* k^v, $v \in -\mathbf{N}$, $g_1 \neq 0$. *Let* $f \in \mathcal{D}'(\mathbf{R})$ *and* $f \overset{q}{\sim} g$ *at* $\pm\infty$ *related to* $k^\alpha L(k)$, $\alpha \in \mathbf{R} \setminus (-\mathbf{N})$, $g \neq 0$. *Then* $T * f \overset{q}{\sim} g_1 * g$ *at* $\pm\infty$ *related to* $k^{\alpha+v+1}L(k)$.

Proof. Let $\phi \in \mathcal{S}(\mathbf{R})$. Using the properties of the Fourier transform, we have

$$\left\langle \frac{(T * f)(kx)}{k^{\alpha+v+1}L(k)}, \widehat{\phi}(x) \right\rangle = \left\langle \frac{\hat{T}(x/k)\hat{f}(x/k)}{k^{\alpha+v+2}L(k)}, \phi(x) \right\rangle$$

$$= \left\langle \frac{\hat{f}(x/k)}{(1/k)^{-\alpha-1}L(1/k)}, \frac{\hat{T}(x/k)}{(1/k)^{-v-1}}\phi(x) \right\rangle. \tag{2.31}$$

Since $\hat{T}$ is an entire function of polynomial growth when $|x| \to \infty$, it must be of the form $\hat{T}(x) = x^{-v-1}T_1(x)$, $x \in \mathbf{R}$, where T_1 is an entire function of polynomial growth such that $T_1(0) = C \neq 0$. All the derivatives of $\hat{T}$ are of polynomial growth when $|x| \to \infty$. So, the same holds for T_1. This implies that for any $\phi \in \mathcal{S}(\mathbf{R})$

$$\frac{1}{k^{v+1}}(x/k)^{-v-1}T_1(x/k)\phi(x) = x^{-v-1}T_1(x/k)\phi(x) \to x^{-v-1}T_1(0)\phi(x),$$

$k \to \infty$, in the sense of convergence in $\mathcal{S}(\mathbf{R})$. Let us note that $\widehat{g}_1(x) = x^{-v-1}T_1(0)$, $x \in \mathbf{R}$.

In the spaces $\mathcal{S}(\mathbf{R})$ and $\mathcal{S}'(\mathbf{R})$ the strong and weak sequential convergence are equivalent. This implies that

$$\left\langle \frac{\hat{f}(x/k)}{(1/k)^{-\alpha-1}L_1(1/k)}, \frac{\hat{T}(x/k)}{(1/k)^{-v-1}}\phi(x) \right\rangle$$

$$\to \langle \hat{g}(x), \hat{g}_1(x)\phi(x) \rangle = \langle (g_1 * g)(x), \hat{\phi}(x) \rangle \quad k \to \infty.$$

By (2.31), the claim follows from Theorem 2.16. $\quad\square$

Theorem 2.18. *Let $f \in \mathcal{D}'(\mathbf{R})$ and $\{f(kx)/k^\alpha L(k); \ k > a\}$, $\alpha \in \mathbf{R}\backslash (-\mathbf{N})$, be a bounded subset of $\mathcal{D}'(\mathbf{R})$. Let $T \in \mathcal{E}'(\mathbf{R})$ and $T \overset{q}{\sim} g_1$ at $\pm\infty$ related to k^{-1}, $g_1 \neq 0$. If $T * f \overset{q}{\sim} g_2$ at $\pm\infty$ related to $k^\alpha L(k)$, $g_2 \neq 0$, then $f \overset{q}{\sim} g$ at $\pm\infty$ related to $k^\alpha L(k)$ and $g_1 * g = g_2$. (Note, $g_1 = C\delta$).*

Proof. The same arguments, as in the proof of Theorem 2.10 yield that $f \in \mathcal{S}'(\mathbf{R})$ and that

$$\{f(k\cdot)/(k^\alpha L(k)); \ k > a\}$$

is a bounded subset of $\mathcal{S}(\mathbf{R})$. With the same arguments as above, we have $(\phi \in \mathcal{S}(\mathbf{R}))$

$$\left\langle \frac{\hat{f}(x/k)}{(1/k)^{-\alpha-1}L_1(1/k)}, \phi(x)\left(1 - \frac{\hat{T}(x/k)}{\hat{T}(0)}\right) \right\rangle \to 0 \quad \text{as} \ k \to \infty.$$

This implies the assertion. $\quad\square$

2.8 Quasi-asymptotics at the origin

Definition 2.5. Let $f \in \mathcal{F}'(\mathbf{R})$ and $c(x)$, $x \in (0,a)$, $a > 0$, be a positive measurable function. It is said that f has quasi-asymptotics at 0 in $\mathcal{F}'(\mathbf{R})$ related to $c(1/k)$ if there is $g \in \mathcal{F}'(\mathbf{R})$ such that

$$\lim_{k\to\infty} \left\langle \frac{f(x/k)}{c(1/k)}, \varphi(x) \right\rangle = \langle g(x), \varphi(x) \rangle, \ \varphi \in \mathcal{F}(\mathbf{R}).$$

We write in short: $f \overset{q}{\sim} g$ at 0 related to $c(1/k)$ in $\mathcal{F}'(\mathbf{R})$.

Remark. The limit in Definition 2.5 may be formulated as

$$\lim_{\varepsilon \to +0} \left\langle \frac{f(\varepsilon x)}{c(\varepsilon)}, \phi(x) \right\rangle = \langle g(x), \phi(x) \rangle, \quad \phi \in \mathcal{F}(\mathbf{R}).$$

Therefore, we often denote also quasi-asymptotics at 0 by

$$f(\varepsilon x) \overset{q}{\sim} c(\varepsilon) g(x), \quad \varepsilon \to 0^+ \text{ in } \mathcal{F}'(\mathbf{R}),$$

or simply by $f \overset{q}{\sim} g$ at 0 related to $c(\varepsilon)$.

If $c = 1$ then this Definition 2.5 is a slight generalization of the well-known Lojasiewicz definition of a "value at 0" of a distribution (see [93] and [105]), and it leads to a notion of jump behavior for distributions [173], [176], [178], [181] and [57]. Note that in the Lojasiewicz definition $\varepsilon \to 0$ from both sides.

We list some properties of the quasi-asymptotics at 0 for distributions, we omit the proofs if they are similar to those of the corresponding properties of the quasi-asymptotics at infinity. For details see [112].

The next proposition shows that we may always assume c is regularly varying at the origin (cf. **0.3**). The proof is essentially the same as that of (i) in Theorem 2.10.

Proposition 2.7. *Let $f \in \mathcal{D}'(\mathbf{R})$ and c satisfy the conditions of Definition 2.5. Assume further that $g \neq 0$. Then, for some real number v and some slowly varying function L at 0^+*

$$c(x) = x^v L(x), \quad x \in (0, a).$$

Moreover, g is homogeneous with the degree of homogeneity v.

Some obvious properties of the quasi-asymptotics at 0 in $\mathcal{S}'(\mathbf{R})$ are given in the following proposition.

Proposition 2.8. *Let $f \in \mathcal{S}'(\mathbf{R})$ and $f \overset{q}{\sim} g$ at 0 related to $\varepsilon^v L(\varepsilon)$ in $\mathcal{S}'(\mathbf{R})$. Then:*

(i) $f^{(m)} \overset{q}{\sim} g^{(m)}$ at 0 related to $\varepsilon^{v-m} L(\varepsilon)$ in $\mathcal{S}'(\mathbf{R})$, $m \in \mathbf{N}$;

(ii) $x^m f(x) \overset{q}{\sim} x^m g(x)$ at 0 related to $\varepsilon^{v+m} L(\varepsilon)$ in $\mathcal{S}'(\mathbf{R})$, $m \in \mathbf{N}$.

The same assertions hold for the quasi-asymptotics at 0 in $\mathcal{D}'(\mathbf{R})$.

We have seen (see remark after Proposition 2.4) that the quasi-asymptotics at infinity is not in general a local property. The next proposition asserts that *the quasi-asymptotics at 0 is a local property*.

Proposition 2.9. *Let $f \in \mathcal{D}'(\mathbf{R})$ and $f \overset{q}{\sim} g$ at 0 related to $\varepsilon^v L(\varepsilon)$ in $\mathcal{D}'(\mathbf{R})$, and let $f_1 \in \mathcal{D}'(\mathbf{R})$ be such that $f = f_1$ in some neighborhood of zero. Then $f_1 \overset{q}{\sim} g$ at 0 related to $\varepsilon^v L(\varepsilon)$, as well.*

Proof. The assertion follows from $\langle f(x), \varphi(x/\varepsilon) \rangle = \langle f_1(x), \varphi(x/\varepsilon) \rangle$ which holds for any $\varphi \in \mathcal{D}(\mathbf{R})$ if $\varepsilon < \varepsilon_0(\varphi)$.

Remark. The same assertion holds for the quasi-asymptotics at 0 in $\mathcal{S}'(\mathbf{R})$. This was proved in [34], Lemma 1.6. This claim also follows directly if we combine Proposition 2.9 with Theorem 2.35 from **2.11.1**.

Proposition 2.10. (Theorem 3, Chapter I, 3.3 in [192]) *Let $f \in \mathcal{S}'(\mathbf{R})$ be a locally integrable function in $(-a, a)$, $a > 0$. Let $c(\varepsilon) = \varepsilon^\alpha L(\varepsilon)$, $\alpha > -1$, where as usual L is slowly varying at 0^+. If*

$$\lim_{x \to \pm 0} f(x)/c(|x|) = C_\pm,$$

then f has the quasi-asymptotics at zero in $\mathcal{S}'(\mathbf{R})$ related to c and

$$\lim_{\varepsilon \to 0^+} f(\varepsilon x)/c(\varepsilon) = C_+ x_+^\alpha + C_- x_-^\alpha \quad in \ \mathcal{S}'(\mathbf{R}).$$

Proof. According to the remark after Proposition 2.9, it is sufficient to prove that for any $\varphi \in \mathcal{S}(\mathbf{R})$ the limits of the following expressions exist, when $\varepsilon \to 0^+$,

$$\int_0^a f(\varepsilon x)/c(\varepsilon)\varphi(x)dx \quad \text{and} \quad \int_{-a}^0 f(\varepsilon x)/c(\varepsilon)\varphi(x)dx.$$

Let us consider the first one:

$$\lim_{k \to \infty} \int_0^a f(\varepsilon x)/c(\varepsilon)\varphi(x)dx = \lim_{k \to \infty} \int_0^a \frac{c(\varepsilon x)}{c(\varepsilon)} \frac{f(\varepsilon x)}{c(\varepsilon x)}\varphi(x)dx.$$

But since the term in the second integral is dominated by an L^1 function $(c(\varepsilon x)/c(\varepsilon) = O(\varepsilon^{\alpha-\sigma})$, where $\alpha > \sigma - 1$, [148]), it is possible to pass the limit under the integral. The second expression can be considered in the same way. $\square$

Proposition 2.10 provides a relation between the asymptotic behavior at zero of a locally integrable function and the quasi-asymptotics at zero of the distribution defined by it.

The next Proposition is a direct consequence of Theorem 2.15 and Theorem 2.16.

Proposition 2.11. *Let* $f \in \mathcal{S}'(\mathbf{R})$ *such that* $xf \overset{q}{\sim} g$ *at* 0 *related to* $\varepsilon^{v+1} L(\varepsilon)$, $v \in \mathbf{R} \setminus (-\mathbf{N})$ *in* $\mathcal{S}'(\mathbf{R})$. *Let* $\varphi_0 \in \mathcal{D}(\mathbf{R})$ *such that* $\int_{\mathbf{R}} \varphi_0(x)dx = 1$ *and*

$$\left\langle \frac{f(\varepsilon x)}{\varepsilon^v L(\varepsilon)}, \varphi_0(x) \right\rangle \to \langle g_0(x), \varphi_0(x) \rangle \quad as \ \varepsilon \to 0^+$$

such that $g_0 \in \mathcal{S}'(\mathbf{R})$ *and* $xg_0(x) = g(x)$, $x \in \mathbf{R}$. *Then,* $f \overset{q}{\sim} g_0$ *at* 0 *related to* $\varepsilon^v L(\varepsilon)$ *in* $\mathcal{S}'(\mathbf{R})$ (g *and* g_0 *are homogeneous of order* $v+1$ *and* v, *respectively*).

Proposition 2.10 and (i) of Proposition 2.8 directly yield

Proposition 2.12. *Let* $f \in \mathcal{S}'(\mathbf{R})$ *and* $f = F^{(m)}$ *in some neighborhood of* 0, *where* $m \in \mathbf{N}_0$ *and* F *is a locally integrable function such that for some* $v > -1$ *and some slowly varying function* L,

$$\lim_{x \to \pm 0} \frac{F(x)}{|x|^v L(|x|)} = C_{\pm} .$$

Then $f \overset{q}{\sim} g$ *at* 0 *related to* $\varepsilon^v L(\varepsilon)$ *in* $\mathcal{S}'(\mathbf{R})$, *where* $g = (C_+ x_+^v + C_- x_-^v)^{(m)}$.

We now discuss an structural theorem. A more complete result will be the subject of **2.10.3–2.10.5.**

Theorem 2.19. *Let* $f \in \mathcal{D}'(\mathbf{R})$ *have quasi-asymptotics at* 0 *in* $\mathcal{D}'(\mathbf{R})$ *related to* $\varepsilon^v L(\varepsilon)$. *If* $v > 0$ *or if* $v > -1$ *but* L *is bounded on some interval* $(0, a)$, $a > 0$, *then there exist a continuous function* F, *defined on* $(-1, 1)$, *an integer* m, *and constants* C_+, C_-, *such that*

$$f = F^{(m)} \quad and \quad \lim_{x \to \pm 0} \frac{F(x)}{|x|^{v+m} L(|x|)} = C_{\pm} . \tag{2.32}$$

Proof. The proof of this theorem is similar to the proof of the Lojasiewicz structural theorem for a distribution having a value at 0, given in [102], pp. 49–52. However, it is necessary to make several non-trivial refinements in the quoted argument.

Let $I = (-2, 2)$. Since $f(\varepsilon x)/(\varepsilon^v L(\varepsilon)) \to g(x)$ in $\mathcal{D}'(\mathbf{R})$, $\varepsilon \to 0^+$, there is a family of continuous functions F_ε, $\varepsilon \in (0, \varepsilon_0]$, defined on I, an $m \in \mathbf{R}_0$ such that

$$F_\varepsilon^{(m)}(x) = \frac{f(\varepsilon x)}{\varepsilon^v L(\varepsilon)}, \quad x \in I, \quad \varepsilon \in (0, \varepsilon_0]$$

and $F_\varepsilon(x) \to g_1(x)$ uniformly on I when $\varepsilon \to 0^+$, where

$$g_1(x) = (C_+ x_+^{v+m} + C_- x_-^{v+m})/m! \ .$$

With no loss of generality, we assume that $\varepsilon_0 \geq 1$. Let us put

$$\tilde{F}_\varepsilon(x) = F_\varepsilon(x)\varepsilon^v L(\varepsilon), \quad x \in I, \quad 0 < \varepsilon \leq 1 \ .$$

From $\tilde{F}_1^{(m)} = f$ it follows $(\varepsilon^{-m} F_1(\varepsilon x))^{(m)} = f(\varepsilon x)$, $x \in I$, $\varepsilon \in (0, 1]$. So, because $\tilde{F}_\varepsilon^{(m)}(x) = f(\varepsilon x)$, we obtain

$$(\tilde{F}_1(\varepsilon x) - \varepsilon^m \tilde{F}_\varepsilon(x))^{(m)} = 0, \quad x \in I, \quad \varepsilon \in (0, 1] \ .$$

This implies that there is a polynomial which depends on ε such that

$$\tilde{F}_\varepsilon(x) = \varepsilon^{-m}(\tilde{F}_1(\varepsilon x) + b_0(\varepsilon) + b_1(\varepsilon)\varepsilon x + \cdots + b_{m-1}(\varepsilon)(\varepsilon x)^{m-1}) \ ,$$

$x \in I$, $\varepsilon \in (0, 1]$. We have

$$\frac{\tilde{F}_\varepsilon(x)}{\varepsilon^v L(\varepsilon)} - g_1(x) = \frac{1}{\varepsilon^{m+v} L(\varepsilon)}[\tilde{F}_1(\varepsilon x) - g_1(\varepsilon x)L(\varepsilon|x|)$$

$$+ b_0(\varepsilon) + \cdots + b_{m-1}(\varepsilon)(\varepsilon x)^{m-1}]$$

$$+ g_1(x)\left(\frac{L(\varepsilon|x|)}{L(\varepsilon)} - 1\right), \quad x \in I, \quad \varepsilon \in (0, 1].$$

We obtain that for any $x \in I \setminus \{0\}$

$$\frac{1}{\varepsilon^{m+v} L(\varepsilon)}(G(\varepsilon x) + b_0(\varepsilon)(\varepsilon x) + \cdots + b_{m-1}(\varepsilon)(\varepsilon x)^{m-1}) \to 0$$

as $\varepsilon \to 0^+$, where we put $G(x) = \tilde{F}_1(x) - g_1(x)L(|x|)$.

Note that the last limit is not uniform in general because we have only the following property of a slowly varying function:

$$\frac{L(\varepsilon|x|)}{L(\varepsilon)} \to 1, \quad \varepsilon \to 0^+, \quad \text{uniformly for} \quad |x| \in [a, b], \ b > a > 0 \ .$$

Let us fix m points $x_1, \ldots, x_m \in I$ such that $x_i \neq 0$, $i = 1, \ldots, m$, and let

$$d = \frac{1}{2} \min\{|x_i|; \ i = 1, \ldots, m\}, \quad J = \{x; \ |x| > d\} \cap I.$$

Because of the quoted property of L, we have

$$\frac{1}{\varepsilon^{m+v} L(\varepsilon)} (G(\varepsilon x) + b_0(\varepsilon) + b_1(\varepsilon)\varepsilon x + \cdots + b_{m-1}(\varepsilon)(\varepsilon x)^{m-1}) \to 0, \ \varepsilon \to 0^+,$$

uniformly on J. This implies that for some monotonously increasing function $\eta(\varepsilon)$, $\varepsilon > 0$, $\eta(\varepsilon) \to 0$ as $\varepsilon \to 0^+$, there holds

$$|G(\varepsilon x) + b_0(\varepsilon) + b_1(\varepsilon)\varepsilon x + \cdots + b_{m-1}(\varepsilon)\varepsilon^{m-1} x^{m-1}| < \varepsilon^{m+v} L(\varepsilon)\eta(\varepsilon),$$

$$x \in J, \ \varepsilon \in (0, 1]. \tag{2.33}$$

Let $\beta > 0$ and $\varepsilon < \beta \leq 2\varepsilon$. If we put in (2.33) β instead of ε and $x_i \varepsilon/\beta$ instead of x, $i = 1, \ldots, m$), (note $x_i \varepsilon/\beta \in J$, $i = 1, \ldots, m$), we obtain

$$|G(\varepsilon x_i) + b_0(\beta) + \varepsilon b_1(\beta)x_i + \cdots + \varepsilon^{m-1} b_{m-1}(\beta)x_i^{m-1}| < \beta^{m+v} L(\beta)\eta(\beta).$$

From (2.33) and the last inequality it follows

$$|b_0(\varepsilon) - b_0(\beta) + \varepsilon(b_1(\varepsilon) - b_1(\beta))x_i$$

$$+ \cdots + \varepsilon^{m-1}(b_{m-1}(\varepsilon) - b_{m-1}(\beta))x_i^{m-1}| < 2\beta^{m+v} L(\beta)\eta(\beta).$$

Now, in the same way as in [102], p. 51 one can prove

$$|b_i(\varepsilon) - b_i(\beta)| < 2^{i+1} K\eta(\beta)L(\beta)\beta^{m+v-i},$$

$$\varepsilon < \beta \leq 2\varepsilon, i = 0, \ldots, m-1, \tag{2.34}$$

where K is a suitable constant. Let $\varepsilon < \beta < 1/2$. Take $r \in \mathbf{N}_0$ such that $\beta/2^{r+1} \leq \varepsilon < \beta/2^r$. (2.34) implies

$$|b_i(\beta/2^{j-1}) - b_i(\beta/2^j)| < 2^{i+1} K\eta(\beta/2^{j-1})L(\beta/2^{j-1})(\beta/2^{j-1})^{m+v-i},$$

$$\tag{2.35}$$

$i = 0, \ldots, m-1; \ j = 1, \ldots, r$, and

$$|b_i(\beta/2^r) - b_i(\varepsilon)| < 2^{i+1} K\eta(\beta/2^r)L(\beta/2^r)(\beta/2^r)^{m+v-i}, \tag{2.36}$$

$i = 0, \ldots, m-1$.

Let $v > 0$ and $C = \sup\{tL(t); \ t \in (0, \beta)\}$. (2.35) and (2.36) imply $(0 < \varepsilon < \beta < 1/2)$

$$|b_i(\varepsilon) - b_i(\beta)| \leq 2^{i+1} K\eta(\beta)C\left(\sum_{j=1}^{r+1}(\beta/2^{j-1})\right)^{m+v-i-1}$$

$$\leq 2^{i+v+m+2} K\eta(\beta)C\beta^{m+v-i-1}, \quad i = 0, 1, \ldots, m-1. \tag{2.37}$$

Note that the assumption $v > 0$ is essential in the above inequality. From (2.37) it follows $b_i = \lim_{\varepsilon \to 0^+} b_i(\varepsilon) < \infty$, $i = 0, 1, \ldots, m-1$, and

$$|b_i - b_i(\varepsilon)| < \overline{K}\eta(\varepsilon)\varepsilon^{m+v-i-1}, \quad i = 0, \ldots, m-1,$$

where $\overline{K}$ is a suitable constant.

From (2.33) and the last inequality it follows

$$|G(\varepsilon x) + b_0 + \varepsilon b_1 x + \cdots + \varepsilon^{m-1} b_{m-1} x^{m-1}| < K_1 \varepsilon^{m+v} L(\varepsilon)\eta(\varepsilon), \quad (2.38)$$

$x \in J$, $0 < \varepsilon \leq 1$ (with suitable K_1).

Let us show that the function

$$F(x) = G(x) + b_0 + b_1 x + \cdots + b_{m-1} x^{m-1} + g_1(x) L(|x|), \quad x \in (-2, 2),$$

satisfies the conditions of the Theorem. Clearly, $F^{(m)} = f$. Put in (2.38) $x = \pm 1$. We obtain

$$|F(\pm\varepsilon) - g(\pm\varepsilon)L(\varepsilon)| < K_1 \varepsilon^{m+v} L(\varepsilon)\eta(\varepsilon), \quad 0 < \varepsilon \leq 1.$$

Let $v > -1$ and $L(x) < C$, $x \in (0, a)$. Then from (2.35) and (2.36) it follows (for $0 < \varepsilon < \beta$, $\beta < a$, $\beta < 1/2$)

$$|b_i(\varepsilon) - b_i(\beta)| \leq 2^{i+1} K C \eta(\beta) \sum_{j=1}^{r+1} (\beta/2^{j-1})^{m+v-i}$$

$$\leq 2^{i+1} K C \eta(\beta) \beta^{m+v-i} \sum_{j=1}^{r+1} (1/2^{j-1})^{m+v-i}, \quad i = 0, \ldots, m-1.$$

Now, the proof follows as in the previous case. The proof is complete. □

The assertion of Theorem 2.19 also holds for $v < 0$. This situation is analyzed in the following theorem. The proof of it has been given in [116]. We omit the proof because more general results will be given in **2.10.3**.

Theorem 2.20. *Let $f \in \mathcal{S}'(\mathbf{R})$ and let f have the quasi-asymptotic behavior at 0 in $\mathcal{S}'(\mathbf{R})$ relate to $\alpha^v L(\alpha)$, where $v \in (-\infty, 0)$, $v \neq -1, -2, \ldots$ and L is bounded on some interval $(0, a)$, $a > 0$. Then there exist a continuous function F defined on $(-1, 1)$, an integer m and $(C_+, C_-) \neq (0, 0)$ such that (2.32) holds.*

Remarks. 1. If $v = 0$ and $L \equiv 1$, then Theorem 2.19 generalizes the well-known Lojasiewicz structural theorem ([93]) for a distribution which has a value at 0.

2. Let $f \in \mathcal{S}'(\mathbf{R})$ and $f \overset{q}{\sim} g$ at 0 related to $c(\varepsilon)$ in $\mathcal{D}'(\mathbf{R})$. The question is: *Does the same hold in* $\mathcal{S}'(\mathbf{R})$? We shall prove in this section that for $c(\varepsilon) = \varepsilon^v L(\varepsilon)$, $\varepsilon < a$, the answer to the question is affirmative if $v > 0$ or if $0 \geq v > -1$ and L is bounded in some interval $(0, \mu)$, $\mu > 0$. Otherwise, this question remained open for quite long time. A complete affirmative answer has been recently obtained in [186], it will be presented below in **2.11** (Theorem 2.35).

3. Theorems 2.19 and 2.20 describes the structure of quasi-asymptotics only under restrictions over ν and L. The complete structural theorems will be discussed in **2.10**.

Proposition 2.12 and Theorem 2.19 directly imply

Theorem 2.21. *Let* $f \in \mathcal{S}'(\mathbf{R})$ *satisfy conditions of Theorem 2.19. Then* $f \overset{q}{\sim} g$ *at 0 related to* $\varepsilon^v L(\varepsilon)$ *in* $\mathcal{S}'(\mathbf{R})$.

Let us denote by $\mathcal{Z}$ the space of Fourier transforms of elements from $\mathcal{D}(\mathbf{R})$ supplied by the convergence structure transferred from $\mathcal{D}(\mathbf{R})$ ($\mathcal{Z} = \mathcal{F}(\mathcal{D})$). Let $f \in \mathcal{S}'(\mathbf{R})$. As usual, we write $f \overset{q}{\sim} g$ at 0 related to $\varepsilon^v L(\varepsilon)$ in $\mathcal{Z}'$ if $g \in \mathcal{Z}'$, and

$$\lim_{\varepsilon \to 0^+} \left\langle \frac{f(\varepsilon x)}{\varepsilon^v L(\varepsilon)}, \varphi(x) \right\rangle = \langle g(x), \varphi(x) \rangle, \quad \varphi \in \mathcal{Z}.$$

Using the Fourier transform and Theorem 2.10 (ii) one can easily obtain that $g \in \mathcal{S}'(\mathbf{R})$.

For $v < 0$, we have the ensuing related result. above ([112]).

Theorem 2.22. *Let* $f \in \mathcal{S}'(\mathbf{R})$ *and* $f \overset{q}{\sim} g$ *at 0 related to* $\varepsilon^v L(\varepsilon)$ *in* $\mathcal{Z}'$, *where* $v < 0$, $v \notin -\mathbf{N}$, *and* $g \neq 0$. *Then* $f \overset{q}{\sim} g$ *at 0 related to* $\varepsilon^v L(\varepsilon)$ *in* $\mathcal{S}'(\mathbf{R})$.

Proof. Apply Fourier transform, (iii) in Theorem 2.10, and then Theorem 2.16.

2.9   Quasi-asymptotic expansions

Chapter III, §10 in [192] and §4 in [32] were devoted to the quasi-asymptotic expansion of tempered distributions with support in $[0, \infty)$. Two kind of expansions were defined therein. We will recall these definitions but in $\mathcal{F}'_+$. We also survey other definitions appearing in the literature.

Let $\{f_\alpha;\ \alpha \in \mathbf{R}\}$ be the family of tempered distributions belonging to $\mathcal{S}'_+$ which is defined in **0.4**.

We denote by Σ_∞ (by Σ_0) the set of all positive slowly varying functions at ∞ (at 0^+).

Let $\alpha \in \mathbf{R}$ and $L \in \Sigma_\infty$ $(L \in \Sigma_0)$. We introduce another family of tempered distributions:

$$(f_L)_{\alpha+1}(t) = \begin{cases} H(t)L(t)t^\alpha/\Gamma(\alpha+1), & \alpha > -1, \\ D^n(f_L)_{\alpha+n+1}(t), & \alpha \leq -1,\ \alpha+n > -1, \end{cases} \tag{2.39}$$

where n is the smallest natural number such that $\alpha + n > -1$.

Obviously, $(f_L)_{\alpha+1} \overset{q}{\sim} f_{\alpha+1}$ at ∞ (at 0^+) related to $k^\alpha L(k)$ (to $(1/k)^\alpha L(1/k)$).

Definition 2.6. ([192]) A distribution $g \in \mathcal{S}'_+$ is said to have a closed quasi-asymptotic expansion of order α and of length ℓ, $0 \leq \ell \leq \infty$ if there exist $N \in \mathbf{N}, \alpha_j \in \mathbf{R}$ and $c_j \in \mathbf{C}$, $j = 1, \ldots, N$, such that

$$\lim_{k \to \infty} \frac{1}{k^{\alpha-\ell}}\left(g(kt) - \sum_{j=1}^N c_j f_{\alpha_j+1}(kt)\right) = 0, \quad \text{in } \mathcal{S}_+.$$

Definition 2.7. ([192]) A distribution $g \in \mathcal{S}'_+$ is said to have an open quasi-asymptotic expansion of α and of length ℓ, $0 < \ell \leq \infty$ if for every $\ell_1 < \ell$, g has closed quasi-asymptotic expansion of order α and of length ℓ_1.

In [101, 117, 122] these two definitions were slightly altered and extended by using the family $\{(f_L)_\alpha;\ \alpha \in \mathbf{R}\}$ instead of the family $\{f_\alpha;\ \alpha \in \mathbf{R}\}$ (cf. (2.39)). In the next definition we assume that $\{(f_L)_\alpha;\ \alpha \in \mathbf{R}\} \subset \mathcal{F}'_+$.

Definition 2.8. We say that an $f \in \mathcal{F}'_+$ has a closed quasi-asymptotic expansion at ∞ (at 0^+) of order $(\alpha, L) \in \mathbf{R} \times \Sigma_\infty$ (of order $(\alpha, L) \in$

$\mathbf{R} \times \Sigma_0$) and of length ℓ, $0 \leq \ell < \infty$, related to $k^{\alpha-\ell}L_0(k)$ (related to $(1/k)^{\alpha+\ell} L_0(1/k)$) if f has quasi-asymptotics at ∞ (at 0^+) related to $k^\alpha L(k)$ $((1/k)^\alpha L(1/k))$ and if there exist $\alpha_i \in \mathbf{R}$, $L_i \in \Sigma_\infty (L_i \in \Sigma_0)$, and $c_i \in \mathbf{C}$, $i = 1,\ldots,N$, $N \in \mathbf{N}$, $\alpha_1 \geq \alpha_2 \geq \cdots \geq \alpha_N$ ($\alpha_1 \leq \alpha_2 \ldots \leq \alpha_N$), and such that f is of the form

$$f(t) = \sum_{i=1}^{N} c_i (f_{L_i})_{\alpha_i+1}(t) + h(t), \tag{2.40}$$

where for every $\phi \in \mathcal{F}(\mathbf{R})$

$$\lim_{k\to\infty} \left\langle \frac{h(kt)}{k^{\alpha-\ell}L_0(k)}, \phi(t) \right\rangle = 0, \quad \left(\lim_{k\to\infty} \left\langle \frac{h(t/k)}{(1/k)^{\alpha+\ell}L_0(1/k)}, \phi(t) \right\rangle = 0 \right).$$

We write in short:

$$f \overset{q.e.}{\sim} \sum_{i=1}^{N} c_i (f_{L_i})_{\alpha_i+1} \quad \text{at } \infty \text{ (at } 0^+) \quad \text{of order } (\alpha, L).$$

Obviously, we shall assume that $c_i \neq 0$ and that $\alpha_N \geq \alpha-\ell$ ($\alpha_N \leq \alpha+\ell$).

We shall always assume that in representation (2.40), $\alpha_1 > \alpha_2 > \cdots > \alpha_N$ ($\alpha_1 < \alpha_2 < \cdots < \alpha_N$), because, $(f_{L_j})_{\beta+1} + (f_{L_k})_{\beta+1} \sim (f_{L_j+L_k})_{\beta+1}$.

Observe $(f_{L_j})_{\beta_1+1}$ and $(f_{L_k})_{\beta_2+1}$ have the same quasi-asymptotics at ∞ (0^+) if and only if $\beta_1 = \beta_2$ and $L_j \sim L_k$. So, we have:

Proposition 2.13. *Let* $f \in \mathcal{F}'_+$ *satisfy the conditions of Definition 2.8 and assume that there are two representations of* f, *with the same length* l,

$$f(t) = \sum_{i=1}^{N} c_i (f_{L_i})_{\alpha_i+1} + h(t), \quad f(t) = \sum_{i=1}^{M} \tilde{c}_i (f_{L_i})_{\tilde{\alpha}_i+1} + \tilde{h}(t)$$

for which all the assumptions given above hold. Then, $M = N$, $\alpha_1 = \tilde{\alpha}_1,\ldots,\alpha_N = \tilde{\alpha}_N$, $L_1 \sim \tilde{L}_1,\ldots,L_N \sim \tilde{L}_N$.

Examples: All examples are in $\mathcal{S}'_+(\mathbf{R})$.

1. We have that $\displaystyle\sum_{r=1}^{\infty} \frac{H(x-1)}{r!\,x^r}, x \in \mathbf{R}$, converges uniformly to $H(x - 1)e^{1/x}$ but

$$H(x-1)e^{1/x} \overset{q.e.}{\sim} H(x) + ((\log x)_+)' \quad \text{at } \infty \quad \text{of order } (0, L \equiv 1)$$

related to $k^{-1}\log k$ and

$$H(x-1)e^{1/x} \overset{q.e.}{\sim} H(x) + ((\log x)_+)' + \left(-1 + \sum_{r=2}^{\infty}\frac{1}{r!}\frac{1}{r-1}\right)\delta(x) \quad \text{at } \infty$$

of order $(0, L \equiv 1)$ related to k^{-1}.

  2. $H(t-1)/t^3 \overset{q.e.}{\sim} (\delta - \delta')/2$ at ∞ of order $(-1, L \equiv 1)$ related to $k^{-3}\log k$. Moreover, let $n > 2$; then for $j \le n-2$

$$H(t-1)/t^n \overset{q.e.}{\sim} \frac{1}{(n-1)}\delta - \frac{1}{(n-2)1!}\delta' + \cdots + \frac{(-1)^{j-1}\delta^{(j-1)}}{(n-1)(j-1)!}$$

at ∞ of order $(-1, L \equiv 1)$ related to k^{-j};

$$H(t-1)/t^n \overset{q.e.}{\sim} \frac{1}{(n-1)}\delta + \frac{1}{(n-2)1!}\delta' + \cdots + \frac{(-1)^{n-2}}{(n-2)!}\delta^{(n-2)}$$

at ∞ of order $(-1, L \equiv 1)$ related to $k^{-n}\log k$.

Following [32], we define the extended open quasi-asymptotic expansion.

Definition 2.9. An $f \in \mathcal{F}'_+$ has open quasi-asymptotic expansion at ∞ (at 0^+) of order $(\alpha, L) \in \mathbf{R} \times \Sigma_\infty$ $((\alpha, L) \in \mathbf{R} \times \Sigma_0)$ and of the length s, $0 < s \le \infty$, if and only if for every ℓ, $0 \le \ell < s$, f has closed quasi-asymptotic expansion of order (α, L) and of length ℓ, related to $k^{\alpha-\ell}L_\ell(k)$ $((1/k)^{\alpha+\ell}L_\ell(1/k))$.

By the same arguments as in Proposition 2.13 one can prove the following proposition:

Proposition 2.14. *Let $f \in \mathcal{F}'_+$ have open quasi-asymptotic expansion at ∞ (at 0^+) of order (α, L) and of length s and let $0 \le \ell_1 < \ell_2 < s$. Suppose that*

$$f \overset{q.e.}{\sim} \sum_{i=1}^{N} a_i(f_{L_i})_{\alpha_i+1} \quad \text{at } \infty \text{ (at } 0^+)$$

related to $k^{\alpha-\ell_1}L_{\ell_1}(k)((1/k)^{\alpha+\ell_1}L_{\ell_1}(1/k))$, and

$$f \overset{q.e.}{\sim} \sum_{i=1}^{M} b_i(f_{L_i})_{\beta_i+1} \quad \text{at } \infty \text{ (at } 0^+)$$

related to $k^{\alpha-\ell_2}L_{\ell_2}(k)((1/k)^{\alpha+\ell_2}L_{\ell_2}(1/k))$.

  Then, $M \ge N$ and $a_i = b_i$, $\alpha_i = \beta_i$, $L_i \sim \tilde{L}_i$, $i = 1, \ldots, N$.

Let us note that if f has the closed quasi-asymptotic expansion at ∞ of order (α, L) related to $k^{\alpha-\ell}L(k)$, then for any $s \le \ell$, f has the open quasi-asymptotic expansion at ∞ of order (α, L) and of length s. A similar conclusion holds for the point 0^+ as well.

We will actually use the notation

$$f \overset{q.e.}{\sim} \sum_{i=1}^{N} c_i (f_{L_i})_{\alpha_i+1} |\{k^{\alpha_i} L_i(k)\} \quad (\varepsilon^{k_i} L_i(\varepsilon))$$

and instead of the "open or closed quasi-asymptotic expansion of order (α, L) and length s", we will just say that f has the *quasi-asymptotic expansion at* ∞ or with respect to the given scale.

Let us redefine the notion of quasi-asymptotic expansion. We state our definition at 0^+, but one can do the same for the quasi-asymptotic expansion at ∞.

We denote by Λ the set $\mathbf{N}$ or a finite set of the form $\{1, 2, \ldots, N\}$, $N \in \mathbf{N}$. In the second case, we shall sometimes use the symbol Λ_N. In the following definition, for $\alpha_i \in \mathbf{R}$ and $L_i \in \Sigma_0$, $i \in \Lambda$, we assume that if $i, j \in \Lambda$, $i < j$, then $\alpha_i \le \alpha_j$ and if $\alpha_i = \alpha_j$, then $L_j(\varepsilon)/L_i(\varepsilon) \to 0$, $\varepsilon \to 0^+$.

Definition 2.10. An $f \in \mathcal{F}'(\mathbf{R})$ has quasi-asymptotic expansion at 0^+ related to $\{\varepsilon^{\alpha_k} L_k(\varepsilon); \ k \in \Lambda\}$ if there are complex numbers $A_k \ne 0$, $k \in \Lambda$, so that for any $m \in \Lambda$

$$\underset{\varepsilon \to 0^+}{w.\lim} \frac{\left(f - \sum\limits_{k=1}^{m} A_k (f_{L_k})_{\alpha_k+1}\right)(\varepsilon x)}{\varepsilon^{\alpha_m} L_m(\varepsilon)} = 0 \text{ in } \mathcal{F}'(\mathbf{R}).$$

We write in short:

$$f \overset{q.e.}{\sim} \sum_{k \in \Lambda} A_k (f_{L_k})_{\alpha_k+1} |\{\varepsilon^{\alpha_k} L_k(\varepsilon)\},$$

or simply $f \overset{q.e.}{\sim} \sum_{k \in \Lambda} A_k (f_{L_k})_{\alpha_k+1}$ related to $\{\varepsilon^{\alpha_k} L_k(\varepsilon); k \in \Lambda\}$. One can easily prove that in this case

$$f' \overset{q.e.}{\sim} \sum_{k \in \Lambda} A_k (f'_{L_k})_{\alpha_k+1} |\{\varepsilon^{\alpha_k} L_k(\varepsilon)\}.$$

Proposition 2.15. *Let* $f \in \mathcal{F}'(\mathbf{R})$. *If*

$$f \overset{q.e.}{\sim} \sum_{k \in \Lambda} A_k (f_{L_k})_{\alpha_k+1} |\{\varepsilon^{\alpha_k} L_k(\varepsilon)\}$$

and

$$f \overset{q.e.}{\sim} \sum_{k \in \Lambda} \tilde{A}_k (f_{\tilde{L}_k})_{\tilde{\alpha}_k + 1} |\{\varepsilon^{\tilde{\alpha}_k} \tilde{L}_k(\varepsilon)\},$$

then $\alpha_k = \tilde{\alpha}_k$, $A_k = \tilde{A}_k$ *and* $L_k(\varepsilon) \sim \tilde{L}_k(\varepsilon)$, $\varepsilon \to 0^+$, $k \in \Lambda$.

Proof. Since $1 \in \Lambda$, by the properties of the quasi-asymptotics at 0^+, $\alpha_1 = \tilde{\alpha}_1$, $A_1 = \tilde{A}_1$ and $L_1(\varepsilon) \sim \tilde{L}_1(\varepsilon)$, $\varepsilon \to 0^+$. If

$$A_1(f_{L_1})_{\alpha_1+1} + A_2(f_{L_2})_{\alpha_2+1} + R_2 = A_1(f_{\tilde{L}_1})_{\tilde{\alpha}_1+1} + \tilde{A}_2(f_{\tilde{L}_2})_{\tilde{\alpha}_2+1} + \tilde{R}_2,$$

where

$$w.\lim_{\varepsilon \to 0^+} \frac{R_2(\varepsilon x)}{\varepsilon^{\alpha_2} L_2(\varepsilon)} = 0 \quad \text{and} \quad w.\lim_{\varepsilon \to 0^+} \frac{\tilde{R}_2(\varepsilon x)}{\varepsilon^{\tilde{\alpha}_2} \tilde{L}_2(\varepsilon)} = 0 \quad \text{in} \quad \mathcal{F}'(\mathbf{R}).$$

Since $\alpha_2 > \alpha_1$ and $\tilde{\alpha}_2 > \tilde{\alpha}_1$, the assumption $\alpha_2 \neq \tilde{\alpha}_2$ gives a contradiction. In the same way $L_2(\varepsilon) \sim \tilde{L}_2(\varepsilon)$, $\varepsilon \to 0^+$ and thus $A_2 = \tilde{A}_2$. The rest follows by induction. $\qquad\square$

We give a structural proposition in $\mathcal{S}'_+$.

Proposition 2.16. *If $f \in \mathcal{S}'_+$,*

$$f \overset{q.e.}{\sim} \sum_{k \in \Lambda} A_k f_{\alpha_k + 1} |\{\varepsilon^{\alpha_k} L_k(\varepsilon)\},$$

then for each $m \in \Lambda$ there is a $p_m \in \mathbf{N}_0$ and a continuous function F_m with supp $F_m \subset [0, \infty)$ *such that*

$$f = \left(F_m + \sum_{k=1}^{m} A_k f_{\alpha_k + p_m + 1} \right)^{(p_m)} \quad \text{and} \quad \lim_{x \to 0^+} \frac{F_m(x)}{x^{\alpha_m + p_m} L_m(x)} = 0.$$

Proof. Observe that a version of Theorem 2.2 holds at 0^+. Now, let $s \leq m$. By Definition 2.10

$$\left\langle \frac{f(\varepsilon x) - \sum\limits_{k=1}^{m} A_k f_{\alpha_k + 1}(\varepsilon x)}{\varepsilon^{\alpha_m} L_m(\varepsilon)}, \varphi(x) \right\rangle \to 0, \text{ as } \varepsilon \to 0^+, \quad \varphi \in \mathcal{S}(\mathbf{R}),$$

and the claim follows. $\qquad\square$

Other definitions and results of the quasi-asymptotic expansions of distributions are given in [52], [53], [55], [102], [118], [194], and [205].

We end this section with an important example of quasi-asymptotic expansion at ∞.

Example 3. *The quasi-asymptotic expansion at* ∞

$$f(kx) \overset{q}{\sim} \sum_{j=0}^{\infty} \frac{(-1)^j \mu_j}{j! \, k^{j+1}} \delta^{(j)}(x) \,,$$

is called the Estrada-Kanwal moment asymptotic expansion.

Estrada and Kanwal have extensively studied this expansion in several distribution spaces as well as its numerous applications. For example, it holds if f has compact support, and actually not just in the space $\mathcal{D}'$ but in $\mathcal{E}'$. The constants μ_j are the moments of f, namely, they are determined by $\mu_j = \langle f(x), x^j \rangle$. Observe that the moment asymptotic expansion gives a clear explanation of why the quasi-asymptotics at infinity is not a local property. For theory and applications of this interesting and important asymptotic expansion, we refer to [47], [52], [53], and [54].

2.10   The structure of quasi-asymptotics. Up-to-date results in one dimension

In this section we present results related to quasi-asymptotics in $\mathcal{D}'(\mathbf{R})$ and $\mathcal{S}'(\mathbf{R})$. We give complete answers to some questions started in the previous sections (cf. **2.6**, **2.7** and **2.8**). We shall follow the exposition from [171], [172], [173] and [186].

Our first aim is to describe the structure of (one-dimensional) quasi-asymptotics by means of complete structural theorems; this will be done in **2.10.3** and **2.10.5**. The key tool for obtaining such results is the concepts of asymptotically and associate asymptotically homogeneous functions; we discuss these classes of functions in detail in **2.10.2** and **2.10.4**.

We should employ a new notation for quasi-asymptotics at 0 and $\pm\infty$, which is more convenient for the purposes of this section. In order to emphasize the role of the slowly varying function, we will use the following notation for quasi-asymptotics,

$$f(\lambda x) = L(\lambda)g(\lambda x) + o(\lambda^{\alpha} L(\lambda)) \quad \text{in } \mathcal{F}'(\mathbf{R}) \,, \tag{2.41}$$

where the parameter λ is taken to either zero or infinity. We call α *the degree* of the quasi-asymptotic behavior. Observe that g may be identically zero, and all the results presented in this section are applicable to this situation as well. Recall that if (2.41) is satisfied and $\mathcal{F}' = \mathcal{D}'$, then g is automatically homogeneous of degree α and it therefore has the form (2.3), depending on whether $\alpha \in \mathbf{R} \setminus (-\mathbf{N})$ or not.

Let $g_j \in \mathcal{F}'(\mathbf{R})$ and let ρ_j be arbitrary measurable functions, $j = 1, 2, \ldots, n$, we write

$$f(\lambda x) = \sum_{j=1}^{n} \rho_j(\lambda) g_j(x) + o(\rho_n(\lambda)) \quad \text{in } \mathcal{F}'(\mathbf{R}),$$

if

$$\langle f(\lambda x), \phi(x) \rangle = \sum_{j=1}^{n} \rho_j(\lambda) \langle g_j(x), \phi(x) \rangle + o(\rho_n(\lambda)), \quad \phi \in \mathcal{F}(\mathbf{R}).$$

2.10.1 *Remarks on slowly varying functions*

In this section we collect some results about slowly varying functions to be used in the future.

Let us assume that L is a slowly varying function at the origin (cf. **0.3**). Similar considerations are applicable for slowly varying functions at infinity.

Our first obvious observation is that only the behavior of L near 0 plays a role in (2.41), and so we may impose to L any behavior we want in intervals of the form $[A, \infty)$. Moreover, if $\tilde{L}$ is any slowly varying function which satisfies

$$\lim_{x \to 0^+} \frac{\tilde{L}(x)}{L(x)} = 1,$$

we may replace L by $\tilde{L}$ in any statement about quasi-asymptotics without loosing generality.

One of the most basic (and most important) results in the theory of slowly varying functions is the representation formula (see first two pages of [148]). Furthermore, the representation formula completely characterizes all the slowly varying functions; L is slowly varying at the origin if and only if there exist measurable functions u and w defined on some interval $(0, B]$, u being bounded and having a finite limit at 0 and w being continuous on $[0, B]$ with $w(0) = 0$, such that

$$L(x) = \exp\left(u(x) + \int_x^B \frac{w(t)}{t} dt \right), \quad x \in (0, B].$$

This formula is important because it enables us to obtain some estimates on L. Since we are looking for suitable modifications of L, our first remark is that we can always assume that L is defined in the whole $(0, \infty)$ and L is everywhere positive. This is shown by extending u and w to $(0, \infty)$ in any way we want.

Given any fixed $\sigma > 0$, then, by modifying u and w, we can assume, when it is convenient, that $B = 1$, u is bounded on all over $(0, \infty)$ and $|w(x)| < \sigma$, $x \in (0, \infty)$. In particular, we obtain the estimate

$$\tilde{M} \min \left\{ x^{-\sigma}, x^{\sigma} \right\} < \frac{L(\lambda x)}{L(\lambda)} < M \max \left\{ x^{-\sigma}, x^{\sigma} \right\}, \quad \forall x, \lambda \in (0, \infty), \quad (2.42)$$

for some positive constants M and $\tilde{M}$. This result is known as *Potter's estimate* [9], p. 25, and will be of vital importance in our investigations of the structural properties for quasi-asymptotics. Under the assumption of the last estimate we can use Lebesgue's dominated convergence theorem in

$$\int_0^\infty \left(\frac{L(\lambda x)}{L(\lambda)} - 1 \right) \phi(x) dx,$$

for $\phi \in \mathcal{S}(\mathbf{R})$, to deduce that

$$L(\lambda x) H(x) = L(\lambda) H(x) + o\left(L(\lambda) \right), \quad \text{in } \mathcal{S}'(\mathbf{R}). \quad (2.43)$$

The reader should keep in mind (2.42) and (2.43), since from now on they will be implicitly used without any further reference, especially for differentiating asymptotic expressions in the future sections. We finally comment a well-known fact [9], [148]: As soon as $L(ax) \sim L(x)$ holds for each $a > 0$, it automatically holds uniformly for a in compact subsets of $(0, \infty)$.

2.10.2 *Asymptotically homogeneous functions*

We study some properties of asymptotically homogeneous functions which will be applied later to the structural study of quasi-asymptotics. Let us proceed to define this class of functions.

Definition 2.11. A function b is said to be asymptotically homogeneous of degree α at the origin (resp. at infinity) with respect to the slowly varying function L, if it is measurable and defined in some interval $(0, A)$ (resp. on (A, ∞)), $A > 0$, and for each $a > 0$,

$$b(ax) = a^\alpha b(x) + o(L(x)), \ x \to 0^+ \quad (resp. \ x \to \infty). \quad (2.44)$$

Obviously, asymptotically homogeneous functions at the origin and at infinity are connected by the change of variables $x \leftrightarrow x^{-1}$; therefore, most of the properties of the class of asymptotically homogeneous functions at infinity can be obtained from those of the corresponding class at the origin.

Observe that no uniformity with respect to a is assumed in Definition 2.11; however, the definition itself forces (2.44) to hold uniformly for a on compact subsets. Indeed, we will show this fact by using a classical argument of J. Korevaar, T. van Aardenne Ehrenfest and N. G. de Bruijn [83], [9], [148] and [186].

Lemma 2.2. *Let b be asymptotically homogeneous of degree α with respect to L. Then, the relation*

$$b(ax) = a^\alpha b(x) + o(L(x)),$$

holds uniformly for a in compact subsets of $(0, \infty)$.

Proof. We show the assertion at the origin, the case at infinity can be obtained by the change of variables $x \leftrightarrow x^{-1}$. So assume that b is asymptotically homogeneous function of degree α at the origin with respect to L. We may assume that b is defined on $(0, 1]$. We rather work with the functions $c(x) = e^{\alpha x} b(e^{-x})$ and $s(x) = L(e^{-x})$, hence c and s are defined in $[0, \infty)$. By using a linear transformation between an arbitrary compact subinterval of $[0, \infty)$ and $[0, 1]$, it is enough to show that

$$c(h + x) - c(x) = o(e^{\alpha x} s(x)), \quad x \to \infty, \tag{2.45}$$

uniformly for $h \in [0, 1]$. Suppose that (2.45) is false. Then, there exist $0 < \varepsilon < 1$, a sequence $\langle h_m \rangle_{m=1}^{\infty} \in [0, 1]^{\mathbf{N}}$ and an increasing sequence of real numbers $\langle x_m \rangle_{m=1}^{\infty}$, $x_m \to \infty$, $m \to \infty$, such that

$$|c(h_m + x_m) - c(x_m)| \geq \varepsilon e^{\alpha x_m} s(x_m), \quad m \in \mathbf{N}. \tag{2.46}$$

Define, for $n \in \mathbf{N}$,

$$A_n = \left\{ h \in [0, 2]; |c(h + x_m) - c(x_m)| < \frac{\varepsilon}{3} e^{\alpha x_m} s(x_m), m \geq n \right\},$$

$$B_n = \left\{ h \in [0, 2]; |c(h + x_m + h_m) - c(h_m + x_m)| < \frac{\varepsilon}{3} e^{\alpha x_m} s(x_m + h_m), m \geq n \right\}.$$

Note that

$$[0, 2] = \bigcup_{n \in \mathbf{N}} A_n = \bigcup_{n \in \mathbf{N}} B_n,$$

so we can select N such that $\mu(A_n)$, $\mu(B_n) > \frac{3}{2}$ (here $\mu(\cdot)$ stands for Lebesgue measure), for all $n \geq N$. For each $n \in \mathbf{N}$, put $C_n = \{h_n\} + B_n$.

Then, we have $\mu(C_n) > \frac{3}{2}, n \geq N$, and $C_n, A_n \subseteq [0,3]$. It follows that $A_n \cap C_n \neq \emptyset, n > N$. For each $n \geq N$, select $u_n \in A_n \cap C_n$. In particular, we have $u_n - h_n \in B_n$, and hence,

$$\left|c\left(u_n + x_n\right) - c\left(x_n\right)\right| < \frac{\varepsilon}{3} e^{\alpha x_n} s\left(x_n\right),$$

$$\left|c\left(u_n + x_n\right) - c\left(x_n + h_n\right)\right| < \frac{\varepsilon}{3} e^{\alpha x_n} s\left(x_n + h_n\right)$$

which implies that for all $n \geq N$,

$$\left|c\left(x_n + h_n\right) - c\left(x_n\right)\right| < \frac{\varepsilon}{3} e^{\alpha x_n} \left(s\left(x_n\right) + s\left(x_n + h_n\right)\right).$$

Using that $s(x + h) - s(x) = o(s(x))$, $x \to \infty$, uniformly for h on compact subsets of $(0, \infty)$, we have that for all n sufficiently large, $s\left(x_n + h_n\right) \leq 2s\left(x_n\right)$, which implies that for n big enough

$$\left|c\left(x_n + h_n\right) - c\left(x_n\right)\right| < \varepsilon e^{\alpha x_n} s\left(x_n\right),$$

in contradiction to (2.46), Therefore, (2.45) must hold uniformly for $h \in [0,1]$. $\square$

Corollary 2.1. *If b is asymptotically homogeneous at the origin (resp. at infinity) with respect to a slowly varying function, then b is locally bounded in some interval of the form $(0, B)$ (resp. (B, ∞)).*

Proof. It follows directly from Lemma 2.2. $\square$

We now obtain the behavior of asymptotically homogeneous functions when the degree is not a negative integer.

Theorem 2.23. *Suppose that b is asymptotically homogeneous at the origin (resp. at infinity) with respect to the slowly varying function L. Assume that its degree is not a negative integer. Then*

(i) *If $\alpha > 0$ (resp. $\alpha < 0$ for the case at infinity), then*

$$b(x) = o(L(x)), \ x \to 0^+ \ (resp. \ x \to \infty). \tag{2.47}$$

(ii) *If $\alpha < 0$ (resp. $\alpha > 0$), then there exists a constant γ such that*

$$b(x) = \gamma x^\alpha + o\left(L(x)\right), \ x \to 0^+ \ (resp. \ x \to \infty). \tag{2.48}$$

Proof. We show only the assertion at the origin, the case at infinity follows again from a change of variables.

Let us first show i). Assume that $\alpha > 0$. Let $0 < \eta$ be any arbitrary number. We keep $\eta < 2^\alpha - 1$. Let $x_0 > 0$ such that

$$\left| b\left(\frac{x}{2}\right) - 2^{-\alpha}b(x) \right| \leq \eta L(x) \quad \text{and} \quad |L(2x) - L(x)| \leq \eta L(x), \ 0 < x < x_0,$$

$$(2.49)$$

and $M = \sup\left\{ |b(x)| / L(x); \frac{1}{2}x_0 \leq x \leq x_0 \right\} < \infty$. Let $x \in [x_0/2, x_0]$. We obtain from (2.49),

$$\left| \frac{b(x/2^n)}{L(x/2^n)} \right| \leq 2^{-\alpha n} \frac{|b(x)|}{L(x/2^n)} + \eta \sum_{j=0}^{n-1} 2^{-\alpha(n-1-j)} \frac{L(x/2^j)}{L(x/2^n)}.$$

Thus, with $t = x/2^n$, and $t \in [x_0/2^{n+1}, x_0/2^n]$,

$$\left| \frac{b(t)}{L(t)} \right| \leq 2^{-n\alpha} M \frac{L(2^n t)}{L(t)} + \eta \sum_{j=0}^{n-1} 2^{-j\alpha} \frac{L(2^{j+1} t)}{L(t)}.$$

By this and

$$L(2^{j+1}t)/L(2^j t) \leq (1+\eta), \ j = 0, \ldots, n-1,$$

we have that if $t \in \left[2^{-(n+1)}x_0, 2^{-n}x_0 \right]$, then

$$\left| \frac{b(t)}{L(t)} \right| \leq M \left(\frac{1+\eta}{2^\alpha} \right)^n + \eta(1+\eta) \sum_{j=0}^{\infty} \left(\frac{1+\eta}{2^\alpha} \right)^j$$

$$= M \left(\frac{1+\eta}{2^\alpha} \right)^n + \eta(1+\eta) \frac{2^\alpha}{2^\alpha - 1 - \eta}.$$

Let us prove that for every $\varepsilon > 0$ there exists a positive σ such that $|b(t)/L(t)| < \varepsilon$, $t \in (0, \sigma)$. First, we have to take η, small enough, such that

$$\eta(1+\eta) \frac{2^\alpha}{2^\alpha - 1 - \eta} < \frac{\varepsilon}{2}$$

and $n_0 \in \mathbf{N}$ such that

$$M \left(\frac{1+\eta}{2^\alpha} \right)^n < \frac{\varepsilon}{2}, \ n \geq n_0.$$

Then, it follows that $|b(t)/L(t)| < \varepsilon$, $t \in (0, \sigma)$, if we take $\sigma = x_0/2^{n_0}$. This completes the first part of the proof.

We now show (ii). Assume that $\alpha < 0$. We rather work with $c(x) = e^{\alpha x}b(e^{-x})$ and $s(x) = L(e^{-x})$. Then c satisfies

$$c(h + x) - c(x) = o\left(e^{\alpha x}s(x)\right), \quad x \to \infty,$$

uniformly for $h \in [0, 1]$. Given $\varepsilon > 0$, we can find $x_0 > 0$ such that for all $x > x_0$ and $h \in [0, 1]$,

$$|c(x + h) - c(x)| \leq \varepsilon e^{\alpha x}s(x) \quad \text{and} \quad |s(h + x) - s(x)| \leq \left(e^{-\frac{\alpha}{2}} - 1\right)s(x).$$

So we have that

$$|c(h + n + x) - c(x)| \leq |c(h + n + x) - c(n + x)| + |c(n + x) - c(x)|$$

$$\leq \varepsilon e^{\alpha(n+x)}s(n + x) + \sum_{j=0}^{n-1}|c(j + 1 + x) - c(j + x)|$$

$$\leq \varepsilon e^{\alpha x}\sum_{j=0}^{n}e^{\alpha j}s(j + x)$$

$$\leq \varepsilon e^{\alpha x}s(x)\frac{1}{1 - e^{\frac{\alpha}{2}}},$$

where the last estimate follows from $s(x + j) \leq s(x)e^{-\alpha j/2}$. Since $s(x) = o\left(e^{-\alpha x}\right)$ as $x \to \infty$, it shows that there exists $\gamma \in \mathbf{R}$ such that

$$\lim_{x \to \infty} c(x) = \gamma.$$

Moreover, the estimate shows that

$$c(x) = \gamma + o\left(e^{\alpha x}s(x)\right), \quad x \to \infty,$$

thus, changing the variables back, we have obtained,

$$b(x) = \gamma x^{\alpha} + o\left(L(x)\right), \quad x \to 0^{+}. \qquad \square$$

We remark that (2.47) and (2.48) trivially imply that b is asymptotically homogeneous of degree α with respect to L.

Asymptotically homogeneous functions of degree zero have a more complex asymptotic behavior. For example if $L \equiv 1$, any asymptotically homogeneous function is the logarithm of a slowly varying function. Instead of attempting to find their behavior in the classical sense, we will study their distributional behavior.

Lemma 2.3. *Let b be asymptotically homogeneous of degree 0 at the origin (respectively at infinity) with respect to the slowly varying function L. If $\sigma < 0$ (resp. $\sigma > 0$) then,*

$$b(x) = o(x^\sigma), \quad x \to 0^+ \ (resp. \ x \to \infty).$$

In particular, $b(x)(L(x))^{-1}$ is integrable near the origin (resp. locally integrable near ∞).

Proof. We know that $L(x) = o(x^\sigma)$. Then for each $a > 0$, $b(ax) = b(x) + o(x^\sigma)$ and this implies that $x^{-\sigma}b(x)$ is asymptotically homogeneous of degree $-\sigma$ with respect to the constant function 1. From (i) of Theorem 2.23, it follows that $b(x) = o(x^\sigma)$. □

We now describe the behavior of asymptotically homogeneous functions of degree zero at the origin. The next theorem will be very important in the next section.

Theorem 2.24. *Let b be asymptotically homogeneous of degree zero at the origin with respect to the slowly varying function L. Suppose that b is integrable on $(0, A]$. Then*

$$b(\varepsilon x)(H(x) - H(\varepsilon x - A)) = b(\varepsilon)H(x) + o(L(\varepsilon)) \quad as \ \varepsilon \to 0^+ \ in \ \mathcal{D}'(\mathbf{R}),$$
$$\tag{2.50}$$

where H is the Heaviside function.

Proof. Let $\phi \in \mathcal{D}(\mathbf{R})$. Find B such that supp $\phi \subseteq [-B, B]$, then there exists $\varepsilon_\phi < 1$ such that

$$\langle b(\varepsilon x), \phi(x) \rangle = \int_0^{\frac{A}{\varepsilon}} b(\varepsilon x)\phi(x)dx = \int_0^B b(\varepsilon x)\phi(x)dx, \ \varepsilon < \varepsilon_\phi. \tag{2.51}$$

Replacing $\phi(x)$ by $B\phi(Bx)$ and ε_ϕ by $B\varepsilon_\phi$, we may assume that $B = 1$. Our aim is to show that for some $\varepsilon_0 < 1$,

$$\frac{b(\varepsilon x) - b(\varepsilon)}{L(\varepsilon)}, \ x \in (0, 1], \ \varepsilon < \varepsilon_0,$$

is dominated by an integrable function in $(0, 1]$ for the use of the Lebesgue theorem. For this goal, we assume that L satisfies the following estimate,

$$\frac{L(\varepsilon x)}{L(\varepsilon)} \leq Mx^{-\frac{1}{2}}, x \in (0, 1], \varepsilon \in (0, \varepsilon_\phi). \tag{2.52}$$

By Lemma 2.2, there exists $0 < \varepsilon_0 < \varepsilon_\phi$ such that

$$|b(\varepsilon x) - b(\varepsilon)| < L(\varepsilon), \ x \in [1/2, 2], \ \varepsilon < \varepsilon_0.$$

We keep $\varepsilon < \varepsilon_0$ and $x \in \left[2^{-n-1}, 2^{-n}\right]$. Then

$$|b(\varepsilon x) - b(\varepsilon)| \leq |b(2\varepsilon x) - b((2x\varepsilon)/2)| + |b(2\varepsilon x) - b(\varepsilon)|$$

$$\leq L(2\varepsilon x) + |b(2\varepsilon x) - b(\varepsilon)| \leq \sum_{i=1}^{n} L\left(2^i \varepsilon x\right) + L(\varepsilon)$$

$$\leq \sum_{i=1}^{n} (2^i x)^{-1/2} L(\varepsilon) + L(\varepsilon),$$

where the last inequality follows from (2.52). Then, if $\varepsilon < \varepsilon_0$ and $x \leq 1$, then

$$\left| \frac{b(\varepsilon x) - b(\varepsilon)}{L(\varepsilon)} \right| \leq M_1 x^{-\frac{1}{2}} + 1,$$

where $M_1 = M(\sqrt{2} + 1)$. Therefore we can apply Lebesgue's dominated convergence theorem to deduce (2.50). $\qquad\square$

We also have a similar result at infinity, this fact is stated in the next theorem. Since its a corollary of Theorem 2.27 , we omit its proof and refer the reader to **2.10.4**.

Theorem 2.25. *Let b be asymptotically homogeneous of degree zero at infinity with respect to the slowly varying function L. Suppose that b is locally integrable on $[A, \infty)$. Then*

$$b(\lambda x) H(\lambda x - A) = b(\lambda) H(x) + o(L(\lambda)) \quad as \ \lambda \to \infty \ in \ \mathcal{S}'(\mathbf{R}).$$

2.10.3 Relation between asymptotically homogeneous functions and quasi-asymptotics

We introduced asymptotically homogeneous functions in order to study the structure of the quasi-asymptotics for Schwartz distributions. We now derive structural theorems for quasi-asymptotics in some cases from the fundamental properties of this class of functions (Theorems 2.23, 2.24 and 2.25).

The technique to be employed here is based on the analysis of the parametric coefficients resulting after performing several integrations of the quasi-asymptotic behavior, these coefficients are naturally connected with the class of asymptotically homogeneous functions. The technique of integration of distributional asymptotic relations goes back to the classical work of Lojasiewicz [93, 116] (cf. Theorem 2.19 in **2.8**). Later on, the properties of the parametric coefficients were singled out and recognized as asymptotically homogeneous functions in [45], [171], [172], [173], [176], [186].

The next proposition provides the intrinsic link between quasi-asymptotics and asymptotically homogeneous functions.

Proposition 2.17. *Let* $f \in \mathcal{D}'(\mathbf{R})$ *have quasi-asymptotic behavior in* $\mathcal{D}'(\mathbf{R})$

$$f(\lambda x) = L(\lambda)g(\lambda x) + o(\lambda^\alpha L(\lambda)) \quad \text{as } \lambda \to \infty \ (\text{resp. } \lambda \to 0^+), \quad (2.53)$$

where L *is a slowly varying function and* g *is a homogeneous distribution of degree* $\alpha \in \mathbf{R}$. *Let* $n \in \mathbf{N}$. *Suppose that* g *admits a primitive of order* n, *that is,* $G_n \in \mathcal{D}'(\mathbf{R})$ *and* $G_n^{(n)} = g$, *which is homogeneous of degree* $n + \alpha$. *Then, for any given* F_n, *an* n-*primitive of* f *in* $\mathcal{D}'(\mathbf{R})$, *there exist functions* $b_0, \ldots, b_{n-1}$, *continuous on* $(0, \infty)$, *such that*

$$F_n(\lambda x) = L(\lambda)G_n(\lambda x) + \sum_{j=0}^{n-1} \lambda^{\alpha+n} b_j(\lambda) \frac{x^{n-1-j}}{(n-1-j)!} + o(\lambda^{\alpha+n} L(\lambda)) \quad (2.54)$$

as $\lambda \to \infty$ *(resp.* $\lambda \to 0^+$*) in* $\mathcal{D}'(\mathbf{R})$, *where each* b_j *is asymptotically homogeneous of degree* $-\alpha - j - 1$.

Proof. Recall that any $\phi \in \mathcal{D}(\mathbf{R})$ is of the form

$$\phi = C_\phi \phi_0 + \theta', \text{ where } C_\phi = \int_{-\infty}^{\infty} \phi(t)dt, \ \theta \in \mathcal{D}(\mathbf{R}), \quad (2.55)$$

and $\phi_0 \in \mathcal{D}(\mathbf{R})$ is chosen so that $\int_{-\infty}^{\infty} \phi_0(t)dt = 1$. The evaluations of primitives F_1 of f and G_1 of g on ϕ are given by

$$\langle F_1, \phi \rangle = C_\phi \langle F_1, \phi_0 \rangle - \langle f, \theta \rangle \quad \text{and} \quad \langle G_1, \phi \rangle = C_\phi \langle G_1, \phi_0 \rangle - \langle g, \theta \rangle.$$

This implies

$$\left\langle \frac{F_1(\lambda x)}{\lambda^{\alpha+1} L(\lambda)}, \phi(x) \right\rangle = C_\phi \left\langle \frac{F_1(\lambda x)}{\lambda^{\alpha+1} L(\lambda)}, \phi_0(x) \right\rangle - \left\langle \frac{f(\lambda x)}{\lambda^\alpha L(\lambda)}, \theta(x) \right\rangle, \quad (2.56)$$

and

$$\left\langle \frac{G_1(\lambda x)}{\lambda^{\alpha+1} L(\lambda)}, \phi(x) \right\rangle = C_\phi \left\langle \frac{G_1(\lambda x)}{\lambda^{\alpha+1} L(\lambda)}, \phi_0(x) \right\rangle - \left\langle \frac{g(\lambda x)}{\lambda^\alpha L(\lambda)}, \theta(x) \right\rangle. \quad (2.57)$$

With $c_0(\lambda) = \langle (F_1 - G_1)(\lambda x), \phi_0(x) \rangle$, $\lambda \in (0, \infty)$, from (2.53), it follows

$$F_1(\lambda x) = L(\lambda)G_1(\lambda x) + c_0(\lambda) + o(\lambda^{\alpha+1} L(\lambda)) \quad \text{in } \mathcal{D}'(\mathbf{R}). \quad (2.58)$$

So relation (2.54) follows by induction from (2.58) and (2.53).

We shall now concentrate in showing the property of the b_j's. We set $F_m = F_n^{(n-m)}$ and $G_m = G_n^{(n-m)}, m \in \{1, \ldots, n\}$. By differentiating relation (2.54) $(n-m)$-times, it follows that

$$F_m(\lambda x) = L(\lambda)G_m(\lambda x) + \sum_{j=0}^{m-1} \lambda^{\alpha+m} b_j(\lambda) \frac{x^{m-1-j}}{(m-1-j)!} + o\left(\lambda^{\alpha+m}L(\lambda)\right)$$

$$(2.59)$$

in $\mathcal{D}'(\mathbf{R})$. Choose $\phi \in \mathcal{D}(\mathbf{R})$ such that $\int_{-\infty}^{\infty} \phi(x)x^j \, dx = 0$ for $j = 1, \ldots, m-1$, and $\int_{-\infty}^{\infty} \phi(x)dx = 1$. Then evaluating (2.59) at ϕ, we have that

$$(a\lambda)^{\alpha+m} b_{m-1}(a\lambda) + L(a\lambda) \langle G_m(a\lambda x), \phi(x) \rangle + o\left(\lambda^{\alpha+m}L(\lambda)\right)$$
$$= \langle F_m(a\lambda x), \phi(x) \rangle = \frac{1}{a} \left\langle F_m(\lambda x), \phi\left(\frac{x}{a}\right) \right\rangle$$
$$= \lambda^{\alpha+m} b_{m-1}(\lambda) + L(\lambda) \langle G_m(a\lambda x), \phi(x) \rangle + o\left(\lambda^{\alpha+m}L(\lambda)\right),$$

and so, with $j = m - 1 \in \{0, \ldots, n-1\}$, for each $a > 0$,

$$b_j(a\lambda) = a^{-\alpha-j-1} b_j(\lambda) + o\left(L(\lambda)\right). \qquad \square$$

With the aid of asymptotically homogeneous functions, we can now obtain our first structural theorem. Observe that for the case at $\pm\infty$ we recover Theorem 2.14, while at 0 we actually extend Theorem 2.19 and Theorem 2.20 (cf. Remark 2 in **2.8**).

Theorem 2.26. *Let $f \in \mathcal{D}'(\mathbf{R})$ have quasi-asymptotic behavior at $\pm\infty$ (resp. at the origin) in $\mathcal{D}'(\mathbf{R})$,*

$$f(\lambda x) = C_- L(\lambda) \frac{(\lambda x)_-^\alpha}{\Gamma(\alpha+1)} + C_+ L(\lambda) \frac{(\lambda x)_+^\alpha}{\Gamma(\alpha+1)} + o\left(\lambda^\alpha L(\lambda)\right). \qquad (2.60)$$

If $\alpha \notin \{-1, -2, \ldots\}$, then there exist a non-negative integer $m > -\alpha - 1$ and an m-primitive F of f such that F is continuous (resp. continuous near 0) and

$$\lim_{x \to \pm\infty} \frac{\Gamma(\alpha+m+1)F(x)}{x^m |x|^\alpha L(|x|)} = C_\pm \quad (resp. \lim_{x \to 0^\pm}). \qquad (2.61)$$

Conversely, if these conditions hold, then (by differentiation) (2.60) follows.

Proof. On combining Proposition 2.17 and Theorem 2.23, one obtains that for each $n \in \mathbf{N}$ and F_n, an n-primitive of f, there exist constants $\gamma_0, \ldots, \gamma_{n-1}$ such that in the sense of convergence in $\mathcal{D}'(\mathbf{R})$,

$$F_n(\lambda x) = \sum_{j=0}^{n-1} \gamma_j \frac{(\lambda x)^j}{j!} + C_- \frac{(-1)^n L(\lambda)(\lambda x)_-^{\alpha+n}}{\Gamma(\alpha+n+1)} + C_+ \frac{L(\lambda)(\lambda x)_+^{\alpha+n}}{\Gamma(\alpha+n+1)}$$
$$+ o\left(\lambda^{\alpha+n} L(\lambda)\right). \tag{2.62}$$

It follows from the convergence $\mathcal{D}'(\mathbf{R})$ that there is $m \in \mathbf{N}$, sufficiently large, such that any m-primitive of f is continuous and (2.62) holds (with $n = m$) uniformly for $x \in [-1, 1]$. Pick a specific m-primitive of f, say F_m, then from (2.62) there is a polynomial p of degree at most $m - 1$ such that

$$F_m(\lambda x) = p(\lambda x) + C_- L(\lambda) \frac{(-1)^m (\lambda x)_-^{\alpha+m}}{\Gamma(\alpha+m+1)} + C_+ L(\lambda) \frac{(\lambda x)_+^{\alpha+m}}{\Gamma(\alpha+m+1)}$$
$$+ o\left(\lambda^{\alpha+m} L(\lambda)\right),$$

uniformly for $x \in [-1, 1]$. Then setting $F = F_m - p$, $x = 1, -1$ and replacing λ by x, relation (2.61) follows at once. The converse follows by differentiation and the properties of regularly varying functions. $\square$

We now start to analyze quasi-asymptotics of negative integral degrees. In this section we will only study the quasi-asymptotics $f(\lambda x) \overset{q}{\sim} \gamma L(\lambda)\delta(\lambda x)$. We postpone the general case for **2.10.5**, after the introduction of associate asymptotically homogeneous function in **2.10.4**.

Proposition 2.18. *Let $f \in \mathcal{D}'(\mathbf{R})$ have quasi-asymptotic behavior at $\pm\infty$ (at the origin) in $\mathcal{D}'(\mathbf{R})$,*

$$f(\lambda x) = \gamma L(\lambda)\delta(\lambda x) + o\left(\lambda^{-1} L(\lambda)\right) \quad \text{as } \lambda \to \infty \text{ (resp. } \lambda \to 0^+). \tag{2.63}$$

Then, there exist $m \in \mathbf{N}$, a function b, being asymptotically homogeneous of degree 0 with respect to L, and an $(m+1)$-primitive F of f such that F is continuous (resp. continuous near 0) and

$$F(x) = \gamma L(|x|) \frac{x^m}{2m!} \operatorname{sgn} x + b(|x|) \frac{x^m}{m!} + o\left(|x|^m L(|x|)\right). \tag{2.64}$$

Conversely, if (2.64) holds, then (2.63) follows by differentiation.

Proof. The existence of m, b, and F satisfying (2.64) follows from the weak convergence of (2.63), Proposition 2.17 and Theorem 2.23, as in the proof of Theorem 2.26. The converse is shown by applying Theorem 2.25 (resp. Theorem 2.24) and differentiating $(m+1)$-times. $\square$

2.10.4 *Associate asymptotically homogeneous functions*

We now introduce the main tool for the study of structural properties of quasi-asymptotics of negative integral degree. Associate asymptotically homogeneous functions are a generalization of asymptotically homogeneous functions. Let us define this class of functions.

Definition 2.12. A function b is said to be *associate asymptotically homogeneous of degree 0 at the origin (resp. at infinity) with respect to the slowly varying function L*, if it is measurable and defined in some interval (A, ∞), $A > 0$, and there exists a constant β such that for each $a > 0$,

$$b(ax) = b(x) + \beta L(x) \log a + o(L(x)) \, , \quad x \to 0^+ \text{ (resp. } x \to \infty) \, . \quad (2.65)$$

Remark. Associate asymptotically homogeneous functions are also known as de Haan functions (cf. [9] and [11]).

We may use the same argument employed in the proof of Lemma 2.2 to show uniform convergence.

Lemma 2.4. *Relation (2.65) holds uniformly in compact subsets of $(0, \infty)$.*

We shall study the distributional asymptotic properties of this class of functions in detail. We first roughly estimate the behavior of associate asymptotically homogeneous functions of degree 0.

Lemma 2.5. *Let b be associate asymptotically homogeneous of degree 0 at the origin (resp. at infinity) with respect to L, then for each $\sigma < 0$ (resp. $\sigma > 0$),*

$$b(x) = o(x^\sigma) \, , \quad x \to 0^+ \text{ (resp. } x \to \infty) \, . \quad (2.66)$$

Hence, b is integrable near the origin (resp. locally integrable near infinity).

Proof. We know that $L(x) = o(x^\sigma)$, for each $\sigma > 0$ [148]. Hence $b(ax) = b(x) + o(x^\sigma)$ and thus $x^{-\sigma} b(x)$ is asymptotically homogeneous of degree $-\sigma$ with respect to $L \equiv 1$, so (2.66) follows from Theorem 2.23.   □

The next two theorems will be crucial in **2.10.5**. They generalize Theorems 2.24 and 2.25. We only give the proof at infinity, the proof at the origin is similar to that of Theorem 2.24.

Theorem 2.27. *Let b be a locally integrable associate asymptotically homogeneous function of degree zero at infinity with respect to the slowly varying function L. Suppose that b is defined on $[A, \infty)$. Then*

$$b(\lambda x)H(\lambda x - A) = b(\lambda)H(x) + L(\lambda)\beta H(x)\log x + o(L(\lambda)) \quad in\ \mathcal{S}'(\mathbf{R})\,, \quad (2.67)$$

$\lambda \to \infty$, *where H is the Heaviside function.*

Proof. Let λ_0 be any positive number. The function b can be decomposed as $b = b_1 + b_2$, where $b_1 \in L^1(\mathbf{R})$ has compact support and $b_2(x) = b(x)H(x - \lambda_0)$ is associate asymptotically homogeneous function of degree zero at infinity. Since b_1 satisfies the *moment asymptotic expansion* (cf. Example **3** in **2.9**), it follows that $b_1(\lambda x) = O(\lambda^{-1}) = o(L(\lambda))$ as $\lambda \to \infty$ in $\mathcal{S}'(\mathbf{R})$. Therefore, we can always assume that $A = \lambda_0$, where λ_0 is selected at our convenience.

Our aim is to show that there is some $\lambda_0 > 1$ such that

$$J(x, \lambda) := \phi(x) \frac{b(\lambda x) - b(\lambda) - \beta L(\lambda)\log x}{L(\lambda)} H(\lambda x - \lambda_0)$$

is dominated by an integrable function, whenever $\phi \in \mathcal{S}(\mathbf{R})$, for the use of the Lebesgue dominated convergence theorem. For this goal, we can always assume that L is positive everywhere and satisfies the following estimate (cf. **2.10.1**)

$$\frac{L(\lambda x)}{L(\lambda)} \leq M \max\left\{x^{-\frac{1}{4}}, x^{\frac{1}{4}}\right\}, \quad x, \lambda \in (0, \infty)\,, \quad (2.68)$$

for some positive constant M. Because of the uniformity of (2.65) on compact sets, there exists a $\lambda_0 > 1$ such that

$$|b(\lambda x) - b(\lambda) - \beta L(\lambda)\log x| < L(\lambda), \quad x \in [1, 2], \ \lambda_0 < \lambda\,.$$

Let n be a positive integer. We keep $\lambda_0 < \lambda$ and $x \in \left[2^n, 2^{n+1}\right]$. Then

$$|b(\lambda x) - b(\lambda) - \beta L(\lambda)\log x| \leq |b(\lambda x) - b(\lambda)| + |\beta|\,L(\lambda)\log x$$

$$\leq |\beta|L(\lambda)\log x + |b(2(\lambda x/2)) - b(\lambda x/2) - \beta L(\lambda x/2)\log 2|$$

$$+ |\beta|L(\lambda x/2)\log 2 + |b(\lambda x/2) - b(\lambda)|$$

$$\leq |\beta|L(\lambda)\log x + (1 + |\beta|\log 2)\,L(\lambda x/2) + |b(\lambda x/2) - b(\lambda)|$$

$$\leq (1 + |\beta|\log 2)\sum_{j=1}^{n} L(2^{-j}\lambda x) + |\beta|L(\lambda)\log 2x + L(\lambda)$$

$$\leq \left(Mx^{\frac{1}{4}}(1 + |\beta|\log 2)\sum_{j=1}^{n}(1/2)^{\frac{j}{4}} + |\beta|\log 2x + 1\right)L(\lambda)\,,$$

where the last inequality follows from (2.68). So if $\lambda_0 < \lambda$ and $1 < x$, then

$$\left| \frac{b(\lambda x) - b(\lambda) - \beta L(\lambda) \log x}{L(\lambda)} \right| \leq M_1 x^{\frac{1}{4}},$$

for some $M_1 > 0$. Now if $\lambda_0/\lambda < x < 1$, we have that

$$\left| \frac{b(\lambda x) - b(\lambda) - \beta L(\lambda) \log x}{L(\lambda)} \right|$$

$$\leq \left(1 + \frac{L(\lambda x)}{L(\lambda)} \right) |\beta \log x| + \left| \frac{b(\lambda) - b(\lambda x) - \beta L(\lambda x) \log x^{-1}}{L(\lambda)} \right|$$

$$\leq \left(1 + M x^{-\frac{1}{4}} \right) |\beta \log x| + \frac{L(\lambda x)}{L(\lambda)} \left| \frac{b(\lambda x(x^{-1})) - b(\lambda x) - \beta L(\lambda x) \log x^{-1}}{L(\lambda x)} \right|$$

$$\leq \left(1 + M x^{-\frac{1}{4}} \right) |\beta \log x| + M M_1 x^{-\frac{1}{2}}.$$

Therefore $J(x, \lambda)$ is dominated by an integrable function for $\lambda > \lambda_0$, so we apply Lebesgue dominated convergence theorem to deduce that

$$\lim_{\lambda \to \infty} \int_0^\infty J(x, \lambda) dx = 0.$$

Finally,

$$\langle b(\lambda x) H(\lambda x - \lambda_0), \phi(x) \rangle - b(\lambda) \int_0^\infty \phi(x) dx - \beta L(\lambda) \int_0^\infty \log x \, \phi(x) dx$$

$$= \int_{\lambda_0/\lambda}^\infty b(\lambda x) \phi(x) dx - b(\lambda) \int_0^\infty \phi(x) dx - \beta L(\lambda) \int_0^\infty \log x \, \phi(x) dx$$

$$= L(\lambda) \int_0^\infty J(x, \lambda) dx + L(\lambda) O \left(\frac{\log \lambda}{\lambda} \right) + O \left(\frac{b(\lambda)}{\lambda} \right)$$

$$= o(L(\lambda)) + L(\lambda) O \left(\frac{b(\lambda)}{\lambda L(\lambda)} \right) = o(L(\lambda)), \quad \lambda \to \infty,$$

where in the last equality we have used Lemma 2.5 and the fact that slowly varying functions are $o(\lambda^\sigma)$ for any $\sigma > 0$. This completes the proof of (2.67). $\qquad \square$

Theorem 2.28. *Let b be a locally integrable associate asymptotically homogeneous function of degree zero at the origin with respect to the slowly varying function L. Suppose that b is defined on $(0, A]$. Then*

$$b(\varepsilon x)(H(x) - H(\varepsilon x - A)) = b(\varepsilon) H(x) + L(\varepsilon) \beta H(x) \log x + o(L(\varepsilon)), \quad (2.69)$$

as $\varepsilon \to 0^+$ in $\mathcal{D}'(\mathbf{R})$.

Corollary 2.2. *Let b be an associate asymptotically homogeneous function of degree 0 with respect to L. Then, there exists an associate asymptotically homogeneous function $c \in C^\infty[0, \infty)$ such that $b(x) = c(x) + o(L(x))$.*

Proof. We may assume that $L \in C^\infty[0, \infty)$ [148], Section 1.4. Find B such that b is locally bounded in $[B, \infty)$, this can be done because of Lemma 2.4. Take $\phi \in \mathcal{D}(\mathbf{R})$ such that $\int_0^\infty \phi(t)dt = 1$ and set $c(x) = \int_{B/x}^\infty b(xt)\phi(t)dt - \beta L(x) \int_0^\infty \phi(t) \log t\, dt$, the corollary now follows from Theorem 2.27 (resp. Theorem 2.28). $\qquad\square$

We may also use Corollary 2.2 to obtain a representation formula for associate asymptotically homogeneous functions, this is the analog to [148], Theorem 1.2 for slowly varying functions.

Theorem 2.29. *The function b is associate asymptotically homogeneous of degree 0 at ∞ satisfying (2.65) if and only if there is a positive number A such that*

$$b(x) = \eta(x) + \int_A^x \frac{\tau(t)}{t}\, dt \, , \quad x \geq A, \qquad (2.70)$$

where η is a locally bounded measurable function on $[A, \infty)$ such that $\eta(x) = M + o(L(x))$ as $x \to \infty$, for some number M, and τ is a C^∞-function such that $\tau(x) \sim \beta L(x)$ as $x \to \infty$.

Proof. Assume first that b_1 is C^∞, defined on $[0, \infty)$ and satisfies that hypothesis of the theorem. We can find $L_1 \sim L$ which is C^∞ and satisfies $xL_1'(x) = o(L(x))$ as $x \to \infty$ [148], p. 7. Let ϕ and c as in the proof of Corollary 2.2 corresponding to b_1 and L_1, additionally assume that $\mathrm{supp}\phi \subseteq (0, \infty)$. From Theorem 2.27, we have that

$$b_1'(\lambda x) = \frac{b_1(\lambda)}{\lambda}\delta(x) + \beta\frac{L(\lambda)}{\lambda}\mathrm{Pf}\left(\frac{H(x)}{x}\right) + o\left(\frac{L(\lambda)}{\lambda}\right) \quad \text{as } \lambda \to \infty$$

in $\mathcal{S}'(\mathbf{R})$, where $\mathrm{Pf}(H(x)/x) = (H(x) \log |x|)'$, since distributional asymptotics can be differentiated. Then, for x positive

$$\begin{aligned}
xc'(x) &= x \int_0^\infty b_1'(xt)t\phi(t)dt - \beta x L_1'(x) \int_0^\infty \phi(t) \log t\, dt \\
&= x \int_0^\infty b_1'(xt)t\phi(t)dt + o(L(x)) \\
&= b_1(x) \cdot 0 + \beta L(x) \int_0^\infty \phi(t)dt + o(L(x)) \\
&= \beta L(x) + o(L(x)) \quad \text{as } x \to \infty .
\end{aligned}$$

Set $\tau(x) = xc'(x)$. Find $A > 0$ such that L is locally integrable on $[A, \infty)$, one has that $b_1(x) = c(A) + \int_A^x (\tau(t)/t)dt + o(L(x))$.

In the general case, let A be a number such that b and L are locally bounded on $[A, \infty)$ and let b_1 the function from Corollary 2.2 such that $b(x) = b_1(x) + o(L(x))$, then we can apply the previous argument to b_1 to find τ as before, so we obtain (2.70) with $\eta(x) = b(x) - \int_A^x (\tau(t)/t)\, dt = c(A) + o(L(x))$.   □

A change of variables $x \leftrightarrow x^{-1}$ in Theorem 2.29 implies the analog result at 0.

Theorem 2.30. *The function b is associate asymptotically homogeneous of degree 0 at the origin satisfying (2.65) if and only if there is a positive number A such that*

$$b(x) = \eta(x) + \int_x^A \frac{\tau(t)}{t}\, dt\,, \quad x \geq A\,, \tag{2.71}$$

where η is a locally bounded measurable function on $[A, \infty)$ such that $\eta(x) = M + o(L(x))$ as $x \to 0^+$, for some number M, and τ is a C^∞-function such that $\tau(x) \sim \beta L(x)$ as $x \to 0^+$.

A slightly different representation formula is given in [148], but, except for the smoothness of τ, both are equivalent.

2.10.5   *Structural theorems for negative integral degrees. The general case*

This section is dedicated to the study of structural properties of quasi-asymptotics with negative integral degree, solving the question posed in Remark 3 of **2.8**.

The next lemma reduces the analysis of negative integral degrees to the case of degree -1.

Lemma 2.6. *Let $f \in \mathcal{D}'(\mathbf{R})$ and k be a positive integer. Then f has the quasi-asymptotic behavior in $\mathcal{D}'(\mathbf{R})$,*

$$f(\lambda x) = \gamma\lambda^{-k}L(\lambda)\,\delta^{(k-1)}(x) + (-1)^{k-1}(k-1)!\,\beta L(\lambda)(\lambda x)^{-k} + o\left(\lambda^{-k}L(\lambda)\right)$$

if and only if there exists a $(k-1)$-primitive g of f satisfying

$$g(\lambda x) = \gamma\lambda^{-1}L(\lambda)\,\delta(x) + \beta L(\lambda)(\lambda x)^{-1} + o\left(\lambda^{-1}L(\lambda)\right) \quad \text{in } \mathcal{D}'(\mathbf{R}).$$

Proof. Apply Proposition 2.17.   □

We should introduce some notation that will be needed. In the following for all $n \in \mathbf{N}$ we denote by l_n the primitive of $\log|x|$ with the property that $l_n(0) = 0$ and $l'_n = l_{n-1}$. We have an explicit formula for them:

$$l_n(x) = \frac{x^n}{n!}\log|x| - \frac{x^n}{n!}\sum_{j=1}^{n}\frac{1}{j}, \quad x \in \mathbf{R},$$

which can be easily verified by direct differentiation. They satisfy

$$l_n(ax) = a^n l_n(x) + \frac{(ax)^n}{n!}\log a, \quad a > 0. \tag{2.72}$$

Theorem 2.31. *Let $f \in \mathcal{D}'(\mathbf{R})$ have quasi-asymptotics at $\pm\infty$ (resp. at the origin) of the form*

$$f(\lambda x) = \gamma\lambda^{-1}L(\lambda)\delta(x) + \beta\lambda^{-1}L(\lambda)x^{-1} + o\left(\lambda^{-1}L(\lambda)\right) \quad in\ \mathcal{D}'(\mathbf{R}). \tag{2.73}$$

Then, there exist an associate asymptotically homogeneous function b satisfying

$$b(ax) = b(x) + \beta\log aL(x) + o(L(x)), \quad x \to \infty\ (resp.\ x \to 0^+), \tag{2.74}$$

an integer m, and a continuous (resp. continuous near 0) $(m+1)$-primitive F of f such that

$$F(x) = b\left(|x|\right)\frac{x^m}{m!} + \gamma\frac{x^m}{2n!}L\left(|x|\right)\mathrm{sgn}x - \beta L\left(|x|\right)\frac{x^m}{m!}\sum_{j=1}^{m}\frac{1}{j} + o\left(|x|^m L\left(|x|\right)\right)$$

$$\tag{2.75}$$

as $x \to \pm\infty$ (resp. $x \to 0$), in the ordinary sense. Conversely, relation (2.75) implies (2.73).

Proof. We will show the assertion only at infinity, the proof at the origin is exactly the same. We shall study, as we have been doing, the coefficients of the integration of (2.73). For each $n \in \mathbf{N}$, choose an n primitive F_n of f satisfying $F'_n = F_{n-1}$. We now proceed to integrate (2.73) once, so we obtain

$$F_1(\lambda x) = b(\lambda) + \frac{\gamma}{2}L(\lambda)\mathrm{sgn}x + \beta L(\lambda)\log|x| + o(L(\lambda)) \quad in\ \mathcal{D}'(\mathbf{R}). \tag{2.76}$$

Now, using the standard trick of evaluating at $\phi \in \mathcal{D}(\mathbf{R})$ with the property $\int_{-\infty}^{\infty}\phi(x)dx = 1$, one obtains that

$$b(\lambda a) + \frac{\gamma}{2}L(\lambda a)\int_{-\infty}^{\infty}\mathrm{sgn}x\,\phi(x)dx + \beta L(\lambda a)\int_{-\infty}^{\infty}\log|x|\phi(x)dx + o(L(\lambda))$$

$$= \langle F_1(\lambda ax), \phi(x)\rangle = \frac{1}{a}\left\langle F_1(\lambda x), \phi\left(\frac{x}{a}\right)\right\rangle$$

$$= b(\lambda) + \frac{\gamma}{2}L(\lambda)\int_{-\infty}^{\infty}\mathrm{sgn}x\phi(x)dx + \beta L(\lambda)\int_{-\infty}^{\infty}\log|ax|\phi(x)dx + o(L(\lambda)),$$

$\lambda \to \infty$, for each $a > 0$. So, we see that b satisfies (2.74) for each $a > 0$. Further integration of (2.76) gives,

$$F_{n+1}(\lambda x) = b(\lambda)\frac{(\lambda x)^n}{n!} + \sum_{j=1}^{n} \lambda^n b_j(\lambda)\frac{x^{n-j}}{(n-j)!} + \gamma L(\lambda)\mathrm{sgn}x\frac{(\lambda x)^n}{2n!}$$
$$+ \beta L(\lambda)\lambda^n l_n(x) + o\left(\lambda^n L(\lambda)\right) \quad \text{as } \lambda \to \infty \text{ in } \mathcal{D}'(\mathbf{R}).$$

As in the proof of Proposition 2.17, one shows that the b_j's are asymptotically homogeneous functions of degree $-j$ with respect to L. Hence if we apply Theorem 2.23 to the b_j's, we obtain that

$$F_{m+1}(\lambda x) = b(\lambda)\frac{(\lambda x)^m}{m!} + \gamma L(\lambda)\frac{(\lambda x)^m}{2m!}\mathrm{sgn}x + \beta L(\lambda)\lambda^m l_m(x) + o\left(\lambda^m L(\lambda)\right)$$
$$(2.77)$$

in the sense of convergence in $\mathcal{D}'(\mathbf{R})$. Moreover, it follows from the definition of convergence in $\mathcal{D}'(\mathbf{R})$ there exists $m_0 \in \mathbf{N}$ such that for all $m \geq m_0$ the distribution F_{m+1} is a continuous function and (2.77) holds uniformly for $x \in [-1, 1]$. Relation (2.75) is shown by making $x = \pm 1$ in (2.77) and then changing $\lambda \leftrightarrow x$.

Conversely, since only the behavior of b at infinity plays a roll in (2.75), we may assume that b is locally integrable, so the converse is obtained after application of Theorem 2.27 and then $(m + 1)$ differentiations. $\quad\square$

Theorem 2.31 is a structural theorem, but we shall give a version free of b.

Theorem 2.32. *Let $f \in \mathcal{D}'(\mathbf{R})$. Then f has quasi-asymptotics at $\pm\infty$ (resp. at the origin) of the form (2.73) if and only if there exists an $(m+1)$-primitive F of f, continuous (resp. continuous near 0), such that for each $a > 0$,*

$$\lim_{x\to\infty} \frac{m!\left(a^{-m}F(ax) - (-1)^m F(-x)\right)}{x^m L(x)} = \gamma + \beta \log a \quad \left(resp. \lim_{x\to 0^+}\right). \quad (2.78)$$

Proof. The limit (2.78) follows from (2.75), (2.74) and (2.72) by direct computation. For the converse, rewrite (2.78) as

$$a^{-m}F(ax) - (-1)^m F(-x) = (\gamma + \beta \log a)\frac{x^m}{m!}L(x) + o\left(x^m L(x)\right),$$

for each $a > 0$. Set

$$b(x) = m!\,x^{-m}F(x) - \left(\frac{\gamma}{2} - \beta\sum_{j=1}^{m}\frac{1}{j}\right)L(x), \quad x > 0.$$

By setting $a = 1$ in (2.78), one sees that for $x < 0$,

$$F(x) = b\left(|x|\right) \frac{x^m}{m!} + \gamma L\left(|x|\right) \frac{x^m}{2m!} \operatorname{sgn} x - \beta L\left(|x|\right) \frac{x^m}{m!} \sum_{j=1}^m \frac{1}{j} + o\left(|x|^m L\left(|x|\right)\right).$$

Since

$$a^{-m} F(ax) - F(x) = \beta \log a \frac{x^m}{m!} L(x) + o\left(x^m L(x)\right),$$

it is clear that for each $a > 0$,

$$b(ax) = b(x) + \beta \log a L(x) + o(L(x)). \qquad \square$$

It is remarkable that, initially, no uniform condition on a is assumed in (2.78). However, the proof of Theorem 2.32 forces this relation to hold uniformly for a in compact subsets.

We are now ready to state the general structural theorem for negative integral degrees which now follows directly from Lemma 2.6, Theorem 2.31 and Theorem 2.32.

Theorem 2.33. *Let $f \in \mathcal{D}'(\mathbf{R})$ and let k be a positive integer. Then f has the quasi-asymptotic behavior in $\mathcal{D}'(\mathbf{R})$ at $\pm\infty$ (resp. at the origin),*

$$f(\lambda x) = \gamma \lambda^{-k} L(\lambda) \delta^{(k-1)}(x) + (-1)^{k-1} (k-1)! \beta L(\lambda)(\lambda x)^{-k} + o\left(\lambda^{-k} L(\lambda)\right)$$

if and only if there exist $m \in \mathbf{N}$, $m \geq k$, an associate asymptotically homogeneous function b of degree 0 at infinity (resp. at the origin) with respect to L satisfying

$$b(ax) = b(x) + \beta \log a L(x) + o(L(x)), \quad x \to \infty \text{ (resp. } x \to 0^+),$$

for each $a > 0$, and an m-primitive F of f which is continuous (resp. continuous near 0) and satisfies

$$F(x) = b\left(|x|\right) \frac{x^{m-k}}{(m-k)!} + \gamma L\left(|x|\right) \frac{x^{m-k}}{2(m-k)!} \operatorname{sgn} x$$

$$- \beta L\left(|x|\right) \frac{x^{m-k}}{(m-k)!} \sum_{j=1}^{m-k} \frac{1}{j} + o(|x|^{m-k} L(|x|))$$

as $x \to \pm\infty$ (resp. $x \to 0$), in the ordinary sense. The last property is equivalent to

$$\lim_{x \to \infty} \frac{(m-k)! \left(a^{k-m} F(ax) - (-1)^{m-k} F(-x)\right)}{x^{m-k} L(x)} = \gamma + \beta \log a \quad \left(\text{resp. } \lim_{x \to 0^+}\right),$$

$$\tag{2.79}$$

for each $a > 0$.

It should be noticed that in (2.79) is not absolutely necessary to assume that the limit is of the form $\gamma + \beta \log a$. Indeed, we have the following corollary.

Theorem 2.34. *Let $f \in \mathcal{D}'(\mathbf{R})$. Then f has quasi-asymptotics at infinity (resp. at the origin) of degree $-k$, $k \in \mathbf{N}$, if and only if there exists a continuous (resp. continuous near 0) m-primitive F of f, $m \geq k$, such that the following limit exists for each $a > 0$,*

$$\lim_{x \to \infty} \frac{\left(a^{k-m} F(ax) - (-1)^{m-k} F(-x)\right)}{x^{m-k} L(x)} = I(a) \quad \left(resp. \lim_{x \to 0^+} \right).$$

Proof. We show that $I(a)$ must be of the form $I(1) + \beta \log a$, for some constant β. We easily see that I is measurable and satisfies

$$I(ab) = I(a) + I(b) - I(1),$$

setting $h(x) = e^{I(x) - I(1)}$, one has that h is positive, measurable and satisfies $h(ab) = h(a)h(b)$, from where it follows [148] that $h(x) = x^\beta$, for some β, and so I has the desired form. $\qquad \square$

2.11 *Quasi-asymptotic extension*

We analyze some problems about the extensions of distributions to other spaces together with their quasi-asymptotic properties, we name this problem *quasi-asymptotic extension problem*.

Let $f \in \mathcal{F}'$ have quasi-asymptotic behavior in $\mathcal{F}'$, that is,

$$\langle f(\lambda x), \phi(x) \rangle \sim \lambda^\alpha L(\lambda) \langle g(x), \phi(x) \rangle, \quad \forall \phi \in \mathcal{F}. \tag{2.80}$$

Suppose that f belongs also to another space $\mathcal{U}'$ such that $\mathcal{F} \subset \mathcal{U}$ (not necessarily densely contained). For various spaces, we investigate in this section the possibility of extending (2.80) to $\mathcal{U}'$, in the sense of obtaining the asymptotic behavior of $\langle f(\lambda x), \varphi(x) \rangle$, for each $\varphi \in \mathcal{U}$. Sometimes the statement "$f \in \mathcal{U}'$" is also part of the problem. The results of the present section were obtained in [171], [172], [173], [186].

2.11.1 *Quasi-asymptotics at the origin in $\mathcal{D}'(\mathbf{R})$ and $\mathcal{S}'(\mathbf{R})$*

In this subsection we conclude the discussion initiated in Remark 2 of **2.8** (cf. Theorem 2.21).

Theorem 2.35. *Let $f \in \mathcal{S}'(\mathbf{R})$. If f has quasi-asymptotic behavior at 0 in $\mathcal{D}'(\mathbf{R})$, then f has the same quasi-asymptotic behavior at 0 in in the space $\mathcal{S}'(\mathbf{R})$.*

Proof. Let α be the degree of the quasi-asymptotic behavior. We shall divide the proof into three cases:

$\alpha \notin \{-1, -2, -3, \ldots\}$,

$\alpha = -1$,

$\alpha = -2, -3, \ldots$

Suppose its degree is $\alpha \notin \{-1, -2, -3, \ldots\}$ and

$$f(\varepsilon x) = C_- L(\varepsilon) \frac{(\varepsilon x)^\alpha_-}{\Gamma(\alpha+1)} + C_+ L(\varepsilon) \frac{(\varepsilon x)^\alpha_+}{\Gamma(\alpha+1)} + o(\varepsilon^\alpha L(\varepsilon))$$

as $\varepsilon \to 0^+$ in $\mathcal{D}'(\mathbf{R})$. Then, by using Theorem 2.26 and the fact $f \in \mathcal{S}'(\mathbf{R})$, we conclude the existence of an integer m, a real number β such that $m > -\alpha$, $\beta > m + \alpha$, and a continuous m-primitive F of f such that

$$F(x) = \frac{|x|^{m+\alpha}}{\Gamma(m+\alpha+1)} L(|x|) ((-1)^m C_- H(-x) + C_+ H(x))$$
$$+ o(|x|^{m+\alpha} L(|x|)), \quad x \to 0^+,$$

and

$$F(x) = O(|x|^\beta), \quad |x| \to \infty. \tag{2.81}$$

We make the usual assumptions over L. Assume (cf. **2.10.1**) that L is positive, defined in $(0, \infty)$ and there exists $M_1 > 0$ such that

$$\frac{L(\varepsilon x)}{L(\varepsilon)} \leq M_1 \max\left\{x^{-\frac{1}{2}}, x^{\frac{1}{2}}\right\}, \quad \varepsilon, x \in (0, \infty). \tag{2.82}$$

Let $\phi \in \mathcal{S}(\mathbf{R})$, then we can decompose $\phi = \phi_1 + \phi_2 + \phi_3$, where supp $\phi_1 \subseteq (-\infty, 1]$, supp ϕ_2 is compact and supp $\phi_3 \subseteq [1, \infty)$. Observe that since $\phi_2 \in \mathcal{D}(\mathbf{R})$ we have that

$$\langle f(\varepsilon x), \phi_2(x) \rangle = (-1)^m C_- \varepsilon^\alpha L(\varepsilon) \left\langle \frac{x^\alpha_-}{\Gamma(\alpha+1)}, \phi_2(x) \right\rangle$$
$$+ C_+ \varepsilon^\alpha L(\varepsilon) \left\langle \frac{x^\alpha_+}{\Gamma(\alpha+1)}, \phi_2(x) \right\rangle + o(\varepsilon^\alpha L(\varepsilon)), \quad \varepsilon \to 0^+. \tag{2.83}$$

So, if we want to show (2.83) for ϕ, it is enough to show it for ϕ_3 placed instead of ϕ_2 in the relation because by symmetry it would follow for ϕ_1 and hence for ϕ. Set

$$G(x) = \frac{F(x)}{x^{\alpha+m}L(x)}, \; x > 0.$$

Then

$$\lim_{x \to 0^+} G(x) = \frac{C_+}{\Gamma(\alpha+m+1)}. \tag{2.84}$$

On combining (2.81), (2.82) and (2.84), we find a constant $M_2 > 0$ such that

$$|G(x)| < M_2(1 + x^{\beta+\frac{1}{2}-m-\alpha}), \; x > 0. \tag{2.85}$$

Relation (2.85) together with (2.82) show that for $\varepsilon \leq 1$,

$$\left| G(\varepsilon x)\frac{L(\varepsilon x)}{L(\varepsilon)}x^{\alpha+m}\phi_3^{(m)}(x) \right| \leq 2M_1 M_2 x^{\beta+1}\left| \phi_3^{(m)}(x) \right| H(x-1).$$

The right hand side of the last estimate belongs to $L^1(\mathbf{R})$ and thus we can use the Lebesgue dominated convergence theorem to obtain,

$$\lim_{\varepsilon \to 0^+}\frac{1}{\varepsilon^\alpha L(\varepsilon)}\langle f(\varepsilon x), \phi_3(x)\rangle$$

$$= \lim_{\varepsilon \to 0^+}(-1)^m \int_0^\infty G(\varepsilon x)\frac{L(\varepsilon x)}{L(\varepsilon)}x^{\alpha+m}\phi_3^{(m)}(x)dx$$

$$= (-1)^m \frac{C_+}{\Gamma(\alpha+m+1)}\int_0^\infty x^{\alpha+m}\phi_3^{(m)}(x)dx$$

$$= C_+ \left\langle \frac{x_+^\alpha}{\Gamma(\alpha+1)}, \phi_3(x) \right\rangle.$$

This shows the result in the case $\alpha \notin \{-1, -2, -3, \ldots\}$.

We now consider the case $\alpha = -1$. Assume that

$$f(\varepsilon x) = \gamma\varepsilon^{-1}L(\varepsilon)\delta(x) + \beta\varepsilon^{-1}L(\varepsilon)x^{-1} + o\left(\varepsilon^{-1}L(\varepsilon)\right) \text{ as } \varepsilon \to 0^+ \text{ in } \mathcal{D}'(\mathbf{R}).$$

As in the last case, it suffices to assume that $\phi \in \mathcal{S}(\mathbf{R})$, supp $\phi \subseteq [1, \infty)$ and show that

$$\lim_{\varepsilon \to 0^+}\frac{\varepsilon}{L(\varepsilon)}\langle f(\varepsilon x), \phi(x)\rangle = \beta\int_1^\infty \frac{\phi(x)}{x}dx.$$

We may proceed as in the previous case to apply the structural theorem, but we rather reduce it to the previous situation. So, set $g(x) = xf(x)$, then

$$g(\varepsilon x) = \beta L(\varepsilon) + o(L(\varepsilon)) \text{ as } \varepsilon \to 0^+ \text{ in } \mathcal{D}'(\mathbf{R}). \tag{2.86}$$

But $g \in \mathcal{S}'(\mathbf{R})$, then since the degree of the quasi-asymptotics is 0, the first case implies that (2.86) is valid in $\mathcal{S}'(\mathbf{R})$. Therefore

$$\lim_{\varepsilon \to 0^+} \frac{\varepsilon}{L(\varepsilon)} \langle f(\varepsilon x), \phi(x) \rangle = \lim_{\varepsilon \to 0^+} \frac{1}{L(\varepsilon)} \left\langle g(\varepsilon x), \frac{\phi(x)}{x} \right\rangle = \beta \int_1^\infty \frac{\phi(x)}{x} dx \,.$$

This shows the case $\alpha = -1$.

It remains to show the theorem when $\alpha \in \{-2, -3, \dots\}$. Suppose the degree is $-k$, $k \in \{2, 3, \dots\}$. It is easy to see that any primitive of degree $(k-1)$ of f has quasi-asymptotics of degree -1 at the origin with respect to L (in fact this is the content of Proposition 2.17 when combined with Theorem 2.23). The $(k-1)$-primitives of f are in $\mathcal{S}'(\mathbf{R})$, so we can apply the case $\alpha = -1$ to them, and then, by differentiation, it follows that f has quasi-asymptotics at the origin in $\mathcal{S}'(\mathbf{R})$.

This completes the proof of Theorem 2.35. $\square$

2.11.2   *Quasi-asymptotic extension problem in* $\mathcal{D}'(0, \infty)$

The purpose of this subsection is to study the following problem. Suppose that a distribution $f \in \mathcal{D}'(\mathbf{R})$ with support in $[0, \infty)$ has quasi-asymptotics of degree α in the space $\mathcal{D}'(0, \infty)$, that is, for each $\phi \in \mathcal{D}(0, \infty)$

$$\langle f(\lambda x), \phi(x) \rangle \sim \lambda^\alpha L(\lambda) \langle g(x), \phi(x) \rangle \,. \tag{2.87}$$

What can we say about the quasi-asymptotic properties of f in $\mathcal{D}'(\mathbf{R})$?

We can apply the techniques from **2.10.3** and **2.10.5** to give a complete answer to this question. The answer depends on α. We start with the quasi-asymptotic behavior at infinity, the same arguments are applicable to quasi-asymptotics at the origin.

Let us start with the case $\alpha > -1$. It is not difficult to show that g must be of the form $C x_+^\alpha / \Gamma(\alpha + 1)$, for some constant C. Next, Proposition 2.17 still holds replacing the space $\mathcal{D}'(\mathbf{R})$ by $\mathcal{D}'(0, \infty)$ (actually this holds without the restriction $\alpha > -1$). Hence, the same argument given in Theorem 2.26 applies here, but this time we only require the uniform convergence on $[1/2, 2]$, and hence we can still conclude the existence of the integer such that (2.61) holds with the limit taken only as $x \to \infty$. Actually, because $\alpha > -1$, relation (2.61) holds for any m-primitive of f. Let F be the m-primitive of f supported on the interval $[0, \infty)$, then we have that

$$F(x) \sim \frac{C x^{\alpha+m} L(x)}{\Gamma(\alpha + m + 1)}, \quad x \to \infty,$$

so we have that $F(\lambda x) = CL(\lambda)(\lambda x)_+^{\alpha+m}/\Gamma(\alpha + m + 1) + o(\lambda^{\alpha+m} L(\lambda))$ in the space $\mathcal{S}'(\mathbf{R})$, differentiating m-times, we obtain the following result.

Theorem 2.36. *Let $f \in \mathcal{D}'(\mathbf{R})$ be supported on $[0, \infty)$. If f has quasi-asymptotic behavior of degree $\alpha > -1$ in $\mathcal{D}'(0, \infty)$, then it is a tempered distribution and has the same quasi-asymptotic behavior in the space $\mathcal{S}'(\mathbf{R})$.*

Suppose now that $\alpha < -1$ and α is not a negative integer. This case differs from the last one essentially in one point, we cannot conclude (2.61) for every m-primitive of f but only for some of them. In any case, denoting again by F the m-primitive (we keep $m > -\alpha - 1$) of f supported on $[0, \infty)$, we have that there exists a polynomial of degree at most $m - 1$ such that

$$F(x) - p(x) \sim \frac{C x^{\alpha+m} L(x)}{\Gamma(\alpha + m + 1)}, \quad x \to \infty \,;$$

therefore,

$$F(\lambda x) = \frac{CL(\lambda)(\lambda x)_+^{\alpha+m}}{\Gamma(\alpha + m + 1)} + \sum_{j=0}^{m-1} a_j (\lambda x)_+^j + o(\lambda^{\alpha+m} L(\lambda)) \quad \text{as } \lambda \to \infty \,,$$

in the space $\mathcal{S}'(\mathbf{R})$, for some constants $a_0, \ldots, a_{m-1}$. Thus, our arguments immediately yield the next theorem.

Theorem 2.37. *Let $f \in \mathcal{D}'(\mathbf{R})$ be supported on $[0, \infty)$. Suppose that*

$$f(\lambda x) = CL(\lambda) \frac{(\lambda x)_+^\alpha}{\Gamma(\alpha + 1)} + o(\lambda^\alpha L(\lambda)) \quad \text{as } \lambda \to \infty \text{ in } \mathcal{D}'(0, \infty).$$

If $\alpha < -1$ and α is not a negative integer, then f is a tempered distribution. Moreover, there exist constants $a_0, a_1, \ldots, a_n$ $(n < -\alpha - 1)$ such that

$$f(\lambda x) = CL(\lambda) \frac{(\lambda x)_+^\alpha}{\Gamma(\alpha + 1)} + \sum_{j=0}^n a_j \frac{\delta^{(j)}(x)}{\lambda^{j+1}} + o(\lambda^\alpha L(\lambda)) \quad \text{as } \lambda \to \infty \text{ in } \mathcal{S}'(\mathbf{R}).$$

When $\alpha = -k$, k being a positive integer, the distribution g in (2.87) must have the form $C x^{-k} \in \mathcal{D}'(0, \infty)$, for some constant C; these distributions are homogeneous as elements of $\mathcal{D}'(0, \infty)$, but they do not have homogeneous extensions to $\mathcal{D}(\mathbf{R})$. The behavior of $f(\lambda x)$ as $\lambda \to \infty$ in $\mathcal{S}'(\mathbf{R})$ is described in the next theorem.

We denote by $\mathrm{Pf}(H(x)/x^k)$ the distribution

$$\left\langle \mathrm{Pf}\left(\frac{H(x)}{x^k}\right), \phi(x) \right\rangle = \mathrm{F.p.} \int_0^\infty \frac{\phi(x)}{x^k} dx, \quad \phi \in \mathcal{D}(\mathbf{R}),$$

where F.p. stands for the Hadamard finite part of the divergent integral [56].

Theorem 2.38. *Let* $f \in \mathcal{D}'(\mathbf{R})$ *be supported on* $[0, \infty)$. *Suppose that*

$$f(\lambda x) = CL(\lambda)\frac{H(x)}{(\lambda x)^k} + o\left(\frac{L(\lambda)}{\lambda^k}\right) \quad \text{as } \lambda \to \infty \text{ in } \mathcal{D}'(0, \infty),$$

where k *is a positive integer. Then* f *is a tempered distribution and there exist an associate asymptotically homogeneous function* b *satisfying*

$$b(ax) = b(x) + \frac{(-1)^{k-1}}{(k-1)!} CL(x) \log a + o(L(x)), \quad x \to \infty, \quad (2.88)$$

for each $a > 0$, *and constants* $a_0, a_1, \dots, a_{k-2}$ *such that*

$$f(\lambda x) = C\frac{L(\lambda)}{\lambda^k}\text{Pf}\left(\frac{H(x)}{x^k}\right) + \frac{b(\lambda)}{\lambda^k}\delta^{(k-1)}(x) + \sum_{j=0}^{k-2} a_j \frac{\delta^{(j)}(x)}{\lambda^{j+1}} + o\left(\frac{L(\lambda)}{\lambda^k}\right) \tag{2.89}$$

as $\lambda \to \infty$ *in* $\mathcal{S}'(\mathbf{R})$.

Proof. For each $n \in \mathbf{N}$, let F_n denote the n-primitive of f with support in $[0, \infty)$. Set $C_1 = (-1)^{k-1}C/(k-1)!$. Adapting the arguments of **2.10.5** and reasoning as in the previous two cases, we obtain the existence of a positive integer $m > k$ such that F_m is continuous and

$$F_m(x) = b_1(x)\frac{x^{m-k}}{(m-k)!} - C_1 L(x)\frac{x^{m-k}}{(m-k)!}\sum_{j=1}^{m-k}\frac{1}{j} + p_{m-1}(x) + o(x^{m-k}L(x)),$$

$x \to \infty$, where b_1 is a locally integrable associate asymptotically homogeneous function satisfying (2.88) and p_{m-1} is a polynomial of degree at most $m - 1$. Throwing away the irrelevant terms of the polynomial p_{m-1} and using Theorem 2.27, we obtain the following asymptotic expansion as $\lambda \to \infty$ in the space $\mathcal{S}'(\mathbf{R})$,

$$F_m(\lambda x) = b_1(\lambda)\frac{(\lambda x)_+^{m-k}}{(m-k)!} + C_1 \lambda^{m-k} L(\lambda) l_{m-k}(x) H(x)$$

$$+ \sum_{j=0}^{k-2} a_j \frac{(\lambda x)_+^{m-j-1}}{(m-j-1)!} + o(\lambda^{m-k}L(\lambda)).$$

Differentiating $(m - k)$-times this expansion, we have that

$$F_k(\lambda x) = b_1(\lambda)H(x) + C_1 L(\lambda)H(x)\log x + \sum_{j=0}^{k-2} a_j \frac{(\lambda x)_+^{k-j-1}}{(k-j-1)!} + o(L(\lambda)). \tag{2.90}$$

The well known formulas [56], p. 68,

$$\frac{d}{dx}(H(x)\log x) = \mathrm{Pf}\left(\frac{H(x)}{x}\right)$$

and

$$\frac{d}{dx}\left(\mathrm{Pf}\left(\frac{H(x)}{x^n}\right)\right) = -n\mathrm{Pf}\left(\frac{H(x)}{x^{n+1}}\right) + \frac{(-1)^n\delta^{(n)}(x)}{n!}$$

imply that

$$\frac{d^{k-1}}{dx^{k-1}}\left(\mathrm{Pf}\left(\frac{H(x)}{x}\right)\right) = (-1)^{k-1}(k-1)!\,\mathrm{Pf}\left(\frac{H(x)}{x^k}\right) - \delta^{(k-1)}(x)\sum_{j=1}^{k-1}\frac{1}{j}.$$

Hence, differentiating (2.90) k-times, one has (2.89) with

$$b(x) = b_1(x) + \frac{(-1)^k C}{(k-1)!}\left(\sum_{j=1}^{k-1}\frac{1}{j}\right)L(x). \qquad \Box$$

The corresponding result at the origin is stated in the next theorem.

Theorem 2.39. *Let $f_0 \in \mathcal{D}'(0,\infty)$. Let L be slowly varying at the origin and $\alpha \in \mathbf{R}$. Suppose that*

$$f_0(\varepsilon x) = \varepsilon^\alpha L(\varepsilon)g_0(x) + o\left(\varepsilon^\alpha L(\varepsilon)\right) \text{ as } \varepsilon \to 0^+ \text{ in } \mathcal{D}'(0,\infty), \qquad (2.91)$$

$g_0 \in \mathcal{D}'(0,\infty)$. Then f_0 admits extensions to $[0,\infty)$. Let f be any of such extensions. Then f has the following asymptotic properties at the origin:

(i) *If $\alpha \notin -\mathbf{N}$, then there exist constants $C, a_0, \ldots, a_n$ such that*

$$f(\varepsilon x) = \sum_{j=0}^{n} a_j\delta^{(j)}(\varepsilon x) + C\varepsilon^\alpha L(\varepsilon)x_+^\alpha + o(\varepsilon^\alpha L(\varepsilon)),$$

as $\varepsilon \to 0^+$ in $\mathcal{D}'(\mathbf{R})$.

(ii) *If $\alpha = -k$, $k \in \mathbf{N}$, then there are constants $C, a_0, \ldots, a_n$ and an associate asymptotically homogeneous function of degree 0 with respect to L satisfying*

$$b(ax) = b(x) + \frac{(-1)^{k-1}}{(k-1)!}CL(x)\log a + o(L(x)), \qquad (2.92)$$

such that

$$f(\varepsilon x) = \sum_{j=0}^{n} a_j\delta^{(j)}(\varepsilon x) + b(\varepsilon)\delta^{(k-1)}(\varepsilon x) + C\varepsilon^{-k}L(\varepsilon)\mathrm{Pf}\left(\frac{H(x)}{x^k}\right)$$

$$+ o\left(\varepsilon^{-k}L(\varepsilon)\right), \qquad (2.93)$$

as $\varepsilon \to 0^+$ in $\mathcal{D}'(\mathbf{R})$.

2.11.3 *Quasi-asymptotics at infinity and spaces* $\mathcal{V}'_\beta(\mathbf{R})$

Sometimes is very useful to have the right of evaluating (2.41) in more test functions than in $\mathcal{S}(\mathbf{R})$, this section is dedicated to give some conditions under the test function which guarantee one can do this for the quasi-asymptotic behavior at $\pm\infty$. We shall now improve Theorem 2.10 and its corollary (cf. **2.6**).

We need the following definition.

Definition 2.13. Let $\phi \in \mathcal{E}(\mathbf{R})$ and $\beta \in \mathbf{R}$. We say that

$$\phi(x) = O(|x|^\beta) \text{ strongly as } |x| \to \infty, \tag{2.94}$$

if for each $m \in \{0, 1, 2, \dots\}$

$$\phi^{(m)}(x) = O(|x|^{\beta-m}) \text{ as } |x| \to \infty. \tag{2.95}$$

The set of ϕ satisfying Definition 2.13 for a particular β forms the space $\mathcal{V}_\beta(\mathbf{R})$ which we topologize in the obvious way [56]. These spaces and their dual spaces are very important in the theory of asymptotic expansions of distributions [56]. In fact, if we set $\mathcal{V}(\mathbf{R}) = \bigcup \mathcal{V}_\beta(\mathbf{R})$ (the union having a topological meaning), we have that $\mathcal{V}'(\mathbf{R})$ is the space of distributional small distributions at infinity [47], [56], they satisfy the moment asymptotic expansion at infinity (cf. Example **3** in **2.9**). We point out that in [56] these spaces are denoted by $= \mathcal{V}'_\beta = \mathcal{K}'_\beta$ and $\mathcal{V}' = \mathcal{K}'$.

The next theorem shows that if f has quasi-asymptotics at $\pm\infty$, then the distributional evaluation of f at $\phi \in \mathcal{V}_\beta(\mathbf{R})$ makes sense under some conditions on β, specifically, we show that f has extensions to some of the spaces $\mathcal{V}_\beta(\mathbf{R})$.

Theorem 2.40. *Let* $f \in \mathcal{D}'(\mathbf{R})$ *have quasi-asymptotic behavior of degree* α *at* $\pm\infty$ *with respect to the slowly varying function* L. *If* $\alpha + \beta < -1$, *then* f *admits extensions to* $\mathcal{V}_\beta(\mathbf{R})$.

Proof. Let $\sigma > 0$ such that $\alpha + \beta + \sigma < -1$, then from Theorem 2.26, Theorem 2.33 and Lemma 2.5 we deduce that there exist $m \in \mathbf{N}$ and a continuous m-primitive of f, say F, such that

$$F(x) = O(|x|^{m+\alpha+\sigma}) \text{ as } |x| \to \infty. \tag{2.96}$$

Notice that here we have used that $L(x) = O(x^\sigma)$ as $x \to \infty$ [148]. So it is evident that the extension of f to $\mathcal{V}_\beta(\mathbf{R})$ is given by

$$\langle f(x), \phi(x) \rangle = (-1)^m \int_{-\infty}^{\infty} F(x)\phi^{(m)}(x)dx, \quad \phi \in \mathcal{V}_\beta(\mathbf{R}), \qquad (2.97)$$

which in view of (2.95) and (2.96) is well-defined and defines an element of $\mathcal{V}'_\beta(\mathbf{R})$. $\square$

We now show that the quasi-asymptotic behavior remains valid in $\mathcal{V}'_\beta(\mathbf{R})$ for one extension of f, with the assumption under β imposed in Theorem 2.40.

Theorem 2.41. *Let $f \in \mathcal{D}'(\mathbf{R})$ have quasi-asymptotic behavior at ∞ of degree α with respect to a slowly varying function L, then f admits an extension to $\mathcal{V}_\beta(\mathbf{R})$ which has the same quasi-asymptotics in $\mathcal{V}'_\beta(\mathbf{R})$, provided that $\alpha + \beta < -1$.*

Proof. The proof is similar to that of Theorem 2.35 with some modifications in the estimates. We use one of the extensions of f found in Theorem 2.40; denote the extension by $\tilde{f}$. We shall divide the proof into two cases: $\alpha \notin \{-1, -2, -3, \dots\}$ and $\alpha \in \{-1, -2, -3, \dots\}$.

Suppose its degree is $\alpha \notin \{-1, -2, -3, \dots\}$ and

$$f(\lambda x) = C_- L(\lambda)\frac{(\lambda x)^\alpha_-}{\Gamma(\alpha + 1)} + C_+ L(\lambda)\frac{(\lambda x)^\alpha_+}{\Gamma(\alpha + 1)} + o\left(\lambda^\alpha L(\lambda)\right) \ as \ \lambda \to \infty,$$

in $\mathcal{D}'(\mathbf{R})$. Find $\sigma > 0$ such that $\alpha + \beta + \sigma < -1$. Then from Theorem 2.26, there are an m such that $m + \alpha > 0$ and a continuous m-primitive F of f such that

$$F(x) = \frac{x^m |x|^\alpha}{\Gamma(m + \alpha + 1)}L(|x|)\left(C_- H(-x) + C_+ H(x)\right) + o\left(|x|^{m+\alpha} L(|x|)\right),$$

$x \to \infty$. We recall that H denotes the Heaviside function. We make the usual assumptions over L (cf. **2.10.1**), assume that L is positive, defined and continuous in $(0, \infty)$ and there exists $M_1 > 0$ such that

$$\frac{L(\lambda x)}{L(\lambda)} \leq M_1 \max\left\{x^\sigma, x^{-\sigma}\right\}, \ \lambda \geq 1, \ x \in (0, \infty). \qquad (2.98)$$

Let $\phi \in \mathcal{V}_\beta(\mathbf{R})$. As in the proof of Theorem 2.35, we may assume that $\mathrm{supp}\, \phi \subseteq [1, \infty)$, and the proof would be complete after we show

$$\left\langle \tilde{f}(\lambda x), \phi(x) \right\rangle \sim C_+ \lambda^\alpha L(\lambda) \left\langle \frac{x^\alpha_+}{\Gamma(\alpha + 1)}, \phi(x) \right\rangle, \qquad (2.99)$$

as $\lambda \to \infty$.

Set

$$G(x) = \frac{F(x)}{x^{\alpha+m}L(x)} \quad \text{for } x \geq 1, \tag{2.100}$$

then

$$\lim_{x \to \infty} G(x) = \frac{C_+}{\Gamma(\alpha+m+1)}. \tag{2.101}$$

So, we can find a constant $M_2 > 0$ such that

$$|G(x)| < M_2, \quad \text{globally.} \tag{2.102}$$

Relation (2.102) together with (2.98) show that for $\lambda \geq 1$,

$$\left| G(\lambda x)\frac{L(\lambda x)}{L(\lambda)}x^{\alpha+m}\phi^{(m)}(x) \right| \leq M_1 M_2 x^{\alpha+m+\sigma}\left| \phi^{(m)}(x) \right| H(x-1).$$

Since $\phi \in \mathcal{V}_\beta(\mathbf{R})$, the right hand side of the last estimate belongs to $L^1(\mathbf{R})$ and thus we can use the Lebesgue dominated convergence theorem to obtain,

$$\begin{aligned}
\lim_{\lambda \to \infty} \frac{\left\langle \tilde{f}(\lambda x), \phi(x) \right\rangle}{\lambda^\alpha L(\lambda)} &= \lim_{\lambda \to \infty} (-1)^m \int_0^\infty G(\lambda x)\frac{L(\lambda x)}{L(\lambda)}x^{\alpha+m}\phi^{(m)}(x)dx \\
&= (-1)^m \frac{C_+}{\Gamma(\alpha+m+1)}\int_0^\infty x^{\alpha+m}\phi^{(m)}(x)dx \\
&= C_+ \left\langle \frac{x_+^\alpha}{\Gamma(\alpha+1)}, \phi(x) \right\rangle.
\end{aligned}$$

This shows the result in the case $\alpha \notin \{-1, -2, -3, \dots\}$.

We now consider the case $\alpha = -k$, $k \in \mathbf{N}$. Assume that

$$f(\lambda x) = \gamma\lambda^{-k}L(\lambda)\delta^{(k-1)}(x) + \beta\lambda^{-k}L(\lambda)x^{-k} + o\left(\lambda^{-k}L(\lambda)\right),$$

as $\lambda \to \infty$ in $\mathcal{D}'(\mathbf{R})$. As in the last case, it suffices to assume that $\phi \in \mathcal{V}_\beta(\mathbf{R})$, supp $\phi \subseteq [1, \infty)$ and show that

$$\lim_{\lambda \to \infty} \frac{\lambda^k}{L(\lambda)}\left\langle \tilde{f}(\lambda x), \phi(x) \right\rangle = \beta\int_1^\infty \frac{\phi(x)}{x^k}dx.$$

So, set $g(x) = x^k f(x)$, then

$$g(\lambda x) = \beta L(\lambda) + o(L(\lambda)) \quad \text{as } \lambda \to \infty \text{ in } \mathcal{D}'(\mathbf{R}). \tag{2.103}$$

But $\phi \in \mathcal{V}_\beta(\mathbf{R})$ implies $\phi(x)/x^k \in \mathcal{V}_{\beta-k}(\mathbf{R})$ then since the degree of the quasi-asymptotic behavior of g is 0, last case implies that (2.103) is valid in $\mathcal{V}'_{\beta-k}(\mathbf{R})$ for a suitable extension $\tilde{g}$ because $\beta - k < -1$, therefore

$$\lim_{\lambda \to \infty} \frac{\lambda^k}{L(\lambda)}\left\langle \tilde{f}(\lambda x), \phi(x) \right\rangle = \lim_{\lambda \to \infty} \frac{1}{L(\lambda)}\left\langle \tilde{g}(\lambda x), \frac{\phi(x)}{x^k} \right\rangle = \beta\int_1^\infty \frac{\phi(x)}{x^k}dx.$$

This completes the proof of Theorem 2.41.   □

The importance Theorem 2.41 lies in the fact that we can relax the growth restrictions over the test functions, this permits to apply quasi-asymptotics to obtain ordinary asymptotics in many interesting situations, for example for certain integral transforms or for solutions to partial differential equations. We discuss a simple example.

Example. Let $f \in \mathcal{D}'(\mathbf{R})$ have quasi-asymptotic behavior at infinity of degree $\alpha < 1$,

$$f(\lambda x) = \lambda^\alpha L(\lambda) g(x) + o\left(\lambda^\alpha L(\lambda)\right) \quad \text{as } \lambda \to \infty \text{ in } \mathcal{D}'(\mathbf{R}).$$

Consider the Poisson kernel,

$$P(t) = \frac{1}{\pi \left(t^2 + 1\right)}.$$

Clearly $P \in \mathcal{V}_2(\mathbf{R})$. By Theorem 2.41, f has an extension $\tilde{f}$ such that the evaluation of $\tilde{f}$ at P is well defined and $\tilde{f}$ preserves the quasi-asymptotic properties of f. Thus

$$U(z) = U(x + yi) = \left\langle f(t), \frac{1}{y} P\left(\frac{x - t}{y}\right) \right\rangle$$

is a solution of the boundary value problem

$$\frac{\partial^2 U}{\partial x^2} + \frac{\partial^2 U}{\partial y^2} = 0, \quad U(x + i0^+) = f(x) \quad (\text{in } \mathcal{D}'(\mathbf{R})).$$

Using Theorem 2.41, we can find the asymptotic behavior of U at infinity over cones. Indeed, let $0 < \sigma < \pi/2$, then Theorem 2.41 implies that as $r \to \infty$

$$U(re^{i\vartheta}) \sim \sin^\alpha(\vartheta) C_\vartheta r^\alpha L(r), \quad \text{uniformly for } \vartheta \in [\sigma, \pi - \sigma],$$

where $C_\theta = g * P(\cot \vartheta)$.

2.12   *Quasi-asymptotic boundedness*

This section is intended to study the structure of the distributional relation

$$f(\lambda x) = O(\rho(\lambda)),$$

where here $\lambda \to \infty$ or $\lambda \to 0^+$ and ρ is a regularly varying function. Our approach to the problem follows the exposition from [172]. Distributions satisfying this relation will be called *quasi-asymptotically bounded distributions*, we make this more precise in the following definition.

Definition 2.14. Let L be a slowly varying function at infinity (respectively at the origin). We say that $f \in \mathcal{D}'$ is *quasi-asymptotically bounded at infinity (at the origin) in* $\mathcal{D}'(\mathbf{R})$ *with respect to* $\lambda^\alpha L(\lambda)$, $\alpha \in \mathbf{R}$, if

$$\langle f(\lambda x), \phi(x) \rangle = O\left(\lambda^\alpha L(\lambda)\right) \quad \text{as } \lambda \to \infty \ \ \forall \phi \in \mathcal{D}(\mathbf{R}), \tag{2.104}$$

(respectively $\lambda \to 0^+$). If (2.104) holds, it is also said that f is *quasi-asymptotically bounded of degree* α *at infinity (at the origin) with respect to the slowly varying function* L. We express (2.104) by

$$f(\lambda x) = O(\lambda^\alpha L(\lambda)) \quad \text{as } \lambda \to \infty \ \text{in } \mathcal{D}'(\mathbf{R}), \tag{2.105}$$

(respectively $\lambda \to 0^+$).

Note that in analogy to the quasi-asymptotic behavior of distributions, we may talk about (2.105) in other spaces of distributions. In the case at infinity, It will follow from our structural theorem that $f \in \mathcal{S}'(\mathbf{R})$ and actually the relation (2.105) holds in $\mathcal{S}'(\mathbf{R})$. The case at the origin is related to the problem of extension of distributions from $\mathbf{R} \setminus \{0\}$ to $\mathbf{R}$. Indeed, if $f \in \mathcal{D}'(\mathbf{R} \setminus \{0\})$ and (2.104) holds for all $\phi \in \mathcal{D}(\mathbf{R} \setminus \{0\})$, we will see later that f admits an extension to $\mathbf{R}$.

We now proceed to obtain the structure of quasi-asymptotically bounded distributions. For this aim, the program established in **2.10** will be followed. We will integrate the relation (2.105) and the coefficients of this integration will satisfy the properties of the following definition.

Definition 2.15. A function b is said to be asymptotically homogeneously bounded of degree α at infinity with respect to the slowly varying function L if it is measurable and defined in some interval (A, ∞), $A > 0$, and for each $a > 0$

$$b(ax) = a^\alpha b(x) + O(L(x)), \ x \to \infty. \tag{2.106}$$

Similarly, one defines asymptotically homogeneously bounded functions at the origin. Our first goal is to study the asymptotic properties of this class of functions. Proceeding as in Lemma 2.2, or using the results of [148], Section 2.4, one has that (2.106) must hold uniformly in compact subsets of $(0, \infty)$. Most of the proofs of the following results are the analog to those for asymptotically homogeneous functions by replacing the o symbol by the O symbol and making obvious modifications to the estimates, therefore they will be omitted.

Proposition 2.19. *Let b be asymptotically homogeneously bounded at infinity (at the origin) with respect to the slowly varying function L. If the degree is negative (respectively positive), then $b(x) = O(L(x))$, as $x \to \infty$ $(x \to 0^+)$.*

Proposition 2.20. *Let b be asymptotically homogeneously bounded at infinity (at the origin) with respect to the slowly varying function L. If the degree α is positive (respectively negative), then there exits a constant γ such that $b(x) = \gamma x^\alpha + O(L(x))$, as $x \to \infty$ $(x \to 0^+)$.*

Note that for the case at infinity since $L(x) = O(x^\sigma)$ as $x \to \infty$, for any $\sigma > 0$, then any asymptotically homogeneously bounded function of degree 0 at infinity satisfies that $b(x)/x^\sigma$ is asymptotically homogeneously bounded of degree $-\sigma$ with respect to the trivial slowly varying function $L \equiv 1$ and hence by Proposition 2.19 it satisfies $b(x) = O(x^\sigma)$ as $x \to \infty$, hence for large argument it is a regular tempered distribution. Similarly, any asymptotically homogeneously bounded function of degree 0 at the origin satisfies $b(x) = O(x^{-\sigma})$ as $x \to 0^+$, for any $\sigma > 0$, consequently it is a distribution for small argument. The proof of the next proposition is totally analogous to those of Theorems 2.24 and 2.27, and therefore it will be omitted.

Proposition 2.21. *Let b be asymptotically homogeneously bounded of degree zero at the infinity (at the origin) with respect to the slowly varying function L. Suppose that b is locally integrable on $[A, \infty)$ (respectively $(0, A]$). Then*

$$b(\lambda x)H(\lambda x - A) = b(\lambda)H(x) + O(L(\lambda)) \text{ as } \lambda \to \infty \text{ in } \mathcal{S}'(\mathbf{R}), \quad (2.107)$$

(resp. $b(\lambda x)(H(x) - H(\lambda x - A)) = b(\lambda)H(x) + O(L(\lambda))$ as $\lambda \to 0^+$ in $\mathcal{D}'(\mathbf{R})$).

Corollary 2.3. *Let b be an asymptotically homogeneously bounded function of degree 0 at infinity (at the origin) with respect to L. Then, there exists $c \in C^\infty[0, \infty)$, being asymptotically homogeneously bounded of degree 0, such that $b(x) = c(x) + O(L(x))$ as $x \to \infty$ (resp. as $x \to 0^+$).*

Proof. We only show the assertion at infinity, the case at the origin is similar. Find B such that b is locally bounded in $[B, \infty)$. Take $\phi \in \mathcal{D}(\mathbf{R})$ supported in $(0, \infty)$ such that $\int_0^\infty \phi(t)dt = 1$ and set $c(x) = \int_{B/x}^\infty b(xt)\phi(t)dt$, the corollary now follows from Proposition 2.21. $\square$

The main connection between quasi-asymptotically bounded distributions and asymptotically homogeneously bounded functions is given in the next proposition, again the proof will be omitted since it is analogous to that of Proposition 2.17.

Proposition 2.22. *Let* $f \in \mathcal{D}'(\mathbf{R})$ *be quasi-asymptotically bounded of degree* α *at infinity (at the origin) with respect to the slowly varying function* L. *Let* $m \in \mathbf{N}$. *Then, for any given* F_m, *an* m-*primitive of* f *in* $\mathcal{D}'(\mathbf{R})$, *there exist functions* $b_0, \ldots, b_{m-1}$, *continuous on* $(0, \infty)$, *such that*

$$F_m(\lambda x) = \sum_{j=0}^{m-1} \lambda^{\alpha+m} b_j(\lambda) \frac{x^{m-1-j}}{(m-1-j)!} + O\left(\lambda^{\alpha+m} L(\lambda)\right) \text{ in } \mathcal{D}'(\mathbf{R}),$$

$$(2.108)$$

as $\lambda \to \infty$ *(respectively* $\lambda \to 0^+$*), where each* b_j *is asymptotically homogeneously bounded of degree* $-\alpha - j - 1$ *with respect to* L.

Thus we obtain from Propositions 2.19–2.22 our first structural theorem.

Theorem 2.42. *Let* $f \in \mathcal{D}'(\mathbf{R})$ *and* $\alpha \notin -\mathbf{N}$. *Then* f *is quasi-asymptotically bounded of degree* α *at infinity (at the origin) with respect to the slowly varying function* L *if and only if there exist* $m \in \mathbf{N}$, $m + \alpha > 0$, *and a continuous (continuous near 0)* m-*primitive* F *of* f *such that*

$$F(x) = O\left(|x|^{m+\alpha} L(|x|)\right),$$

$$(2.109)$$

as $|x| \to \infty$ *(respectively* $x \to 0$*) in the ordinary sense. Moreover, in the case at infinity,* f *belongs to* $\mathcal{S}'(\mathbf{R})$ *and is quasi-asymptotically bounded of degree* α *with respect to* L *in* $\mathcal{S}'(\mathbf{R})$.

Proof. We only discuss the case at infinity, the proof of the assertion at the origin is similar to this case. It follows from Proposition 2.22, Proposition 2.19 and Proposition 2.20 that given $m \in \mathbf{N}$ and an m-primitive F_m, there is a polynomial p_{m-1} of degree at most $m - 1$ such that

$$F_m(\lambda x) = p_{m-1}(\lambda x) + O(\lambda^{\alpha+m} L(\lambda)) \quad \text{as } \lambda \to \infty \text{ in } \mathcal{D}'(\mathbf{R}), \quad (2.110)$$

from the definition of boundedness in $\mathcal{D}'(\mathbf{R})$ it follows that there is an $m > -\alpha$ such that (2.110) holds uniformly for $x \in [-1, 1]$. We let $F = F_m - p_{m-1}$, so by taking $x = -1$, $x = 1$ and replacing λ by x in (2.110) we obtain (2.109). The converse follows by observing that (2.109) implies that $F(\lambda x) = O(\lambda^{\alpha+m} L(\lambda))$ in $\mathcal{S}'(\mathbf{R})$ which gives the result after differentiating m-times. $\qquad \square$

We now analyze the case of negative integral degree.

Theorem 2.43. *Let $f \in \mathcal{D}'(\mathbf{R})$ and let k be a positive integer. Then f is quasi-asymptotically bounded of degree $-k$ at infinity (at the origin) with respect to L if and only if there exist $m > k$, an asymptotically homogeneously bounded function b of degree 0 at infinity (at the origin) with respect to L and a continuous (continuous near 0) m-primitive F of f such that*

$$F(x) = b\left(|x|\right) x^{m-k} + O\left(|x|^{m-k} L\left(|x|\right)\right), \tag{2.111}$$

as $|x| \to \infty$ ($x \to 0$). Moreover (2.111) is equivalent to have

$$a^{k-m}F(ax) - (-1)^{m-k}F(-x) = O\left(x^{m-k}L(x)\right), \tag{2.112}$$

as $x \to \infty$ ($x \to 0^+$), for each $a > 0$. In the case at infinity, it follows that f is quasi-asymptotically bounded of degree $-k$ with respect to L in $\mathcal{S}'(\mathbf{R})$.

Proof. Again we only give the proof of the assertion at infinity, the case at the origin is similar. If $f(\lambda x) = O(\lambda^{-k}L(\lambda))$ in $\mathcal{D}'(\mathbf{R})$, then after $k-1$ integrations Proposition 2.22 and Proposition 2.20 provide us of a $(k-1)$-primitive of f which is quasi-asymptotically bounded of degree -1 at infinity with respect to L, hence we may assume that $k = 1$. Next, Proposition 2.22, Proposition 2.19 and the definition of boundedness in $\mathcal{D}'(\mathbf{R})$ give to us the existence of an $m > 1$, an asymptotically homogeneously bounded function of degree -1 with respect to L and an m-primitive F of f such that $F(\lambda x)$ is continuous for $x \in [-1, 1]$ (hence F is continuous on $\mathbf{R}$) and $F(\lambda x) = \lambda^{m-1}b(\lambda)x^{m-1} + O(\lambda^{m-1}L(\lambda))$ as $\lambda \to \infty$ uniformly for $x \in [-1, 1]$. By taking $x = -1$, $x = 1$ and replacing λ by x one gets (2.111). Assume now (2.111), by using Corollary 2.3, we may assume that b is locally integrable on $[0, \infty)$. This allows the application of Proposition 2.21 to deduce that $F(\lambda x) = \lambda^{m-1}b(\lambda)x^{m-1} + O(\lambda^{m-1}L(\lambda))$ as $\lambda \to \infty$ in $\mathcal{S}'(\mathbf{R})$ and hence the converse follows by differentiating m-times. That (2.111) implies (2.112) is a simple calculation; conversely, setting $b(x) = x^{k-m}F(x)$ for $x > 0$, one obtains (2.111). $\square$

Remark I.1. Even if not assumed initially, the proof of Theorem 2.43 forces (2.112) to hold uniformly on compact subsets of $(0, \infty)$.

We remark that the results from **2.11** are also true in the context of quasi-asymptotic boundedness. Indeed, if one proceeds as in **2.11** but now using the structural theorems of the present section, then one obtains the proofs for the following theorems.

Theorem 2.44. *Let $f \in \mathcal{S}'(\mathbf{R})$. If f is quasi-asymptotically bounded at 0, with respect to a slowly varying function L, in $\mathcal{D}'(\mathbf{R})$, then f is quasi-asymptotically bounded at 0 of the same degree with respect to L in the space $\mathcal{S}'(\mathbf{R})$.*

Theorem 2.45. *Let $f \in \mathcal{D}'(\mathbf{R})$ satisfy $f(\lambda x) = O(\lambda^\alpha L(\lambda))$ as $\lambda \to \infty$ in the space $\mathcal{D}'(\mathbf{R})$. If $\alpha + \beta < -1$, then f admits an extension to $\mathcal{V}_\beta(\mathbf{R})$ which is equally quasi-asymptotic bounded in the space $\mathcal{V}'_\beta(\mathbf{R})$.*

Theorem 2.46. *Let $f_0 \in \mathcal{D}'(0, \infty)$. Let L be slowly varying at the origin and $\alpha \in \mathbf{R}$. Suppose that*

$$f_0(\varepsilon x) = O\left(\varepsilon^\alpha L(\varepsilon)\right) \text{ as } \varepsilon \to 0^+ \text{ in } \mathcal{D}'(0, \infty). \tag{2.113}$$

Then f_0 admits extensions to $[0, \infty)$. Let f be any of such extensions. Then f has the following asymptotic properties at the origin:

(i) If $\alpha \notin -\mathbf{N}$, then there exist constants $a_0, \ldots, a_n$ such that

$$f(\varepsilon x) = \sum_{j=0}^{n} a_j \delta^{(j)}(\varepsilon x) + O(\varepsilon^\alpha L(\varepsilon)),$$

as $\varepsilon \to 0^+$ in $\mathcal{D}'(\mathbf{R})$.

(ii) If $\alpha = -k$, $k \in \mathbf{N}$, then there are constants $a_0, \ldots, a_n$ and an associate asymptotically homogeneously bounded function of degree 0 with respect to L,

$$b(ax) = b(x) + O(L(x)), \tag{2.114}$$

such that

$$f(\varepsilon x) = \sum_{j=0}^{n} a_j \delta^{(j)}(\varepsilon x) + b(\varepsilon)\delta^{(k-1)}(\varepsilon x) + O\left(\varepsilon^{-k} L(\varepsilon)\right), \tag{2.115}$$

as $\varepsilon \to 0^+$ in $\mathcal{D}'(\mathbf{R})$.

2.13   Relation between the S-asymptotics and
quasi-asymptotics at ∞

It is not easy to find conditions for a $T \in \mathcal{F}'$ which imply that T has both the S-asymptotics and quasi-asymptotics. The S-asymptotics is a local property (Theorem 1.4), whereas the quasi-asymptotics has in general a global character (Proposition 2.4, Example 4 in **2.3**, and Example **3** in **2.9**). Also, we have to find a subspace of $\mathcal{F}'$ in which we can compare these two definitions of the asymptotic behavior and choose the class of functions which will measure the asymptotic behavior following these definitions.

We have seen that the space of tempered distributions is a "natural" one for the quasi-asymptotics (Theorem 2.3, Theorem 2.10 and **2.11.2**), while for the S-asymptotics the space $\mathcal{K}'_1$ has this role (Theorem 1.17). The following examples illustrate the problem of comparison of these two types of asymptotic behavior in $\mathcal{S}'(\mathbf{R})$.

i) The regular distribution $T(t) = H(t)e^{iat}$, $t \in \mathbf{R}$, $a \neq 0$, has the quasi-asymptotics $\dfrac{i}{a}\delta$ in $\mathcal{S}'(\mathbf{R})$ related to $c(k) = k^{-1}$ (cf. [32]):

$$k\langle H(kt)e^{ikat}, \varphi(t)\rangle = k \int\limits_0^\infty e^{ikat}\varphi(t)dt = \frac{1}{ia} \int\limits_0^\infty \varphi\left(\frac{x}{k}\right) d(e^{iax})$$

$$= \frac{-1}{ia}\varphi(0) - \frac{1}{k} \int\limits_0^\infty e^{iax}\varphi'\left(\frac{x}{k}\right) dx \to \frac{i}{a}\varphi(0), \quad k \to \infty.$$

But the distribution T has no S-asymptotics related to h^α with a $U \neq 0$ for any $\alpha \in \mathbf{R}$. We start with

$$\langle H(t+h)e^{ia(t+h)}, \varphi(t)\rangle = e^{iah} \int\limits_{-h}^\infty e^{iat}\varphi(t)dt$$

$$\sim e^{iah} \int\limits_{-\infty}^\infty e^{iat}\varphi(t)dt, \quad h \in \mathbf{R}_+.$$

This distribution has the S-asymptotics, but related to the oscillatory function $c(h) = e^{iah}$.

ii) The regular distribution $T(t) = H(t)\sin t, t \in \mathbf{R}$, has the quasi-asymptotics related to $c(k) = k^{-1}$, but it has no S-asymptotics at all:

$$k\langle H(kt)\sin kt, \varphi(t)\rangle = \int_0^\infty \sin u\varphi\left(\frac{u}{k}\right) du = \varphi(0) + \frac{1}{k}\int_0^\infty \varphi'\left(\frac{u}{k}\right) \cos u \, du.$$

For the S-asymptotics we have, $h \to \infty$,

$$\langle H(t+h)\sin(t+h), \varphi(t)\rangle = \cos h \int_{-\infty}^\infty \sin t\varphi(t)dt + \sin h \int_{-\infty}^\infty \cos t\varphi(t)dt + o(1).$$

iii) For the regular distribution $T = H(t)\sin\sqrt{t}, t \in \mathbf{R}$, we cannot find an $\alpha \in \mathbf{R}$ and a distribution $U_\alpha \neq 0$ such that

$$\lim_{k\to\infty} \int_0^\infty k^\alpha \sin\sqrt{kt}\, \varphi(t)dt = \langle U_\alpha, \varphi\rangle, \quad \varphi \in \mathcal{S}(\mathbf{R}).$$

Suppose on the contrary that such α and U_α do exist. We choose $\varphi \in \mathcal{S}_+$ such that, $\varphi(t) = e^{-pt}$, $t > 0$, where $\mathrm{Re}p > 0$. Then, we have

$$\lim_{k\to\infty} k^\alpha \int_0^\infty \sin\sqrt{kt}e^{-pt}dt = \langle U_\alpha(t), e^{-pt}\rangle.$$

The value of the last integral is $\sqrt{\pi k}/\sqrt{4p^3}\exp(-k/4p)$ and $\langle U_\alpha(y), e^{-pt}\rangle$ is the Laplace transform of U_α. Thus the last relation says that the Laplace transform of U_α equals zero for $\mathrm{Re}p > 0$, hence $U_\alpha = 0$.

If we change the basic space, the conclusion can be quite different. Suppose that the basic space is $\mathcal{K}_1(\mathbf{R})$.

The function $H(x)\cosh x = \dfrac{H(x)}{2}(e^x + e^{-x}), x \in \mathbf{R}$ defines an element in $\mathcal{K}'_1$ and $\{\exp(-px^2); \ p > 0\}$ is in $\mathcal{K}_1$.

In order to find the quasi-asymptotics at infinity in $\mathcal{K}'_1$ of $H(x)\cosh(x)$, $x \in \mathbb{R}$, we use

$$\int_0^\infty \cosh(kx)e^{-px^2}dx = \frac{1}{2}\int_0^\infty \cosh(k\sqrt{x})e^{-px}\frac{dx}{\sqrt{x}} = \frac{\sqrt{\pi}}{\sqrt{4p}}\exp(k^2/4p).$$

This shows that there exists no function $c(k)$ of the form $e^{ak^r}k^b L(k)$, where $a, b \in \mathbf{R}$, $0 \le r < 2$, $L(k)$ is a slowly varying function, such that $H(x)\cosh x$ has the quasi-asymptotics related to $c(k)$ with a limit $U \neq 0$.

For the S-asymptotics, we have

$$e^{-h} \int_{-h}^{\infty} \cosh(x+h)\varphi(x)dx = \frac{1}{2} \int_{-h}^{\infty} e^x \varphi(x)dx + \frac{1}{2} e^{-2h} \int_{-h}^{\infty} \varphi(x)dx$$

$$\to \frac{1}{2} \int_{-\infty}^{\infty} e^x \varphi(x)dx, \quad k \to \infty, \quad \varphi \in \mathcal{K}_1 \,.$$

Therefore, $H(x)\cosh x, x \in \mathbf{R}$, has the S-asymptotics related to $c(h) = e^h$ with the limit $U = \frac{1}{2}e^x$.

The common space in which we can compare the two types of asymptotic behavior is the space of tempered distributions and the common class of functions for the comparison is the class of regularly varying functions $\{h^\alpha L(h); \ \alpha \in \mathbf{R}\}$, where L is a slowly varying function (cf. **0.3**).

The first result of this kind is the following one.

Proposition 2.23. ([32]). *Let $g \in \mathcal{S}'_+$, $\alpha > -1$, and let $\varphi_0 \in \mathcal{S}(\mathbf{R})$ be such that $(\mathcal{F}\varphi_0)(x) = 1$ on a neighborhood of zero. If there exists*

$$\lim_{h \to \infty} \langle g(x+h)/h^\alpha, \varphi_0(x) \rangle = C \int \varphi_0(x)dx, \quad C \neq 0,$$

then $g(kx) \overset{q}{\sim} k^\alpha C f_{\alpha+1}(x)$, $k \to \infty$, in $\mathcal{S}'(\mathbf{R})$. (For the function f_α see **0.4***).*

Later on, this result has been improved (see Theorem 6, Chapter I, §3.3 in [192]) in such a way that, instead of h^α, $h^\alpha L(h)$ was used, where L is a slowly varying function. In case $\alpha \leq -1$, the situation is quite different. In [120], one can find precise results for $\alpha = -1$, but for $\alpha < -1$ the obtained results must include the quasi-asymptotic behavior when the limit equals zero, as well.

We quote some results of this kind proved in [159]. Denote $\widehat{L}(x) = \int_a^x \frac{L(t)}{t}dt$, $x > a$. If $\widehat{L}(x) \to \infty$ and L is slowly varying, then $\widehat{L}$ is slowly varying, as well.

Theorem 2.47. *Suppose that $T \in \mathcal{S}'_+$ has S-asymptotics related to $h^\alpha L(h)$. Then T has also the quasi-asymptotics related to $h^\beta L'(h)$, where β and L' are determined as follows: If $\alpha > -1$, then $\alpha = \beta$, $L' = L$; if*

$\alpha = -1$ *and* $\widehat{L}(x) \to \infty$, $x \to \infty$, *then* $\beta = -1$, $L' = \widehat{L}$; *if* $\alpha < -1$ *or* $\alpha = -1$ *and* $\widehat{L}(x) < \infty$, *then* $\beta = -1, L' = const.$, *but the limit can be zero.*

Finally, as an illustration of the case $\alpha < -1$ in Theorem 2.47, we can use the distribution $T = C_1 \delta^{(r)} + C_2 (x^\alpha L(x))_+$; the quasi-asymptotics depends not only on α but also on r.

We can compare the quasi-asymptotics with the S-asymptotics not only for the elements belonging to $\mathcal{S}'_+$ but also when they belong to $\mathcal{D}'$. In this case, we shall use Definition 2.4 instead of Definition 2.2. We know that the general form of the function c, related to which we can measure the S-asymptotics, is $c(h) = \exp(\alpha h) L(\exp h)$, $\alpha \in \mathbf{R}$.

Theorem 2.48. *If* $T \in \mathcal{D}'(\mathbf{R})$ *has S-asymptotics related to* $c(h) = \exp(\alpha h) \cdot L(\exp h)$ *with* $\alpha \neq 0$ *and with non-zero limit, then* T *cannot have quasi-asymptotics.*

At the end of this part dealing with the quasi-asymptotics, we shall cite additionally the following papers: [8], [46], [50], [55], [113], [121], [124], [159], [174], [199] and [201].

Chapter II

Applications of the Asymptotic Behavior of Generalized Functions

3 Asymptotic behavior of solutions to partial differential equations

3.1 S-asymptotics of solutions

We refer to [152], [154], [69] and [160] for the S-asymptotics of solutions of partial differential equations.

Let $P(y)$, $y \in \mathbf{R}^n$ be a polynomial. By $\mathrm{reg}(1/P(y))$ we denote a solution, belonging to $\mathcal{S}'$, of the equation $P(y) \cdot X = 1$. It is well known that Hörmander [70] proved that the last equation can always be solved in $\mathcal{S}'$ if $P \not\equiv 0$ (cf. [70], [190]).

Proposition 3.1. *Let*

$$P(D) = \sum_{|\alpha| \geq 0}^{m} a_\alpha D^\alpha, \ a_\alpha \in \mathbf{R}, \ \alpha \in \mathbf{N}_0^n. \tag{3.1}$$

A necessary and sufficient condition that there exists a solution to the equation

$$P(D)E = \delta \,,$$

such that

$$E(t + h) \overset{s}{\sim} c(h)U(t), \ h \in \Gamma \quad in \ \mathcal{S}' \tag{3.2}$$

is that there exists

$$\lim_{h \in \Gamma, \|h\| \to \infty} \frac{1}{c(h)} \exp(-it \cdot h) \mathrm{reg}\frac{1}{P(-it)} = \mathcal{F}[U](t) \quad in \ \mathcal{S}'. \tag{3.3}$$

($\mathcal{F}$ denotes the Fourier transform).

Proof. We know that $E \in \mathcal{S}'$ is a *fundamental solution* of the operator $P(D)$ if and only if $\mathcal{F}[E]$ is a solution to the equation $P(-ix)\mathcal{F}[E] = 1$ (see [190], p. 192]) where, $\mathcal{F}[E] = reg(1/P(-it)) = f \in \mathcal{S}'$.

We have $E(\cdot + h) = \mathcal{F}^{-1}[\exp(-it \cdot h)f(t)]$ and

$$\langle E(x + h)/c(h), \phi(t) \rangle$$

$$= \left\langle \frac{1}{c(h)} \mathcal{F}^{-1}[\exp(-it \cdot h)f(t)](x), \phi(x) \right\rangle. \qquad (3.4)$$

Let us suppose now that the limit (3.2) exists in $\mathcal{S}'$. Then by (3.4) there exists the limit

$$\lim_{h \in \Gamma, \|h\| \to \infty} \frac{1}{c(h)} \mathcal{F}^{-1}[\exp(-it \cdot h)f(t)](x) \quad \text{in} \ \mathcal{S}'.$$

We know that

$$\exp(-it \cdot h)f(t) = \mathcal{F}[\mathcal{F}^{-1}[\exp(-iy \cdot h)f(y)](x)](t).$$

The continuity of $\mathcal{F}$ implies (3.3). $\qquad \square$

Theorem 3.1. *Let $P(D)$ be of the form (3.1). A sufficient condition that there exists a solution X to the equation*

$$P(D)X = G, \quad G \in \mathcal{E}', \ G \neq 0, \qquad (3.5)$$

such that $X(t + h) \overset{s}{\sim} c(h) \, (G * U)(t), \ h \in \Gamma$ in $\mathcal{S}'$ is that there exists the limit (3.3).

Proof. We proved that the limit (3.3) is necessary and sufficient for $E(t + h) \overset{s}{\sim} c(h)U(t), \ h \in \Gamma$ in $\mathcal{S}'$, where E is a solution to equation (3.1). In order to find the S-asymptotics of $X = E * G$, which is a solution to (3.5), we have only to apply Theorem 1.2 b). $\qquad \square$

Remark. Theorem 3.1 also holds if $G \in \mathcal{O}'_C \subset \mathcal{S}'$. The convolution $E * G$ exists and it is a separately continuous mapping of $(\mathcal{O}'_C, \mathcal{S}') \to \mathcal{O}'_C * \mathcal{S}' \subset \mathcal{S}'$ (see Chapter VII, 5 in [146]). Moreover, a solution to equation (3.5) is of the form $X = E * G$. This solution is unique in the class of distributions for which the convolution with E exists (see Chapter III, §11 in [190]).

Let $P_0(i\frac{\partial}{\partial x})$ be the principal part of $P(i\frac{\partial}{\partial x})$, where P is given by (3.1). If $P_0(y) \neq 0$ for any real $y \neq 0$, then $P(i\frac{\partial}{\partial x})$ is called an *elliptic operator*.

Assume that the coefficients of P are real numbers. If the elliptic operator is also homogeneous (i.e., if it coincides with its principal part), then the fundamental solution $G(x), x \in \mathbf{R}^n$, has the form [61]:

$$G(x) = \begin{cases} A\left(\dfrac{x}{r}\right)r^{m-n}, \ x \in \mathbf{R}^n, & \text{for } n \text{ odd, or } n \text{ even and } m < n \\[2ex] B\left(\dfrac{x}{r}\right)r^{m-n} + C(x)\log r, \ x \in \mathbf{R}^n, & \text{for } n \text{ even and } m \geq n, \end{cases}$$

$$(3.6)$$

where A and B are analytic functions on $\|y\|= 1$ and $C(x), x \in \mathbf{R}^n$, is a polynomial in x of degree $m-n$; m is the degree of the principal polynomial $P_0(y)$ (m is then necessarily an even number); $r^2 = \|x\|^2 = \sum\limits_{i=1}^{n} x_i^2$.

Our aim is to find the S-asymptotics of the fundamental solution G given by (3.6).

First, we shall treat the function $A(\frac{x}{r})r^{m-n}$, $x \in \mathbf{R}^n$, $m > n$; A is analytic on the sphere $\{y; \ \|y\|= 1\}$. Γ will be the ray $\{\rho w; \rho > 0\}$ for a fixed $w \in \mathbf{R}^n$, $\|w\|= 1$. In this case the S-asymptotics of G is given by

$$J \equiv \lim_{\rho \to \infty} \langle \rho^{n-m} A((x + \rho w)/\|x + \rho w\|)\|x + \rho w\|^{m-n}, \varphi(x)\rangle$$

$$= \lim_{\rho \to \infty} \int_{\mathbf{R}^n} A\left(\left(\frac{x}{\rho} + w\right) \Big/ \left\|\frac{x}{\rho} + w\right\|\right)\left\|\frac{x}{\rho} + w\right\|^{m-n} \varphi(x)dx, \ \varphi \in \mathcal{S}.$$

Since

$$A\left(\left(\frac{x}{\rho} + w\right) \Big/ \left\|\frac{x}{\rho} + w\right\|\right)\left\|\frac{x}{\rho} + w\right\|^{m-n}$$

$$\leq M\left[\sum_{i=1}^{n}(|x_i|+|w_i|)^2\right]^{(m-n)/2}, \ \rho > 1,$$

where $M = \sup\{|A(y)|, \ \|y\|= 1\}$, we can use Lebesgue's theorem which gives

$$J = A(w/\|w\|)\|w\|^{m-n}\int_{\mathbf{R}^n}\varphi(x)dx = A(w)\int_{\mathbf{R}^n}\varphi(x)dx, \ \varphi \in \mathcal{S}.$$

We prove that $A\left(\dfrac{x}{r}\right)r^{m-n}$, $x \in \mathbf{R}^n$, $m > n$, has the S-asymptotics related to ρ^{m-n} on the ray $\{\rho w, \rho > 0\}$, $\|w\|= 1$ and with the limit $U = A$.

The next step is to find the S-asymptotics of the function $A\left(\dfrac{x}{r}\right)r^{m-n}$, $x \in \mathbf{R}^n$, but in case when $m < n$, $m > 0$.

Suppose that in the expression for J the function φ belongs to $\mathcal{D}(\mathbf{R}^n)$ and $\operatorname{supp}\varphi \subset I^n$, $I = (-a, a)$, $a \in \mathbf{R}_+$. Then, for $\rho \geq a/\{\max\|w_i\|;\ i = 1, \ldots, n\}$

$$\lim_{\rho \to \infty} \rho^{n-m} \int\limits_{\mathbf{R}^n} A((x + \rho w)/\|x + \rho w\|)\|x + \rho w\|^{m-n}\varphi(x)dx$$

$$= A(w) \int\limits_{\mathbf{R}^n} \varphi(x)dx, \quad \varphi \in \mathcal{D}.$$

Hence, $A\left(\dfrac{x}{r}\right)r^{m-n}$, $x \in \mathbf{R}^n$, $m < n$, has the S-asymptotics in $\mathcal{D}'$ with the limit A. Since $\mathcal{D}$ is dense in $\mathcal{S}$, to prove that this function has the same S-asymptotics in $\mathcal{S}'$ it suffices to prove that $\rho^{n-m}J(\rho)$, $\rho > 0$, is bounded (cf. the Banach–Steinhaus theorem), where

$$J(\rho) = \int\limits_{\mathbf{R}^n} A\left(\frac{x + \rho w}{\|x + \rho w\|}\right)\|x + \rho w\|^{m-n}\varphi(x)dx, \quad \rho > 0,\ \varphi \in \mathcal{S}.$$

Therefore, we shall split the integral $J(\rho)$ into three parts and use the following inequalities:

For $\|x\| \leq \rho/2$ and $\|x\| \geq 3\rho/2$, we have $\|x + \rho w\| \geq |\ \|x\| - \rho\ | \geq \rho/2$ and

$$\rho^{n-m}\left|\ \int\limits_{\|x\|\leq\frac{\rho}{2}} + \int\limits_{\|x\|\geq\frac{3\rho}{2}} A\left(\frac{x + \rho w}{\|x + \rho w\|}\right)\|x + \rho w\|^{m-n}\varphi(x)dx\ \right|$$

$$\leq 2\left(\frac{2}{\rho}\right)^{n-m} M \int\limits_{\mathbf{R}^n} |\varphi(x)|dx, \quad \varphi \in \mathcal{S},\ 0 < m < n,$$

where $M = \sup|A(w)|$, $\|w\| = 1$. But

$$\left|\ \int\limits_{\|x\|\geq\frac{\rho}{2}}^{\|x\|\leq\frac{3\rho}{2}} A\left(\frac{x + \rho w}{\|x + \rho w\|}\right)\|x + \rho w\|^{m-n}\varphi(x)dx\ \right|$$

$$\leq K\frac{M}{(1 + \frac{\rho^2}{4})^k} \int\limits_{\|x\|\geq\frac{\rho}{2}}^{\|x\|\leq\frac{3\rho}{2}} \|x + \rho w\|^{m-n}dx \to 0, \quad \rho \to \infty,$$

where k is any positive number.

Consequently, $A\left(\dfrac{x}{r}\right)r^{m-n}$, $x \in \mathbf{R}^n$, $m < n$, has the S-asymptotics in $\mathcal{S}'$, as well. Let us note that we omitted the case $m = n$, because m is an even number.

Now, we have to find the S-asymptotics of G for n even and $n \geq m$. After this, it is enough to treat only $C(x)\log r$, $x \in \mathbf{R}^n$.

Assume now that $C(x) = \displaystyle\sum_{|i|\leq m-n} \alpha_i x^i, x \in \mathbf{R}^n, m \geq n$, where $\alpha_i \in \mathbf{R}$, $i = (i_1,\ldots,i_n)$, $x^i = x_1^{i_1}\ldots x_n^{i_n}$; $|i| = i_1 + \cdots + i_n$. Then,

$$\lim_{\rho\to\infty} \rho^{n-m}(\log\rho)^{-1}\langle C(x+\rho w)\log\|x+\rho w\|,\ \varphi(x)\rangle$$

$$= \lim_{\rho\to\infty} \rho^{n-m}(\log\rho)^{-1} \int_{\mathbf{R}^n} \sum_{|i|\leq m-n} \alpha_i(x+\rho w)^i(\log\rho$$

$$+ \log\|\frac{x}{\rho}+w\|)\varphi(x)dx = \sum_{|i|=m-n} \alpha_i w^i \int_{\mathbf{R}^n} \varphi(x)dx$$

by the same reason as in the first case.

The results of this paragraph can be expressed in the following proposition:

Proposition 3.2. *If $G \in \mathcal{S}'$ is a fundamental solution to an elliptic homogeneous operator P of degree m (see (3.6)), then on the ray $\{\rho w;\ \rho > 0\}$, $\|w\| = 1$, G has the S-asymptotics:*

$$G(x + \rho w) \stackrel{s}{\sim} \rho^{m-n} \cdot G(w),\ \rho \to \infty, \qquad \text{for } n \text{ odd, or } n \text{ even}$$
$$\text{and } m < n;$$
$$G(x + \rho w) \stackrel{s}{\sim} \rho^{m-n}\log\rho \cdot D(w),\ \rho \to \infty,\ \text{for } n \text{ even}$$
$$\text{and } m \geq n, \tag{3.7}$$

where $D(w) = \displaystyle\lim_{\rho\to\infty} G(\rho w)/(\rho^{m-n}\log\rho)$.

Now, we can give the S-asymptotic behavior of the solutions to

$$P\left(i\frac{\partial}{\partial x}\right)u(x) = f(x),\ f \in \mathcal{O}'_C, \tag{3.8}$$

where P is an elliptic homogeneous operator of degree m in n dimensions. We recall that a solution to equation (3.8) is of the form $u = G * f$ (G

being the fundamental solution to the operator P given by (3.6) and belongs to $\mathcal{S}'$. Namely, since G belongs to $\mathcal{S}'$ and f to $\mathcal{O}'_C$, it follows that $G * f$ exists and belongs to $\mathcal{S}'$ [190]. Moreover, the mapping: $(G, f) \to G * f$ is separately continuous.

Theorem 3.2. *Let G be the fundamental solution to the elliptic differential equation*

$$P\left(i\frac{\partial}{\partial x}\right)G(x) = \delta(x)$$

and

$$D(w) = \lim_{\rho \to \infty} G(\rho w)/(\rho^{m-n} \log \rho);$$

P is a homogeneous operator of degree m in n dimensions. Equation (3.8) has a solution $u = G * f$ belonging to $\mathcal{S}'$. The solution u has the S-asymptotics on the ray $\{\rho w; \ \rho > 0\}$, $\|w\| = 1$, $\rho \to \infty$:

$$u(x + \rho w) \overset{s}{\sim} \rho^{m-n} \cdot (G(w) * f), \qquad \text{for } n \text{ odd, or } n \text{ even}$$
$$\text{and} \quad m < n;$$
$$u(x + \rho w) \overset{s}{\sim} \rho^{m-n} \log \rho \cdot (D(w) * f), \quad \text{for } n \text{ even}$$
$$\text{and} \quad m \geq n.$$

Proof. Since G has the S-asymptotics given by (3.7), $u = G * f$ and the assertion of the Theorem follows from (3.7). $\qquad \square$

The next proposition shows the use of the S-asymptotic expansion in $\mathcal{D}'$.

Proposition 3.3. *Suppose that E is a fundamental solution of the operator*

$$P(D) = \sum_{|\alpha| \geq 0} a_\alpha D^\alpha, \ a_\alpha \in \mathbf{R}, \ \alpha \in \mathbf{N}_0^n$$

such that

$$E(t + h) \overset{s}{\sim} \sum_{n=1}^{\infty} U_n(t, h) \mid \{c_n(h)\}, \ \|h\| \to \infty, \ h \in \Gamma.$$

Then, there exists a solution X to equation

$$P(D)X = G, \quad G \in \mathcal{E}' \tag{3.9}$$

which has the S-asymptotic expansion

$$X(t + h) \overset{s}{\sim} \sum_{n=1}^{\infty} (G * U_n(t, h)) \mid \{c_n(h)\}, \ \|h\| \to \infty, \ h \in \Gamma.$$

Proof. The well-known Malgrange-Ehrenpreis theorem (cf. for Example [190], p. 212) asserts that there exists a fundamental solution to the operator $P(D)$ which belongs to $\mathcal{D}'$. The solution to equation (3.9) exists and can be expressed by the formula $X = E * G$. To find the S-asymptotics of X, we have only to apply Theorem 1.2 b). $\qquad \square$

Remark. The solution $X = E * G$ is unique in $\mathcal{D}'$ (cf. [190], p. 87).

3.2 Quasi-asymptotics of solutions

Chapter IV in [192] is devoted to asymptotic properties of solutions to systems of convolution equations. We shall quote only two theorems.

Let Γ be a closed convex and regular cone, and Γ^* be its dual cone. Let $\{U_k;\ k \in I\}$ be a family of linear nonsingular transforms of $\mathbf{R}^n$ which leaves the cone Γ invariant. Denote by U_k^T the transposed operator for U_k and $V_k = (U_k^T)^{-1}$. Then, $V_k \Gamma^* = \Gamma^*$, $k \in I$. Let $J_k = \det U_k > 0$. We denote by $\mathcal{L}(f)(z)$, $z \in C = \operatorname{int} \Gamma^*$, the Laplace transform of $f \in \mathcal{S}'(\Gamma)$. (We refer for the definition and the properties of the Laplace transform [192], Chapter I, §2).

Theorem 3.3. ([192]) *Suppose that $K \in \mathcal{S}'(\Gamma)$ has the quasi-asymptotics in the cone Γ over the family $\{U_k;\ k \in I\}$ related to ρ_k with the limit K_0 and that $\mathcal{L}K(z)$ has a bounded argument for $z \in \mathbf{R}^n + iC$. If $f \in \mathcal{S}'(\Gamma)$ has the quasi-asymptotics in Γ over $\{U_k;\ k \in I\}$ related to r_k and with the limit g_0, then the equation $K * u = f$ has a solution in $\mathcal{S}'(\Gamma)$ which has the quasi-asymptotics in Γ over $\{U_k;\ k \in I\}$ related to $(J_k \rho_k)^{-1} r_k$ with the limit u_0 which satisfies $K_0 * u_0 = g_0$.*

Theorem 3.4. ([192]) *Let $P(-iz)$ be a polynomial which is different from zero for $z \in \mathbf{R}^n + iC$, $C = \operatorname{int} \Gamma^*$. Let $\Omega \subset C$ be a domain. If there exists ρ_k such that*

$$J_k \rho_k P(V_k y) \to h(iy), \quad k \to \infty, \ k \in I, \ y \in \Omega,$$

then the fundamental solution E of the hyperbolic operator $P(D)$ exists and has the quasi-asymptotics in Γ over $\{U_k;\ k \in I\}$ related to ρ_k and with the limit E_0, where E_0 has the property that $\mathcal{L}(E_0)(z) = \dfrac{1}{h(z)}$, $z \in \mathbf{R}^n + iC$, and $h(z)$ has a bounded argument $\mathbf{R}^n + iC$.

Now, we will consider in the one-dimensional case applications of the *B-transform* to equation

$$((xu)'' + ru) * h = g, \ r \le 0, \ h, g \in \mathcal{S}'_+ \tag{3.10}$$

(cf. [136]). First we give the definition of the *B*-transform.

We need an equivalent definition of $\mathcal{S}'_+$.

Let $l_n = e^{-x/2} L_n(x)$, $x > 0$, $n \in \mathbf{N}_0$, be the *Laguerre orthonormal system* in $L^2(\mathbf{R}_+)$, where

$$L_n = \sum_{m=0}^{n} \binom{n}{n-m} \frac{(-x)^m}{m!}, \ x > 0, \ n \in \mathbf{N}_0,$$

are the *Laguerre polynomials*, and l_n are the eigenfunctions of the operator $\mathcal{R} = e^{x/2} D x e^{-x} D e^{x/2}$, for which $\mathcal{R}(l_n) = -n l_n$, $n \in \mathbf{N}_0$.

Then $\mathcal{S}_+$, the space of functions $\phi \in C^\infty([0, \infty))$, $|x^k \varphi^{(\ell)}(x)| < \infty$, $k, \ell \in \mathbf{N}_0$ can be defined also through the finite norms

$$\|\phi\|_k = (\int_0^\infty |\mathcal{R}^k \phi(k)|^2 dx)^{1/2}, \ k \in \mathbf{N}_0$$

and the following property:

$$(\mathcal{R}^k \phi, l_n) = (\phi, \mathcal{R}^k l_n), \ k, n \in \mathbf{N}_0, \ (\mathcal{R}^{k+1} = \mathcal{R}(\mathcal{R}^k)) .$$

(cf. [43] and References in [136]).

Then *b-transform* on $\mathcal{S}_+$ is defined as follows: If $\phi = \sum_{n=0}^{\infty} a_n l_n \in \mathcal{S}_+$, then

$$b[\phi](t) = \phi(0) + 1/2 \langle \phi(\tau), \sqrt{t/\tau} J_0'(\sqrt{t\tau}) \rangle$$

$$= -2 \sum_{n=0}^{\infty} (-1)^n (2 \sum_{i=n+1}^{\infty} a_i + a_n) l_n(t), \ t > 0.$$

J_0 is the Bessel function:

$$J_0(z) = \sum_{j=0}^{\infty} \left(\frac{iz}{2}\right)^{2j} \frac{1}{(j!)^2} .$$

The *B*-transform on $\mathcal{S}'_+$ is defined by dualizing the *b*-transform on $\mathcal{S}_+$:

$$\langle B[f], \phi \rangle = \langle f, b[\phi], \rangle, \ \phi \in \mathcal{S}_+ .$$

Thus, for $f = \sum_{n=0}^{\infty} b_n l_n$,

$$B[f] = \sum_{n=0}^{\infty} [2 \sum_{m=0}^{n-1} (-1)^m b_m + (-1)^n b_n] l_n .$$

Recall that $B[f_\alpha] = 4^\alpha f_{-\alpha}$, $\alpha \in \mathbf{R}$, where $\{f_\alpha; \alpha \in \mathbf{R}\}$ is given in **0.4**. We shall use the identity

$$B[f(\varepsilon x)](t) = k^2 B[f(x)](kt), \ t > 0, \ \varepsilon = 1/k, \ k > 0.$$

Proposition 3.4. *Let $f \in \mathcal{S}'_+$, and let $L(\varepsilon)$, $\varepsilon \in (0, \varepsilon_0)$, be slowly varying at 0^+. Then the following condition are equivalent:*

1. f has the quasi-asymptotics at 0^+ (at ∞) related to $\varepsilon^\sigma L(\varepsilon)$ (to $k^\sigma L(1/k)$).

2. Bf has the quasi-asymptotics at ∞ (at 0^+) related to $k^{-\sigma-2} L(1/k)$ (to $\varepsilon^{-\sigma-2} L(\varepsilon)$).

Proof. Since $B : \mathcal{S}'_+ \to \mathcal{S}'_+$ is an isomorphism (cf. [136]), we have to prove only the part of this assertion which corresponds to 0^+.

1. $\Rightarrow$ 2. Let $\phi \in \mathcal{S}_+$. Then, with $\varepsilon = 1/k$, $k \to \infty$,

$$\left\langle \frac{(Bf)(kt)}{k^{-\sigma-2} L(1/k)}, \phi(t) \right\rangle = \left\langle \frac{B(f(x/k))(t)}{k^{-\sigma-2+2} L(1/k)}, \phi(t) \right\rangle = \left\langle \frac{f(x/k)}{k^{-\sigma} L(1/k)}, b[\phi](x) \right\rangle$$

$$= \left\langle \frac{f(\varepsilon x)}{\varepsilon^\sigma L(\varepsilon)}, b[\phi](x) \right\rangle \to \langle C f_{\sigma+1}(x), b[\phi](x) \rangle = \langle CB(f_{\sigma+1}), \phi \rangle \quad (3.11)$$

$$= \langle \tilde{C} f_{-\sigma-1}, \phi \rangle, \quad \text{where } \tilde{C} = C 4^{\sigma+1} .$$

2. $\Rightarrow$ 1. For given $\phi \in \mathcal{S}_+$ let $\phi = b[\psi]$, $\psi \in \mathcal{S}_+$. From $b[\phi] = b[b[\psi]] = \psi$ (cf. [136]) and (3.11), we have

$$\left\langle \frac{f(x/k)}{k^{\sigma+2} L(k)}, \phi(t) \right\rangle = \left\langle \frac{(Bf)(kt)}{k^\sigma L(k)}, b[b[\psi]](x) \right\rangle \to \langle C f_{-\sigma-2+1}, \psi \rangle, \ k \to \infty,$$

and this implies the assertion. $\square$

Now, we shall use the B-transform and the quasi-asymptotics for the analysis of (3.10).
Since

$$B[x u_{xx} + 2 u_x](t) = (-t/4) B[u](t), \ t > 0, \quad (3.12)$$

it is equivalent to the equation

$$(-t/4 + r)\tilde{u} = \tilde{g}, \quad \text{where } \tilde{g} = B[g] \in \mathcal{S}'_+, \text{ and } \tilde{u} = B[u]. \qquad (3.13)$$

Let us remark that the equation $xp = q$, $q \in \mathcal{S}'_+$ has solutions in $\mathcal{S}'_+$ uniquely determined up to $C\delta$, $C \in \mathbf{C}$, and that the equation

$$(x + r)p = q, \quad r > 0, \quad q \in \mathcal{S}'_+,$$

has the unique solution in $\mathcal{S}'_+$.

We need the following assertion

Proposition 3.5. *(i) Let $r = 0$ in (3.10) and let g have the quasi-asymptotics at 0^+ related to $\varepsilon^\sigma L(\varepsilon)$, where $\sigma \neq -2, -1, 0, 1, \ldots$ Then:*

1. If $\sigma < -2$, then the solution u has the quasi-asymptotics at 0^+ related to $\varepsilon^{\sigma+1} L(\varepsilon)$;

2. If $\sigma > -2$, then there exist numbers b_j, $j = 0, 1, \ldots, p$ such that $u + \sum_{j=0}^{p} b_j f_j$ has the quasi-asymptotics related to $\varepsilon^{\sigma+1} L(\varepsilon)$;

(ii) Let $r < 0$ in (3.12) and let g have the quasi-asymptotics at 0^+ related to $\varepsilon^\sigma L(\varepsilon)$, $\sigma \in \mathbf{R}$. Then the solution u to equation (3.12) has the quasi-asymptotics at 0^+ related to $\varepsilon^{\sigma+1} L(\varepsilon)$ if $\sigma < -2$. If $\sigma > -2$ and $\sigma \neq -1, 0, 1, \ldots$, then there exist $a_j \in C$, $j = 0, \ldots, r$ such that $u + \sum_{j=0}^{r} a_j (x^j)_+$ has the quasi-asymptotics at 0^+ with respect to $\varepsilon^{\sigma+1} L(\varepsilon)$.

Proof. (i) 1. Since $\tilde{g}$ has the quasi-asymptotics at ∞ related to $k^{-\sigma-2} L(1/k)$ and $\sigma < -2$, by the first part of Lemma 2.1, it follows that $\tilde{u}$ has the quasi-asymptotics at ∞ related to $k^{-\sigma-3} L(1/k)$, because $-\sigma - 3 > -1$. Proposition 3.4 implies that u has the quasi-asymptotics at 0^+ related to $\varepsilon^{\sigma+1} L(\varepsilon)$.

2. If $\sigma > -2$ then there exist $p \in \mathbf{N}$ and numbers a_j, $j = 0, 1, \ldots, p$, such that $\tilde{u} + \sum_{j=0}^{p} a_j \delta^{(j)}$ has the quasi-asymptotics at ∞ related to $k^{-\sigma-3} L(1/k)$. This follows from the second part of Lemma 2.1.

Thus $u + \sum_{j=0}^{p} b_j x_+^{j-1} / \Gamma(j)$, where $b_j = 4^{-j} a_j$, has the quasi-asymptotics at 0^+ related to $\varepsilon^{\sigma+1} L(\varepsilon)$ because $B[\delta^{(j)}] = 4^{-j} f_j$, $j \in \mathbf{N}$.

(ii) Let $d > 0$ and $f \in \mathcal{S}'_+$. We will prove that if

$$\frac{(kt+d)f(kt)}{k^\alpha L(k)} \to C \text{ in } \mathcal{S}'(\mathbf{R}) \ (k \to \infty), \text{ then } \frac{(kt)f(kt)}{k^\alpha L(k)} \to C \text{ in } \mathcal{S}'(\mathbf{R})$$

$(k \to \infty)$. This follows from

$$\frac{f(kt)}{k^\alpha L(k)} \to 0 \text{ in } \mathcal{S}'(\mathbf{R}), \ k \to \infty.$$

Let us prove this fact.

Our assumption implies that there exist $s \in \mathbf{N}$ and a continuous function F, $\operatorname{supp} F \subset [0, \infty)$, such that

$$(t+d)f(t) = F^{(s)}(t) \text{ and } \frac{F(t)}{t^{\alpha+s}L(t)} \to C \text{ as } t \to \infty.$$

Since

$$f(t) = \frac{F^{(s)}(t)}{t+d} = \sum_{i=0}^{s} (-1)^i \left(F(t) \left(\frac{1}{t+d} \right)^{(i)} \right)^{(s-i)}$$

and

$$\left(\frac{\frac{F(t)}{(t+d)^{1+i}}}{k^\alpha L(k)} \right)^{(s-i)} (kt) \to 0 \text{ in } \mathcal{S}'(\mathbf{R}) \text{ as } k \to \infty, \ i = 0, \ldots, s,$$

we have $\dfrac{f(kt)}{k^\alpha L(k)} \to 0$ in $\mathcal{S}'(\mathbf{R})$ as $k \to \infty$.

Thus (3.13) implies that $(t/4-r)\tilde{u}$ and thus $t\tilde{u}$ has the quasi-asymptotics $k^{-\sigma-2}L(1/k)$. By Lemma 2.1, we have

$$\frac{\tilde{u}(kt)}{k^{-\sigma-3}L(1/k)} \to C \text{ in } \mathcal{S}'(\mathbf{R}) \text{ as } k \to \infty \text{ if } -\sigma - 3 > -1,$$

$$\frac{(\tilde{u} + \sum_{j=1}^{r} a_j \delta^j)(kt)}{k^{-\sigma-3}L(1/k)} \to C \text{ in } \mathcal{S}'(\mathbf{R}) \text{ as } k \to \infty \text{ if } -\sigma - 3 < -1.$$

Hence, by Proposition 3.4, we have that the solution u has the quasi-asymptotics at 0^+ with respect to $\varepsilon^{\sigma+1}L(\varepsilon)$ if $\sigma < -2$ or $u + \sum_{j=0}^{r} a_j(x^j)_+$ has the quasi-asymptotics at 0^+ with respect to $\varepsilon^{\sigma+1}L(\varepsilon)$ if $\sigma > -2$ and $\sigma \neq -1, 0, 1, \ldots$ $\qquad\square$

The main part of previous examinations will be used for the qualitative analysis of equation (3.10). For example, the equation

$$\sum_{k=0}^{j} a_k((xu)'' + ru)^{(k)} = g \quad \text{(with } h = \sum_{k=0}^{j} a_k \delta^{(k)}\text{)}$$

is of this form.

Proposition 3.6. *Assume that $g, h \in \mathcal{S}'_+$ and that the Laplace transform of h, $(\mathcal{L}h)(x + iy)$, $x \in \mathbf{R}$, $y > 0$, has a bounded argument. Let h have the quasi-asymptotics at 0^+ related to $\varepsilon^{\sigma_1} L_1(\varepsilon)$ and g have the quasi-asymptotics at 0^+ related to $\varepsilon^{\sigma_2} L_2(\varepsilon)$.*

1. Let $r = 0$ in (3.10). If $\sigma_1 - \sigma_2 > 1$, then (3.10) has the solution u with the quasi-asymptotics at 0^+ related to $\varepsilon^{\sigma_1 - \sigma_2} L_1(\varepsilon)$.

If $\sigma_1 - \sigma_2 < 1$ and $\sigma_1 - \sigma_2 \neq 1, 0, -1, -2, \ldots$, then there are numbers $b_j \in \mathbf{C}$, $j = 0, 1, \ldots, p$, such that (3.10) has the solution u so that $u + \sum_{j=0}^{p} b_j f_j$ has the quasi-asymptotics at 0^+ related to $\varepsilon^{\sigma_2 - \sigma_1} L_2(\varepsilon)/L_1(\varepsilon)$.

2. Let $r < 0$ in (3.10). Then (3.10) has the solution u with the quasi-asymptotics at 0^+ related to $\varepsilon^{\sigma_2 - \sigma_1} L_2(\varepsilon)/L_1(\varepsilon)$ if $\sigma_2 - \sigma_1 < -2$. If $\sigma_2 - \sigma_1 > -2$ and $\sigma_2 - \sigma_1 \neq -1, 0, 1, \ldots$, then there exist $a_j \in \mathbf{C}$, $j = 0, \ldots, p$ such that $u + \sum_{j=0}^{p} a_j (x^j)_+$ has the quasi-asymptotics at 0^+ related to $\varepsilon^{\sigma_2 - \sigma_1} L_2(\varepsilon)/L_1(\varepsilon)$.

Proof. First, we prove: *If the Laplace transform of h has a bounded argument, then the same holds for $B[h]$.* Namely, by

$$b[e^{iz\tau}](t) = e^{-\frac{ti}{4z}}, \quad t > 0, \ \operatorname{Im} z > 0 \quad \text{([192], Chapter 1, §2.3)},$$

we have

$$\mathcal{L}(B[f])(z) = \langle (B[f](\tau), e^{i\tau z} \rangle = \langle f(t), b[e^{iz\tau}] \rangle = \mathcal{L}f\left(-\frac{1}{4z}\right), \ \operatorname{Im} z > 0,$$

which implies the assertion.

By (3.13) and $B[f * g] = B[f] * B[g]$, (see [136]), (3.10) becomes $(-t/4 + r)\tilde{u} * \tilde{h} = \tilde{g}$.

We shall prove only part 1 of Proposition 3.6. Let $r = 0$. We shall use Theorem 1, Chapter IV, §1.1 in [192]. In the one-dimensional case this theorem reads as follows.

"*Let $K \in \mathcal{S}'_+$, has the quasi-asymptotics at ∞ related to $k^\alpha L_1(k)$ with the limit $C_1 f_{\alpha+1}, C_1 \neq 0$, and $f \in \mathcal{S}'_+$ has the quasi-asymptotics at ∞ related to $k^\beta L_2(k)$ with the limit $C_2 f_{\beta+1}$, $C_2 \neq 0$. Let the Laplace transform of K, $\mathcal{L}K(x+iy)$, has a bounded argument in $\mathbf{R}+i\mathbf{R}_+$. Then the convolution equation $K * u = f$ has the solution $u \in \mathcal{S}'_+$ which has the quasi-asymptotics at ∞ with respect to $k^{\beta-\alpha-1}L_2(k)/L_1(k)$ with the limit $(C_2/C_1)f_{\beta-\alpha}$*" (for $f_{\beta-\alpha}$ see **0.4**).

Now, we can continue with the proof of Proposition 3.6. Since the Laplace transform of h has a bounded argument, it follows that there exists $\tilde{s} \in \mathcal{S}'_+$ such that $\tilde{s} * \tilde{h} = \tilde{g}$, and

$$\frac{\tilde{s}(kx)}{k^{-\sigma_2-2-(-\sigma_1-2)-1}L_2(1/k)/L_1(1/k)} \to const. f_{\sigma_2-\sigma_1}, \ k \to \infty \ \text{in } \mathcal{S}'_+\,,$$

because $\tilde{g}$ has the quasi-asymptotics related to $k^{-\sigma_2-2}L_2(1/k)$ and $\tilde{h}$ has the quasi-asymptotics related to $k^{\sigma_1-2}L_1(1/k)$.

Let u be a solution to

$$xu'' + 2u' = s \iff (-t/4)\tilde{u} = \tilde{s}\,.$$

As in the proof of Proposition 3.5, we have the following situations:

1. Since $\tilde{s}$ has the quasi-asymptotics at ∞ related to

$$k^{-(\sigma_2-\sigma_1-1)-2}L_2(1/k)/L_1(1/k)\,,$$

if $\sigma_2 - \sigma_1 < -1$ it follows that $\tilde{u}$ has the quasi-asymptotics at ∞ related to

$$k^{-(\sigma_2-\sigma_1-1)-3}L_2(1/k)/L_1(1/k)$$

and by Proposition 3.4, h has the quasi-asymptotics at 0^+ related to

$$\varepsilon^{\sigma_2-\sigma_1}L_2(\varepsilon)/L_1(\varepsilon)\,.$$

2. If $\sigma_2 - \sigma_1 > -1$ and $\sigma_1 - \sigma_2 \neq 1, 0, -1, -2, \ldots$, then as in the proof of Proposition 3.5, we conclude that there are numbers a_j, $j = 0, \ldots, p$, such that $\tilde{u} + \sum_{j=0}^{p} a_j \delta^j$ has the quasi-asymptotics at ∞ related to

$$k^{(\sigma_2-\sigma_1-1)-3}L_2(1/k)/L_1(1/k)$$

which implies that

$$u + \sum_{j=0}^{p} b_j x_+^{j-1}/\Gamma(j), \ (b_j = 4^j a_j, \ j = 0, \ldots, p),$$

has the quasi-asymptotics at 0^+ related to $\varepsilon^{\sigma_2-\sigma_1} L_2(\varepsilon)/L_1(\varepsilon)$.   □

As an application to hyperbolic problems, we shall quote the results of P. Wagner [194] in which explicit formulas for the open quasi-asymptotic expansion of the causal fundamental solution of a class of hyperbolic equations are given.

Let $\lambda_0 < \lambda_1 < \ldots$ be a sequence of real numbers and T_j be distributions in $\mathcal{S}'(\Gamma)$, homogeneous of degree $-\lambda_j$. Recall,

$$E(kx) \sim \sum_{j=0}^{\infty} k^{-\lambda_j} T_j(x), \quad k \to \infty$$

means that for every $N \in \mathbf{N}_0$

$$\lim_{k\to\infty} k^{\lambda_N} \left[E(kx) - \sum_{j=0}^{N} k^{-\lambda_j} T_j(x) \right] = 0 \quad \text{in } \mathcal{S}'(\Gamma).$$

(cf. Definition 2.7).

A linear partial differential operator (with constant coefficients) $P(\partial) = \sum_{j=\ell}^{m} P_j(\partial)$, P_j homogeneous of degree j, $P_\ell \not\equiv 0$, $P_m \not\equiv 0$, is called hyperbolic with respect to the direction $N \in \mathbf{R}^n \setminus 0$ if $P_m(N) \neq 0$ and $P(-ix + \tau N) \neq 0$ for all $x \in \mathbf{R}^n$ and $\tau > \tau_0$. In this case $\tau_0 = 0$.

Proposition 3.7. ([194]) *Let $P(\partial) = \sum_{j=\ell}^{m} P_j(\partial)$ and Γ be as above. Denote the causal fundamental solutions (cf. [70]) of $P(\partial)$ and of $P_\ell(\partial)^i$ by E and $E_{l,i}$ respectively. Then E has the following quasi-asymptotic expansion in $\mathcal{S}'(\Gamma)$ for $k \to \infty$:*

$$E \sim \sum_{r=0}^{\infty} k^{l-n-r} T_r,$$

where T_r consists of homogeneous distributions of degree $l - n - r$ given by

$$T_r = \sum_{\substack{\alpha_1 + 2\alpha_2 + \cdots + (m-\ell)\alpha_{m-1} = r \\ \alpha_i \geq 0}} \binom{|\alpha|}{\alpha} (-1)^{|\alpha|} \prod_{j=1}^{m-1} P_{\ell+j}(\partial)^{\alpha_j} E_{l,1+|\alpha|} .$$

Let $Z(\partial)$ be a system of N linear partial differential equations with constant coefficients containing N unknown functions, i.e., $Z(\partial)$ is an $N \times N$-matrix of operators. Assume that $P(\partial) := \det Z(\partial)$ has the property that $P(-iz) \neq 0$ for $z \in \mathbf{R}^n + iC$. The unique (again called causal) two-sided fundamental solution E_Z of $Z(\partial)$ having its support in $\Gamma := C^*$ is given by the formula $E_Z = Z^{adj}(\partial)E$, wherein Z^{adj} denotes the adjoint matrix to Z and E is the causal fundamental solution of $P(\partial)$. Decompose Z^{adj} into its homogeneous components:

$$Z^{adj} = \sum_{s=l_1}^{m_1} Z_s^{adj}, \text{ where } Z_{s,ij}^{adj} \text{ are homogeneous of degree } s, \ 1 \leq i,j \leq N.$$

Proposition 3.8. ([194]) *Let* $Z(\partial) = \sum\limits_{s=0}^{m} Z_s(\partial)$, *where* $Z_{s,ij}$ *are homogeneous of degree* s, *and* E_Z *has the property that* $\det Z(-iz) \neq 0$, $z \in \mathbf{R}^n + iC$. *Assume that* $\det Z_0 = \det Z(0) \neq 0$. *Then the quasi-asymptotic expansion of* $E_Z(kx)$ *in* $\mathcal{S}'(\Gamma)^{N \times N}$ *for* $k \to \infty$ *is given by the formula*

$$E_Z(kx) \sim \sum_{r=0}^{\infty} k^{-n-r} \sum_{\substack{r_1+\cdots+r_j=r \\ j \geq 0, r_j \geq 0}} (-1)^j Z_0^{-1} Z_{r_1}(\partial) Z_0^{-1} \ldots Z_0^{-1} Z_{r_j}(\partial) Z_0^{-1} \delta.$$

3.3 S-asymptotics of solutions to equations with ultra-differential or local operators

We refer to [132] for the material of this section.

We shall cite the following theorem which is a simple consequence of Theorem 1.2e) in order to illustrate possible applications of the S-asymptotics to

$$P(D)X = G, \quad G \in \mathcal{D}'^*, \tag{3.14}$$

where $P(D)$ is an ultradifferential operator of $(*)$-class.

Theorem 3.5. *A necessary condition that (3.14) has a solution* χ_0, *with the property* $\chi_0 \overset{s}{\sim} c(h) \cdot U$, $h \in \Gamma$ *in* $\mathcal{D}'^*$ *is that* $G \overset{s}{\sim} c(h) \cdot P(D)U$, $h \in \Gamma$ *in* $\mathcal{D}'^*$.

The proof will be given with the proof of Theorem 3.6.

Theorem 3.6. *Let* M_p *satisfy (M.1), (M.2) and (M.3) and* $G \in \mathcal{O}'^*_C$ ($\mathcal{O}'^*_C$ *is the space of convolutors of* $\mathcal{S}^{*'}$).

Suppose that there exists a solution $E \in \mathcal{S}'^*$ to equation $P(D)Y = \delta$ such that:

a) the set $\{E(x+h)/(1+\|h\|^2)^{q/2}; \ h \in \mathbf{R}^n\}$, $q \in \mathbf{R}$ is bounded in $\mathcal{D}'^*$;

b) $E \overset{s}{\sim} (1+\|h\|^2)^{q/2} \cdot U$, $h \in \Gamma$ in $\mathcal{D}'^*$.

Then, there exists a solution $\chi_0 \in \mathcal{S}'^*$ of the equation $P(D)\chi = G$ such that $\chi_0 \overset{s}{\sim} (1+\|h\|^2)^{q/2} \cdot (U * G)$, $h \in \Gamma$ in $\mathcal{D}'^*$.

Proof. If we apply Theorem 1.10 to E with the property a), or if we use Theorem 1 in [123], we obtain that $E = Q(D)F_1 + F_2$, where $Q(D)$ is an ultradifferential operator of class $*$ and F_i, $i = 1, 2$, are continuous functions such that

$$|F_i(x)/(1+\|x\|^2)^{q/2}| \leq M_i, \quad x \in \mathbf{R}^n, \ i = 1, 2.$$

Since for every $\phi \in \mathcal{S}^*$, $Q(-D)\phi \in \mathcal{S}^*$, it follows that $\{E(x+h)/(1+\|h\|^2)^{q/2}; \ h \in \Gamma\}$ is bounded in $\mathcal{S}'^*$.

Since $\mathcal{D}^*$ is dense in $\mathcal{S}^*$, by the Banach–Steinhaus theorem

$$\lim_{h \in \Gamma, \|h\| \to \infty} \left\langle \frac{E(x+h)}{(1+\|h\|^2)^{q/2}}, \phi(x) \right\rangle = \langle U, \phi \rangle, \quad \phi \in \mathcal{S}^*. \tag{3.15}$$

Clearly, $\chi_0 = E * G$ is a solution of equation $P(D)\chi = G$. We have to prove that

$$\lim_{h \in \Gamma, \|h\| \to \infty} \langle (E * G)(x+h)/(1+\|h\|^2)^{q/2}, \phi(x) \rangle = \langle U * G, \phi \rangle, \quad \phi \in \mathcal{D}^*.$$

This follows from (3.15) and

$$\langle (E * G)(x+h), \varphi(x) \rangle = \langle E(x+h), (\check{G} * \varphi)(x) \rangle, \quad \check{G}(x) = G(-x), \ x \in \mathbf{R}^n,$$

because $G \in \mathcal{O}'^*_C$ is a convolutor on $\mathcal{S}^{*\prime}$. $\qquad\square$

Let $r_0 > 0$. Denote by $\mathcal{S}_{r_0, h}$, $h > 0$, the space of smooth functions ϕ on $\mathbf{R}^n$ such that

$$\sup_{x \in \mathbf{R}^n, |\alpha| \in \mathbf{N}_0} \frac{(1+\|x\|)^{r_0} \|\phi^{(\alpha)}(x)\|_{L_\infty}}{h^{|\alpha|} M_{|\alpha|}} < \infty.$$

Define

$$\mathcal{S}_{r_0}^{(M_p)} = \operatorname{proj} \lim_{h \to 0} \mathcal{S}_{r_0, h}, \ \mathcal{S}_{r_0}^{\{M_p\}} = \operatorname{ind} \lim_{h \to \infty} \mathcal{S}_{r_0, h}.$$

Clearly, $\mathcal{D}^*$ is dense in $\mathcal{S}_{r_0}^*$. We shall use the definition of the convolution of two ultradistributions as given in [79]: $\langle f * g, \varphi \rangle = \langle g, \check{f} * \varphi \rangle$, whenever this defines a continuous linear functional in φ.

Theorem 3.7. *Suppose:*

1. $g \in S'^*_{r_0}$, $r_0 > 0$, and

$$\lim_{h \in \Gamma, \|h\| \to \infty} \langle g(x + h)/c(h), \phi \rangle = \langle u, \phi \rangle, \ \phi \in S^*_{r_0}. \tag{3.16}$$

2. For the differential operator $L(D) = \sum_{|\alpha| \leq m} a_\alpha D^\alpha$, $a_\alpha \in \mathbf{R}$, there exists a fundamental solution $f \in S'$ of the form

$$f(x) = \sum_{\beta=0}^{m} (F_\beta(x)/(1 + \|x\|)^{r_\beta})^{(\beta)}, \ r_\beta \geq r_0, \ F_\beta \in L^\infty, \ \beta = 0, \ldots, m.$$

Then, the partial differential equation $L(D)X = g$ has a solution $X = f * g$ and $X \overset{s}{\sim} c(h) \cdot f * u$, $h \in \Gamma$ in $\mathcal{D}'^*$.

Proof. One can easily prove that for $F \in L^\infty$, $\phi \in \mathcal{D}^*$ and $r \geq r_0$, the functions:

$$x \mapsto ((F(t)/(1 + \|t\|)^r) * \phi^{(\beta)}(t))(x), \ \beta = 0, \ldots, m,$$

belong to $S^*_{r_0}$. This, in accordance with (3.16), implies

$$\lim_{h \in \Gamma, \|h\| \to \infty} \left\langle \frac{(f * g)(x + h)}{c(h)}, \phi(x) \right\rangle = \langle u, \check{f} * \phi \rangle = \langle u * f, \phi \rangle, \ \phi \in \mathcal{D}^*.$$

Since $f * g = X$ is a solution of the equation $L(D)X = g$, the proof is completed. □

Similar theorems can be proved for hyperfunctions. We quote only the following.

Proposition 3.9. *Let P be a local operator. Then a necessary condition that a solution to equation*

$$P(D)x = f$$

has the S-asymptotics related to c and to the cone Γ with the limit u, is that $f(t + h) \overset{s}{\sim} c(h)u(t)$, $h \in \Gamma$.

Proof. Since $\mathcal{Q}(\mathbf{D}^n)$ is a Montel space, from

$$\underset{h \to \infty, h \in \Gamma}{w.\lim} x(t + h)/c(h) = u(t) \text{ in } \mathcal{Q}(\mathbf{D}^n)$$

it follows that the same holds for the strong convergence. Since a local operator maps continuously $\mathcal{Q}(\mathbf{D}^n)$ into $\mathcal{Q}(\mathbf{D}^n)$, it follows

$$P(D)x(t + h) \overset{s}{\sim} c(h)P(D)u(t), \ h \in \Gamma,$$

and thus

$$f(t + h) \overset{s}{\sim} c(h)P(D)u(t), \ h \in \Gamma. \qquad \square$$

4   Asymptotics and integral transforms

Integral transforms of distributions have been elaborated in the last forty years, or so and they appeared as one of powerful tools, especially in mathematical physics. The book by Zemanian [202] was the first systematic monograph which gave different integral transforms of generalized functions. Brychkov and Prudnikov [19] collected results on the most important integral transforms of generalized functions. The results of Abelian and Tauberian type have been elaborated only for some special integral transforms of distributions.

We shall apply two defined asymptotics to integral transforms with general kernels to obtain results which contain those proved for special kernels. It turned out that the S-asymptotics is well appropriate to integral transforms of convolution type and the quasi-asymptotics to integral transforms of the Melline convolution type.

Integral transforms of generalized functions can be defined in various ways. In the direct approach, one constructs a basic space $\mathcal{A} \subset \mathcal{D}(\mathbf{R})$ of smooth functions to which belongs the kernel of the integral transform, $K(s, \cdot), s \in S_0 \subset \mathbf{C}$. Then, for T belonging to the dual space $\mathcal{A}'$, the transform $\mathcal{K}$ is defined by the expression: $\mathcal{K}(T)(s) = \langle T(t), K(s,t) \rangle, s \in S_0$. We shall prove the theorems of Abelian type for the integral transforms defined in this way. The book by Estrada and Kanwal [56] gives another approach to this matter.

We refer to [129], [130], [133], [134], [135], [158] and [192] for the material on Abelian and Tauberian type theorems.

4.1   *Abelian type theorems*

4.1.1   *Transforms with general kernels*

Let L be a slowly varying function (see **0.3**). One can introduce the function $L^* \in C^\infty(\mathbf{R}_+)$ in the following way: Let $w \in C^\infty, \operatorname{supp} w \subset [-1, 1], w > 0$ and

$$\int_{-1}^{1} w(t)dt = 1; \quad L^*(x) = \int_{\mathbf{R}} w(x - t)L(\exp t)dt, \quad x > 0.$$

Then, L^* has the following properties:

$$L^{*(k)}(x+h)/L(\exp h) \to \begin{cases} 1, & k = 0 \\ 0, & k \in \mathbf{N} \end{cases} \quad h \to \infty, \ x > 0. \quad (4.1)$$

For every $k \in \mathbf{N}_0$ there are $C_k > 0$ and $C_k^* > 0$ such that

$$|L^{*(k)}(x)| \le C_k L(\exp x) \le C_k^* L^*(x), \ x > 0. \quad (4.2)$$

By (4.2), we have that for every $k \in \mathbf{N}_0$ there exists $C^* > 0$ such that

$$\left| \left(\frac{1}{L^*(x)} \right)^{(k)} \right| \le \frac{C^*}{L^*(x)}, \ x > 0. \quad (4.3)$$

We shall use the following functions:

$$c(x) = \begin{cases} \exp(ax)L(\exp x), & x \ge 0 \\ \exp(bx), & x < 0 \end{cases}, \ a > b, \quad (4.4)$$

$$e(x) = \begin{cases} \exp(ax)L^*(x), & x > 1 \\ \exp(bx), & x < 0 \end{cases}, \ a > b, \quad (4.5)$$

where e is extended on $[0,1]$ to be smooth and positive on $\mathbf{R}$ and

$$e_\delta(x) = \begin{cases} \exp((a+\delta)x), & x > 1 \\ \exp(bx), & x < 0 \end{cases}, \ a > b, \ \delta > 0. \quad (4.6)$$

Function e_δ is also extended on $[0,1]$ to be smooth and to satisfy the inequality $e_\delta^{(k)}(x) \ge e^{(k)}(x)$, $x \in \mathbf{R}$, $k \in \mathbf{N}_0$.

We need the following two properties of the function e:

For every compact set $B \subset \mathbf{R}$ and every $k \in \mathbf{N}_0$ there is $C_{B,k} > 0$ such that

$$\left| \left(\frac{c(h)}{e(x+h)} \right)^{(k)} (x) \right| \le C_{B,k}, \ h \in \mathbf{R}, \ x \in B. \quad (4.7)$$

For every $k \in \mathbf{N}_0$ and $\delta > 0$ there is a $C_k > 0$ such that for $h > 0$

$$|(e(x+h)/c(h) - \exp(ax))^{(k)}| \le \begin{cases} C_k \exp(ax + \delta|x|), & x + h \ge 0, \\ C_k \exp(bx), & x + h < 0. \end{cases} \quad (4.8)$$

Note, (4.3) implies that for every $k \in \mathbf{N}_0$ there exists $c_k > 0$ such that

$$|e^{(k)}(x)| \le \begin{cases} c_k L(e^t)e^{at}, & t > 1, \\ c_k, & 0 \le t \le 1, \\ c_k e^{bt}, & t < 0. \end{cases} \quad (4.9)$$

Thus (4.7) follows.

Let us prove (4.8). If $x + h > 1$, by Leibnitz's formula, it is enough to prove that for every $j \in \mathbf{N}_0$ there exists $c_j > 0$ such that

$$\left| \left(\frac{e(\cdot + h)}{L(\exp h) \exp(a(\cdot + h))} \right)^{(j)} (x) \right| \leq c_j \exp(\delta|x|) \quad \text{for any } \delta > 0. \quad (4.10)$$

This easily follows from (4.9) and (4.3). Then it implies (4.8).

Let $x + h \in [0, 1]$. We shall prove (4.10) because it implies (4.8). Since

$$\frac{e(x + h)}{\exp(a(x + h))}, \quad x + h \in [0, 1],$$

is bounded, (4.10) follows from the estimate

$$\frac{1}{L(\exp h)} = \frac{1}{L(\exp(x - t))} \leq e^{\delta|x|}, \quad x + h = t \in [0, 1] \quad \text{for any } \delta > 0,$$

which is a consequence of (4.3).

If $x + h < 0$, then (4.8) follows from (4.9) and the assumption $a > b$.

We shall introduce the space $\mathcal{D}'_{L^1, e}(\mathbf{R})$ *weighted distributions* in a similar way as it has been done by Ortner and Wagner [103] for the space $\mathcal{D}_{L^1, e}(\mathbf{R}) = \{\psi \in C^\infty(\mathbf{R}); e\psi \in \mathcal{D}_{L^1}(\mathbf{R})\}$.

The function $f : f(e, \psi) = e\psi$, which maps $\{e\} \times \mathcal{D}_{L^1, e}(\mathbf{R})$ onto $\mathcal{D}_{L^1}(\mathbf{R})$ is continuous. The same holds for $g : e^{-1} \times (\mathcal{D}_{L^1, e})'(\mathbf{R})$, onto $\mathcal{B}'(\mathbf{R})$, given by $g(e^{-1}, S) = e^{-1}S$.

A topology in $\mathcal{D}_{L^1, e}(\mathbf{R})$ is defined by means of the isomorphism $\mathcal{D}_{L^1, e}(\mathbf{R}) \to \mathcal{D}_{L^1}(\mathbf{R}) : \psi \to e\psi$, through seminorms

$$\|\psi\|_p = \max \left(\int_{\mathbf{R}} |(e\psi)^{(k)}(t)| dt, \ 0 \leq k \leq p \right), \quad p \in \mathbf{N}_0.$$

By the properties of e and e_δ, we have $\mathcal{D}_{L^1, e_\delta}(\mathbf{R}) \subset \mathcal{D}_{L^1, e}(\mathbf{R})$. The topology of $\mathcal{D}_{L^1, e_\delta}(\mathbf{R})$ is finer than the topology induced from $\mathcal{D}_{L^1, e}(\mathbf{R})$. The strong dual space $(\mathcal{D}_{L^1, e})'(\mathbf{R}) = \{S \in \mathcal{D}'(\mathbf{R}); \ e^{-1}S \in \mathcal{B}'(\mathbf{R})\}$ and $(\mathcal{D}_{L^1, e})'(\mathbf{R}) \subset (\mathcal{D}_{L^1, e_\delta})'(\mathbf{R})$, $\delta > 0$ (see **0.5** for $\mathcal{B}'(\mathbf{R})$).

Proposition 4.1. *A necessary and sufficient condition that T belongs to $(\mathcal{D}_{L^1, e})'(\mathbf{R})$ is the existence of continuous and bounded functions F_i, $i = 0, \ldots, p$ such that*

$$T = \sum_{j=0}^{p} D^j (F_j e),$$

where D is the derivative in the sense of distributions.

Proof. Suppose that T is given by the sum of derivatives of $(F_j e)$, then

$$T = e \sum_{k=0}^{p} e^{-1} e^{(k)} \sum_{j=k}^{p} \binom{j}{k} D^{j-k} F_j = eS.$$

Since $e^{-1} e^{(k)} \in C^{\infty}(\mathbf{R}), 0 \le k \le p$, and by (4.2) these functions are bounded, then by the structural theorem for bounded distributions, it follows that S is a bounded distribution. Consequently, $T \in (\mathcal{D}_{L^1,e})'$.

Suppose now that $T \in (\mathcal{D}_{L^1,e})'$, then T is of the form

$$T = e \sum_{j=0}^{p} D^j E_j = \sum_{j=0}^{p} \sum_{k=0}^{j} a_{j,k} D^k (e^{-1} e^{(j-k)} e E_j), \quad a_{j,k} \in \mathbf{R},$$

where E_j, $j = 0, \ldots, p$ are continuous and bounded functions. Since $e^{(j-k)} e^{-1}$, $j = 0, \ldots, p$; $k = 0, \ldots, j$, are smooth and bounded functions, T has the required form. □

Proposition 4.2. *Suppose that $T \in \mathcal{D}'(\mathbf{R})$ and $T(x+h) \overset{s}{\sim} c(h) \cdot A e^{ax}$, $h \in \mathbf{R}_+$, in $\mathcal{D}'(\mathbf{R})$. If $T \in (\mathcal{D}_{L^1,e})'(\mathbf{R})$, then $T_0 = e^{-1} T \in \mathcal{B}'(\mathbf{R})$ has the property $T_0(x+h) \overset{s}{\sim} 1 \cdot A$, $h > 0$ in $\mathcal{D}'(\mathbf{R})$ (and in $\mathcal{B}'(\mathbf{R})$ as well).*

Proof. For h large enough and $\varphi \in \mathcal{D}(\mathbf{R})$

$$\langle T_0(x+h), \varphi(x) \rangle = \left\langle \frac{T(x+h)}{c(h)}, \frac{c(h)}{e(x+h)} \varphi(x) \right\rangle$$

$$= \left\langle \frac{T(x+h)}{c(h)}, \frac{L(\exp h)}{L^*(x+h)} \exp(-ax) \varphi(x) \right\rangle.$$

By relations (4.1) and (4.2) the function $x \mapsto \varphi(x) \exp(-ax) L(\exp h)/L^*(x+h)$ converges to $x \mapsto \varphi(x) \exp(-ax)$ in $\mathcal{D}(\mathbf{R})$ when $x \to \infty$. Now, we can use Theorem XI Chapter III in [146] to obtain

$$\lim_{h \to \infty} \langle T_0(x+h), \varphi(x) \rangle = \langle A, \varphi \rangle \quad \text{for every } \varphi \in \mathcal{D}(\mathbf{R}).$$

The last is valid for $\varphi \in \mathcal{D}_{L^1}(\mathbf{R})$, as well. This follows from Theorem 1.17. □

Definition 4.1. Suppose that $T \in (\mathcal{D}_{L^1,e})(\mathbf{R})$ and $K(s+x) \in \mathcal{D}_{L^1,e_\delta}(\mathbf{R})$ for $s \in S_0 \subset \mathbf{C}$, and $\delta > 0$, where $S_0 - \mathbf{R}_+ \subset S_0$ (i.e., for every $s \in S_0$ and $x > 0$, $S_0 - x \in S_0$.). Then the $\mathcal{K}$-transform of T is defined by

$$\mathcal{K}(T)(s) = \langle T(x), K(s+x) \rangle = \langle T(x) e^{-1}(x), e(x) K(s+x) \rangle, \qquad (4.11)$$

for $s \in S_0$.

Theorem 4.1. *Suppose:*

1. $T \in (\mathcal{D}_{L^1,e})'(\mathbf{R})$ and $K(s+x) \in \mathcal{D}_{L^1,e_\delta}(\mathbf{R})$, $s \in S_0$, $S_0 - \mathbf{R}_+ \subset S_0$, for a $\delta > 0$.

2. $T(x+h) \overset{s}{\sim} Ac(h)e^{ax}$, $h \in \mathbf{R}_+$ in $\mathcal{D}'(\mathbf{R})$, where c is given by (4.4).

Then for the $\mathcal{K}$-transform of T, we have

$$\lim_{h\to\infty} \mathcal{K}(T)(s-h)/c(h) = \langle A, K(s+x)\exp(ax)\rangle, \quad s \in S_0. \tag{4.12}$$

Proof. Note $K(s+x)\exp(ax) \in \mathcal{D}_{L^1}(\mathbf{R})$, $s \in S_0$ because of $a > b$. By Proposition 4.2, if $s \in S_0$, then

$$\lim_{h\to\infty} \langle T(x+h)/e(x+h), K(s+x)\exp(ax)\rangle = \langle A, K(s+x)\exp(ax)\rangle.$$

To prove relation (4.12), it is enough to show

$$\lim_{h\to\infty} \left\langle \frac{T(x+h)}{e(x+h)}, \left(\frac{e(x+h)}{c(h)} - \exp(ax)\right)K(s+x)\right\rangle = 0, \quad s \in S_0. \tag{4.13}$$

For every $h > 0$ the distribution $T(\cdot + h)/e(\cdot + h)$ is in $\mathcal{B}'(\mathbf{R})$ and thus T/e can be given in the form (cf. Theorem 1.22)

$$T/e = \sum_{i=0}^{2} D^{2ik} F_i,$$

where F_i, $i = 0, 1, 2$ are bounded and continuous functions. The left side in (4.13) becomes,

$$\lim_{h\to\infty} \sum_{i=0}^{2} (-1)^{2ik} \int_{\mathbf{R}} F_i(x+h)\left(\left(\frac{e(x+h)}{c(h)} - \exp(ax)\right)K(s+x)\right)^{(2ik)} dx. \tag{4.14}$$

If we prove that for every $j \in \mathbf{N}_0$ and $s \in S_0$ the function

$$x \mapsto |((e(x+h)/c(h) - \exp(ax))K(s+x))^{(j)}|, \quad x \in \mathbf{R},$$

is bounded by a function integrable on $\mathbf{R}$ which does not depend on h, then by the use Lebesgue's theorem in (4.14) and by (4.1), it follows (4.13).

Let $x > 0$, $s \in S_0$ and $h > 0$. By (4.8), we have for $s \in S_0$

$$\left|\left(\frac{e(x+h)}{c(h)} - \exp(ax)\right)^{(i)} K^{(j)}(x+s)\right| \leq C_i |K^{(j)}(x+s)|\exp((a+\delta)x).$$

Let $x < 0$ and $h > 0$. Then (4.8) gives

$$\left| \left(\frac{e(x+h)}{c(h)} - \exp(ax) \right)^{(i)} K^{(j)}(x+s) \right|$$

$$\leq C_i (\exp(ax + \delta|x|) + \exp(bx)) |K^{(j)}(x+s)|$$

$$\leq 2C_i |K^{(j)}(x+s)| \exp(bx), \quad s \in S_0.$$

This completes the proof. $\qquad\square$

If $K(\cdot + s) \in \mathcal{D}_{L^1}(\mathbf{R})$, $s \in S_0$, we can not apply Theorem 4.1 because of supposition that $a > b$ and $a_0 > 0$. In this case, we have

Theorem 4.2. *Suppose:*

1. $T \in \mathcal{B}'(\mathbf{R})$ *and* $K(x+s) \in \mathcal{D}_{L^1}(\mathbf{R})$, $s \in S_0$;
2. $T(\cdot + h) \overset{s}{\sim} 1 \cdot A$, $h \in \mathbf{R}_+$ *in* $\mathcal{D}'(\mathbf{R})$.

Then for the $\mathcal{K}$-transform of $T, \mathcal{K}(T)(s) = \langle T(x), \mathcal{K}(x+s)) \rangle$, $s \in S_0$, we have

$$\lim_{h \to \infty} \mathcal{K}(T)(s-h) = \langle A, K(x+s) \rangle .$$

Proof. We have only to apply assertions a) and d) in Theorem 1.19, and to use $\mathcal{K}(T)(s-h) = \langle T(x+h), K(x+s) \rangle$. $\qquad\square$

Here arises a natural question: If $(1+x^2)^k T(x) \in \mathcal{B}'(\mathbf{R})$ for some $k > 0$, how does the asymptotic behavior of $K(T)(-h)$, $h \to \infty$, depend on k? The following example shows that this dependence is very limited.

The following formula is known

$$\int\limits_0^\infty (is+t)^{-r} \exp(-t)dt = (is)^{-r/2} \exp(is/2) W_{-r/2,(1-r)/2}(is)$$

$$\sim (is)^{-r}, \quad s \to \infty, \quad s \in \mathbf{R}_+, \quad r > 1,$$

where $W_{p,q}$ is the Whittaker function. We have $t \mapsto H(t)(1 + t^2)^k \exp(-t)$, $t \in \mathbf{R}$ defines a distribution which belongs to $\mathcal{B}'$ for any $k > 0$, but the dependence on k in the asymptotic behavior of the K-transform of $H(t)e^{-tx}, t \in \mathbf{R}$, with $K(s+t) = (is+t)^{-r}$ is limited by r (H is the Heaviside function).

We shall give several theorems which correspond to Theorem 4.1 in $\mathcal{D}'(\mathbf{R}_+)$ or in $\mathcal{S}'_+$. By the use of the Mellin convolution type transforms (see Zemanian [202]).

The function $x \to \log x$ is a C^∞-diffeomorphism of $\mathbf{R}_+$ onto $\mathbf{R}$. Its inverse is the function: $y \mapsto \exp y$. To an $f \in \mathcal{D}'(\mathbf{R})$, we assign $f(\log x) \in \mathcal{D}'(\mathbf{R}_+)$ by the relation

$$\langle f(\log x), \varphi(x) \rangle = \langle f(y), \varphi(\exp y) \exp y \rangle, \quad \varphi \in \mathcal{D}(\mathbf{R}_+)$$

(see Chapter 1, Section 1 in [192]) or,

$$\langle f(\log x), \psi(\log x)/x \rangle = \langle f(y), \psi(y) \rangle, \quad \psi \in \mathcal{D}(\mathbf{R})$$

and

$$\langle f(\log xu)/c(\log u), \psi(\log x)/x \rangle = \langle f(y+h)/c(h), \psi(y) \rangle, \qquad (4.15)$$

where $u = \exp h$. The mapping: $\psi(y) \to \psi(\log x)/x$ is an isomorphism of the space $\mathcal{D}(\mathbf{R})$ onto $\mathcal{D}(\mathbf{R}_+)$. Note, $\mathcal{D}(\mathbf{R}_+) \subset \mathcal{S}_+$ and the topology of $\mathcal{D}(\mathbf{R}_+)$ is finer then the topology induced by $\mathcal{S}_+$, $\mathcal{D}(\mathbf{R}_+)$ is not dense in $\mathcal{S}'_+$. Thus, we have to consider the restrictions of $f \in \mathcal{S}_+$ onto $\mathcal{D}(\mathbf{R}_+)$. In this sense $\mathcal{S}'_+ \subset \mathcal{D}'(\mathbf{R}_+)$ (for the space $\mathcal{S}_+$ and $\mathcal{S}'_+$ see **0.5.1**).

Definition 4.2. $\mathcal{E}_{a+\delta,b}(\mathbf{R}_+)$ is the vector space of all smooth functions w on $(0,\infty)$such that for each $k \in \mathbf{N}_0$, $\alpha_k(w) = \int\limits_0^\infty q_\delta(x) x^{k-1} |w^{(k)}(x)| dx < \infty$, where by q_δ is denoted a positive and smooth function such that

$$q_\delta(x) = \begin{cases} x^{-(a+\delta)}, \ 0 < x < e^{-1}, \ \delta > 0 \\ x^{-b}, \quad\quad x > 1 \end{cases}, a > b.$$

Proposition 4.3. *If $w \in \mathcal{E}_{a+\delta,b}(\mathbf{R}_+)$, then $w(\exp(-x)) \in \mathcal{D}_{L^1,e_\delta}(\mathbf{R})$.*

Proof. For $t > 1$ and $m \in \mathbf{N}_0$

$$(e_\delta(t) w(\exp(-t)))^{(m)} = \sum_{k=0}^m a_{k,m} \exp((a+\delta-1)t) w^{(k)}(\exp(-t)),$$

where $a_{k,m}$ are constants. Hence,

$$\int\limits_1^\infty (e_\delta(t) w(\exp(-t)))^{(m)} dt = \sum_{k=0}^m a_{k,m} \int\limits_0^{e^{-1}} q_\delta(x) x^{k-1} w^{(k)}(x) dx.$$

The proof for $t < 0$ is similar. $\square$

Theorem 4.3. *Suppose that:*

1.$\{F(u\cdot)/c(\log u);\ u \in \mathbf{R}_+\}$ is a bounded set in $\mathcal{D}'(\mathbf{R}_+)$, where c is given by (4.4).

2. Let $a > b$ and $V_0 \subset C$ be such that $\mathbf{R}_+ V_0 \subset V_0$ ($\mathbf{R}_+ V_0 = \{xV; x \in \mathbf{R}_+,\ v \in V_0\}$). Let $k(\cdot/v) \in \mathcal{E}_{a+\delta,b}(\mathbf{R}_+)$ for every $v \in V_0$ and for a $\delta > 0$. If for every $\psi \in \mathcal{D}(\mathbf{R}_+)$

$$\lim_{h\to\infty} \langle F(xu)/c(\log u), \varphi(x)\rangle = \langle Ax^a, \varphi(x)\rangle, \qquad (4.16)$$

then there exists a Mellin convolution type transform $\mathcal{M}_k(F)$ (see Zemanian [202])

$$\mathcal{M}_k(F)(v) = (F \underset{M}{*} k)(v) = \langle F(x), k(1/vx)/x\rangle, \quad v \in V_0$$

$$= \langle F(x)/e(\log x), e(\log x)k(1/xv)/x\rangle \qquad (4.17)$$

and

$$\lim_{u\to 0^+} \mathcal{M}_k(F)(vu)/c(-\log u) = \langle A, x^{a-1}k(1/xv)\rangle, \quad v \in V_0.$$

Proof. We shall prove first that the distribution $F(\exp\cdot)$ belongs to $(\mathcal{D}_{L^1,e})'(\mathbf{R})$. We have

$$\left\langle \frac{F(ux)}{c(\log u)}, \psi(x)\right\rangle = \left\langle \frac{F(\exp(t+h))}{c(h)}, \psi(\exp t)\exp t\right\rangle, \quad u > 0,\ h \in \mathbf{R},\ \psi \in \mathcal{D}_+.$$

By the isomorphism $\mathcal{D}(\mathbf{R}_+)$ onto $\mathcal{D}(\mathbf{R}) : \psi(x) \to \psi(\exp t)\cdot\exp t$, and by assumption 1. of Theorem 4.3 it follows that the set $\{F(\exp(\cdot+h))/c(h);\ h \in \mathbf{R}\}$ is bounded in $\mathcal{D}'(\mathbf{R})$. To prove that $F(\exp\cdot) \in (\mathcal{D}_{L^1,e})'(\mathbf{R})$ and $F(\exp\cdot)e^{-1} \in \mathcal{B}'(\mathbf{R})$, it is enough to show that for every $\varphi \in \mathcal{D}(\mathbf{R})$ we have

$$((F(\exp\cdot)/e) * \varphi)(h) \in L^\infty(\mathbf{R}), \quad h \in \mathbf{R}. \qquad (4.18)$$

Let us consider relation:

$$((F(\exp\cdot)/e) * \varphi)(h) = \langle F(\exp(x+h))/c(h), \varphi(-x)c(h)/e(x+h)\rangle.$$

The set $\{\varphi(-\cdot)c(h)/e(\cdot+h);\ h \in \mathbf{R}\}$ is bounded in $\mathcal{D}(\mathbf{R})$ because of (4.7). Now, (4.18) follows from the fact that $\{F(\exp(\cdot+h))/c(h);\ h \in \mathbf{R}\}$ is a bounded set in $\mathcal{D}'(\mathbf{R})$. Hence, the distribution $T = F(\exp\cdot)$ and the function $K(s+\cdot) = k(\exp(-s-\cdot))$ satisfy condition 1 of Theorem 4.1. From (4.15) and (4.16) it follows that the distribution $T = F(\exp\cdot)$ satisfies also condition 2 of Theorem 4.1.

By the change of variables: $x \to \exp t$, $v \to \exp s$, we have

$$\left\langle \frac{F(x)}{e(\log x)}, e(\log x)k(1/xv)/x \right\rangle = \left\langle \frac{F(\exp t)}{e(t)}, e(t)k(\exp(-t-s)) \right\rangle, \quad v \in V_0.$$
(4.19)

Now, from the assertion of Theorem 4.1 and from (4.19) the assertion of Theorem 4.3 follows. $\qquad\square$

The next theorem is a modification of Theorem 4.3 which is better adapted to some special cases. In this theorem, we use the family of tempered distributions $\{f_\alpha; \ \alpha \in \mathbf{R}\}$ (see **0.4**).

Theorem 4.4. *Suppose that $F \in \mathcal{S}'(\mathbf{R}_+)$ and $\mathrm{supp} F \subset [a, \infty), a > 0$, and:*

1. The set $\{F(u)/c(\log u); \ u \in \mathbf{R}_+\}$ is bounded in $\mathcal{S}'(\mathbf{R}_+)$, where c is given by (4.4)

2. Let $a > b$ and $V_0 \subset C$ be such that $\mathbf{R}_+ V_0 \subset V_0$. Let $k(\cdot/v) \in \mathcal{E}_{a+\delta,b}(\mathbf{R}_+)$ for every $v \in V_0$ and for a $\delta > 0$.

3. The function $\hat{k}$, defined by $\hat{k}(1/t)/t = (-1)^m D_t^m (k(1/t)/t), \ t > 0$ has the property that $\hat{k}(\cdot/v) \in \mathcal{E}_{a+m,b+m'}(\mathbf{R}_+)$, where $m \geq m' \geq 0$; $a+m-m' > b$, $a + m > -1$; $v \in V_0$.

4. $F \overset{q}{\sim} c(\log u) A f_{r+1}, \ u \to \infty$ in $\mathcal{S}'(\mathbf{R})$.

Then, there exists

$$\mathcal{M}_k(F)(v) = \langle F(x), k(1/xv)/x \rangle = v^m \langle (f_m * F)(x), \hat{k}(1/xv)/x \rangle, \quad v \in V_0$$

and

$$\lim_{u \to 0+} \mathcal{M}_k(F)(vu)/c(-\log u) = v^m \langle A f_{r+m+1}(x), \hat{k}(1/xv)/x \rangle$$
$$= \langle A f_{r+1}(x), k(1/xv)/x \rangle, \quad v \in V_0.$$

Proof. We denote by $\hat{c}$ the following function:

$$\hat{c} = \begin{cases} \exp(ax)L(\exp x), & x \geq 0 \\ \exp((b+m'-m)x), & x < 0. \end{cases}$$

We have

$$\left\langle \frac{(f_m * F)(ux)}{u^m \hat{c}(\log u)}, \varphi(x) \right\rangle = \frac{c(\log u)}{\hat{c}(\log u)} \left\langle \frac{F(ux)}{c(\log u)}, \langle f_m(t), \varphi(x+t) \rangle \right\rangle, \quad u > 0.$$

Since $\varphi(x) \to \langle f_m(t), \varphi(x+t) \rangle$ is an automorphism of $\mathcal{S}_+$ (see Chapter 1, Section 3 in [192]), it follows that $\{(f_m * F)(u\cdot)/(u^m \hat{c}(\log u)); \ u \in \mathbf{R}_+\}$

is a bounded set in $\mathcal{S}'(\mathbf{R}_+)$ and in $\mathcal{D}'(\mathbf{R}_+)$, as well. Hence, conditions 1 and 2 of Theorem 4.3 are satisfied, where instead of F, k, a and b, we have $f_m * F$, $\hat{k}, a + m$ and $b + m'$, respectively. Now, from assumptions 1,2 and 3 in Theorem 4.4 it follows that for $0 < u < 1$ and $v \in V_0$

$$\mathcal{M}_k(F)(uv) = \langle F(x), k(1/xuv)/x\rangle = (uv)^m \langle (f_m * F)(x), \hat{k}(1/xuv)/x\rangle$$
$$= v^m \langle (f_m * F)(t/u)/(u^{-m}), \hat{k}(1/tv)/t\rangle.$$

By assumption 4 in Theorem 4.4 and by Theorem 1, Chapter 1, Section 3 in [192], we have $f_m * F \overset{q}{\sim} k^m c(\log k) \cdot A f_{r+m+1}$, $k \to \infty$ in $\mathcal{S}'(\mathbf{R})$. Hence,

$$\lim_{u \to 0^+} \frac{\mathcal{M}_k(F)(vu)}{c(-\log u)} = v^m \langle A f_{r+m+1}(x), \hat{k}(1/xv)/x\rangle$$
$$= \langle A f_{r+1}, k(1/xv)/x\rangle, \quad v \in V_0. \qquad \square$$

4.1.2 Special integral transforms

Now we apply results of the previous two sections to some special integral transforms (cf. [158]).

Note that Zemanian [202] introduced the spaces $L_{c,d}$ and $M_{c,d}$ in order to analyze the asymptotic behavior of integral transforms of distributions:

$$L_{c,d} = \{f \in C^\infty(\mathbf{R}); \ \gamma_n(f) = \sup_{t \in \mathbf{R}} |g_{c,d}(t) f^{(n)}(t)| < \infty, \ n \in \mathbf{N}_0\},$$

where $g_{c,d}(t) = \exp(ct)$ if $t \ge 0$ and $g_{c,d}(t) = \exp(dt)$ if $t < 0$; $\{\gamma_n\}$ is a multi-norm in $L_{c,d}$. It is easily seen that $L_{c,d} \subset \mathcal{D}_{L^1, e_\delta}$ if $c > a + \delta$ and $d < b$, and the topology in $L_{c,d}$ is finer than the topology induced by $\mathcal{D}_{L^1, e_\delta}$. This is a consequence of the relation

$$\int_{\mathbf{R}} |(e_\delta f)^{(m)}(t)| dt \le \sum_{k=0}^m a_k \gamma_k(f),$$

where a_k, $k = 0, 1, \ldots, m$ are positive numbers. Hence, for strong duals, we have $\mathcal{D}'_{L^1, e_\delta} \subset L'_{c,d}$ if $c > a + \delta$ and $d < b$. Therefore, Theorem 4.1 can be applied to all integral transforms with the kernels in $L_{c,d}$.

The space $M_{c,d}$ is isomorphic to $L_{c,d}$ via the mapping: $f(x) \to f(e^{-t})e^{-t} = g(t)$, where $x = \exp(-t)$. It can be defined in the following way. Let $G_{c,d}$ be the function: $G_{c,d}(x) = x^{-c}$ if $0 < x \le 1$; $G_{c,d}(x) = x^{-d}$ if $1 < x < \infty$. Then

$$M_{c,d} = \{w \in C^\infty(\mathbf{R}_+); H_k(w) = \sup_{0 < x < \infty} |G_{c,d}(x) x^{k+1} w^{(k)}(x)| < \infty, k \in \mathbf{N}_0\}.$$

$\{H_k\}_{k\in\mathbf{N}_0}$ is a multi norm in $M_{c,d}$.

Proposition 4.4. *If $w \in M_{c,d}$ and $c > a + \delta + 1$, for a $\delta > 0$, $d < b + 1$, then $w \in \mathcal{E}_{a+\delta,b}(\mathbf{R}_+)$, as well.*

Proof. For $0 < x < e^{-1}$,

$$|x^{k-1}q_\delta(x)w^{(k)}(x)| \le x^{-(a-c+\delta+2)}|x^{k+1-c}w^{(k)}(x)| \le C'x^{-(a-c+\delta+2)}.$$

For $x > 1$,

$$|x^{k-1}q_\delta(x)w^{(k)}(x)| \le x^{-(b-d+2)}|x^{k+1-d}w^{(k)}(x)| \le C''x^{-(b-d+2)}.$$

These two inequalities imply the assertion of Proposition 4.4. □

From Proposition 4.4 it follows that Theorem 4.3 and Theorem 4.4 can be applied to all integral transforms with the kernels in $M_{c,d}$. Such kernels appeared in [24], and [104].

Weierstrass transform. (See [202]). The kernel $x \mapsto K(s + x) = (4\pi)^{-1/2} \cdot \exp(-(x + s)^2/4)$, $x \in \mathbf{R}$ belongs to $\mathcal{D}_{L^1,e_\delta}(\mathbf{R})$ for every $a, b \in \mathbf{R}$ and $s \in \mathbf{C}$. We can apply Theorem 4.1 and have the same result as in [108].

Poisson transform. (See [189]. The kernel

$$x \mapsto K(s + x) = y((x + s)^2 + y^2)^{-1}, \; x \in \mathbf{R},$$

belongs to $\mathcal{D}_{L^1}(\mathbf{R})$ for $y > 0$ and $s \in \mathbf{C}$. We can apply Theorem 4.2.

Laplace transform. (See [192]) $\mathcal{L}[f](z) = \langle f(x), e^{ixz}\rangle$, $z \in \mathbf{R} + i\mathbf{R}_+$. The kernel of the Laplace transform is $x \mapsto \exp(-vx)$, $Re v > 0$. Hence, for $t > 0$, $k(1/t)/t = \exp(-t)$ and $\hat{k}(1/t)/t = \exp(-t)$, as well. By the relation

$$D_x^n\left(\frac{1}{x}\exp\left(-\frac{1}{x}\right)\right) = \left(\sum_{k=0}^{n}a_{n,k}x^{-k}\right)\left(x^{-n-1}\exp\left(-\frac{1}{x}\right)\right)$$

it follows that $k(\cdot/v) \in \mathcal{E}_{a,b}(\mathbf{R}_+)$ for any $a \in \mathbf{R}$, $b > -1$ and $Re v > 0$; also $\hat{k}(\cdot/v) \in \mathcal{E}_{a+m,b}(\mathbf{R}_+)$ for any $m \ge 0$; we can choose m in such a way that $a + m > b$ and $a + m > -1$. A consequence of Theorem 4.4 is now:

Theorem 4.5. *Suppose that $F \in \mathcal{S}'_+$ and $b > -1$. Let:*

1. *$\{F(\cdot u)/c(\log u); u \in \mathbf{R}_+\}$ be a bounded set in $\mathcal{S}'(\mathbf{R})$;*

2. *$F \overset{q}{\sim} c(\log u) \cdot Af_{\alpha+1}$, $u \to \infty$ in $\mathcal{S}'(\mathbf{R})$.*

Then

$$\lim_{u\to 0^+} u/c(-\log u)\langle F(x), \exp(-xvu)\rangle = Av^{-\alpha-1}, \; Re v > 0.$$

Remark. Theorem 4.5 asserts the same as Theorem 1, Section 4.2, in [192] in the part concerning the results of Abelian type.

Stieltjes transform. Lavoine and Misra [89] defined the Stieltjes transform by introducing a space $\mathcal{J}'(r)$, $r \in \mathbf{R} \setminus (-\mathbf{N})$; it consists of all distributions $f \in \mathcal{S}'_+$ such that there exists $k \in \mathbf{N}_0$ and a locally integrable function F, supp $F \subset [0, \infty)$ such that $f = D^k F$ (D is the derivative in the sense of distributions) and

$$\int_{\mathbf{R}_+} |F(t)|(t + 1)^{-(r+n+1)} dt < \infty.$$

The Stieltjes transform of $f \in \mathbf{J}'(r)$ is defined by

$$\mathcal{S}_r[f](s) = (r + 1)_n \int_0^\infty F(t)(t + s)^{-(r+n+1)} dt, \quad s \in \mathbf{C} \setminus \overline{\mathbf{R}}_-,$$

where $(r + 1)_n = (r + 1) \ldots (r + n)$, $n \geq 1$, $(r + 1)_0 = 1$. $\mathcal{S}_r[f]$ can be written in the form:

$$\mathcal{S}_r[f](s) = \langle (f_n * f)(t), (r + 1)_n (t + s)^{-(r+n+1)} \rangle, \quad s \in \mathbf{C} \setminus \overline{\mathbf{R}}_-.$$

Or, if we replace s by $1/v$, $v \in \mathbf{C} \setminus \overline{\mathbf{R}}_-$

$$v^{-(r+n+1)} \mathcal{S}_r[f](1/v) = \langle (f_n * f)(t), (r + 1)_n (tv + 1)^{-(r+n+1)} \rangle. \tag{4.20}$$

For other definitions of the Stieltjes transform, we refer to [100], [110], [135], [150] and [157]. In order to apply Theorem 4.4 to the Stieltjes transform, we have to remark that $k(1/t)/t = (t + 1)^{-r-1}$, $t > 0$, and

$$\hat{k}(1/t)/t = (-1)^n D_t^n (k(1/t)/t) = (r + 1)_n (t + 1)^{-(r+n+1)}, \quad t > 0.$$

Hence, $k(x) = x^r (x+1)^{-r-1}$ and $\hat{k}(x) = (r+1)_n x^{r+n} (x+1)^{-(r+n+1)}$, $x > 0$. If we compare (4.17) and (4.20), we find that $\mathcal{M}_k(f)(v) = v^{-r} \mathcal{S}_r[f](1/v)$.

Using the following relation

$$k^{(p)}(x) = \sum_{i=1}^p a_{p,i} x^{r-p+i} (x + 1)^{-(r+i+1)}, \quad x > 0,$$

where $a_{p,i}$ are real numbers, we can see that $k(\cdot/v) \in \mathcal{E}_{a,b}$ if $r > a, b > -1$ and $v \in \mathbf{C} \setminus \overline{\mathbf{R}}_-$. In the same way, we have that $\hat{k}(\cdot/v) \in \mathcal{E}_{a+m,b}(\mathbf{R}_+)$. Now, by Theorem 4.4, we have

Theorem 4.6. *Suppose that $f \in \mathcal{S}'_+$. Moreover, suppose that a, b are real numbers such that $r > a > b > -1$, $r \notin -\mathbf{N}$. Let:*

1. The set $\{f(\cdot u)/c(\log u); \ u \in \mathbf{R}_+\}$ be a bounded set in $\mathcal{S}'(\mathbf{R})$;

2. $f \overset{q}{\sim} c(\log u) \cdot Af_{\alpha+1}, \ u \to \infty$ in $\mathcal{S}'(\mathbf{R})$.

Then, there exist

$$S_r[f](s) = \langle (f_n * f)(x), (r+1)_n(x+s)^{-(r+n+1)} \rangle$$

and

$$\lim_{h \to \infty} S_r[f](sh)/(h^{-r}c(\log h)) = A\Gamma(r-a)/\Gamma(r+1)s^{a-r},$$

where $s \in \mathbf{C} \setminus \overline{\mathbf{R}}_-$.

This is the well-known result (see, for example, [167]).

We shall now apply asymptotics expansions at infinity to obtain Abelian type theorems.

The *Weierstrass transform* with parameter t of an $f \in \mathcal{K}'_1$ is defined by

$$\mathcal{W}_t[f](s) = \langle f(x), k(s-x,t) \rangle,$$

where $k(s,t) = (4\pi t)^{-1/2}\exp(s^{-2}/(4t)), \ s \in \mathbf{C}, \ t > 0$.

Proposition 4.5. *Let $f \in \mathcal{K}'_1(\mathbf{R})$, $u_m \in \mathcal{K}'_1(\mathbf{R})$ and let the slowly varying functions L_m be monotonous for sufficiently large arguments, $m = 1,\ldots,p < \infty$ or $m \in \mathbf{N}$. Denote by $c_m(h) = \exp(a_m h)L_m(\exp h), \ a_m \in \mathbf{R}$. If f has the S-asymptotic expansion of the first type, related to the sequence $\{c_m(h)\}$ (see Remark 3) after Definition 1.4):*

$$f(x+h) \overset{s}{\sim} \sum_{m=1}^{\infty} u_m(x)|\{c_m(h)\}, \ h \in \mathbf{R}_+, \ h \to \infty,$$

then, for any $s \in \mathbf{C}$ in the sense of the ordinary asymptotics

$$\mathcal{W}_t[f](s+h) \sim \sum_{i=1}^{m} A(s+h)_{t,i}|\{c_i(h)\}, \ h \to \infty,$$

where $A(s)_{t,m} = \mathcal{W}_t[u_m](s), \ m = 1,\ldots,p \ or \ m \in \mathbf{N}$.

See [118] for the proof.

Now, we present the Abelian type results for the Stieltjes and Laplace transforms of distributions at zero by using asymptotic expansion at zero.

Suppose that $f \in \mathcal{S}'_+$ and that there exist $m \in \mathbf{N}$ and a locally integrable function F, $\operatorname{supp} F \subset [0, \infty)$ such that $f = D^m F$ and that $F(x)(x+s)^{-r-m-1} \in L^1, \ s \in \mathbf{R}_+$.

Proposition 4.6. *Let $\langle L_k \rangle_{k \in \mathbf{N}}$, be a sequence of slowly varying functions at zero, $\langle \alpha_k \rangle_{k \in \mathbf{N}}$, be a strictly increasing sequence of real numbers, $\langle A_k \rangle_{k \in \mathbf{N}}$, be a sequence of real numbers and let $r \in \mathbf{R} \backslash (-\mathbf{N})$, $r > \alpha_k$, $k \in \mathbf{N}$. Assume that $f \in \mathcal{S}'_+$ has the quasi-asymptotic expansion at zero of the first kind:*

$$f \overset{q.e.}{\sim} \sum_{k \in \mathbf{N}} A_k f_{\alpha_k + 1} | \{ \varepsilon^{\alpha_k} L_k(\varepsilon) \}, \ \varepsilon \to 0^+ .$$

Then, in the sense of ordinary asymptotic expansion, we have

$$\mathcal{S}_r[f](\varepsilon) \sim \sum_{k=1}^{\infty} A_k \frac{\varepsilon^{\alpha_k - 1}}{\Gamma(r+1)} \Gamma(r - \alpha_k) | \{ \varepsilon^{\alpha_m} L_m(\varepsilon) \}, \ \varepsilon \to 0^+. \tag{4.21}$$

Proof. Since for every $m \in \mathbf{N}$,

$$\left(f(\varepsilon x) - \sum_{k=1}^{m} A_k f_{\alpha_k + 1}(\varepsilon x) \right/ (\varepsilon^{\alpha_m} L_m(\varepsilon)) \to 0, \ \varepsilon \to 0^+, \ \text{in } \mathcal{S}'(\mathbf{R}),$$

Theorem 2.17 (at zero) implies that there exists $p_m \in \mathbf{N}_0$ and a continuous function F, $\operatorname{supp} F \in [0, \infty)$, such that $f = D^{p_m} F$ and

$$\left(F(\varepsilon x) - \sum_{k=1}^{m} A_k f_{\alpha_k + p_m + 1}(\varepsilon x) \right/ (\varepsilon^{\alpha_m + p_m} L_m(\varepsilon)) \to 0, \ \varepsilon \to 0^+, \ \text{in } \mathcal{S}'(\mathbf{R}).$$

First, assume that $r - 1 \leq \alpha_m < r$. Then, as in ([[135], Ch. 4]), with $\eta \in C^{\infty}$, $\eta = 1$, $x > -\delta$, $(\delta > 0)$, $\eta = 0$, $x < -2\delta$, we have

$$\frac{1}{\varepsilon^{\alpha_m - 1 - r} L_m(\varepsilon)} \mathcal{S}_{r+1} \left[f - \sum_{k=1}^{m} A_k f_{\alpha_k + 1} \right] (\varepsilon s)$$

$$= \frac{(r+2)_{p_m}}{\varepsilon^{\alpha_m - r - 1} L_m(\varepsilon)} \left\langle \left(F - \sum_{k=1}^{m} A_k f_{\alpha_k + p_m + 1} \right)(x), \frac{\eta(x)}{(\varepsilon s + x)^{r + p_m + 2}} \right\rangle$$

$$= (r+2)_{p_m} \left\langle \frac{F(\varepsilon x) - \sum\limits_{k=1}^{m} A_k f_{\alpha_k + p_m + 1}(\varepsilon x)}{\varepsilon^{\alpha_m + p_m} L_m(\varepsilon)}, \frac{\eta(x)}{(s + x)^{r + p_m + 2}} \right\rangle,$$

where $\langle \cdot, \cdot \rangle$ is the dual pairing of $\mathcal{S}'_{p_m + r + 1}(\mathbf{R})$ and $\mathcal{S}_{p_m + r + 1}(\mathbf{R})$. Because of $\alpha_m > r - 1$, it follows that $\eta(x)/(s + x)^{r + p_m + 2} \in \mathcal{S}_{p_m + r + 1}$ and, hence, for $s > 0$,

$$\frac{1}{\varepsilon^{\alpha_m + p_m} L_m(\varepsilon)} \left(F(\varepsilon s) - \sum_{k=1}^{m} A_k f_{\alpha_k + p_m + 1}(\varepsilon s) \right) \to 0 \ \text{in } \mathcal{S}'_{p_m + r + 1}(\mathbf{R}) \ \text{as } \varepsilon \to 0^+.$$

This implies that for every $s > 0$,

$$\frac{1}{\varepsilon^{\alpha_m - r - 1} L_m(\varepsilon)} S_{r+1}\left(f - \sum_{k=1}^{m} A_k f_{\alpha_k+1}\right)(\varepsilon s) \to 0, \ \varepsilon \to 0^+.$$

Since

$$S_r\left[f - \sum_{k=1}^{m} A_k f_{\alpha_k+1}\right](\varepsilon) = (r+1)\int_\varepsilon^\infty S_{r+1}\left[f - \sum_{k=1}^{m} A_k f_{\alpha_k+1}\right](u)du$$

$$= \varepsilon(r+1)\int_1^\infty S_{r+1}\left[f - \sum_{k=1}^{m} A_k f_{\alpha_k+1}\right](\varepsilon x)dx.$$

By Lebesgue's theorem, we have (4.21) for $\alpha_m \geq r - 1$. If $\alpha_m < r - 1$, then we consider the S_r transform of $f - \sum_{k=1}^{m} A_k f_{\alpha_k+1}$ by noting that $\frac{\eta(x)}{(s+x)^{r+p_m+1}} \in S_{r+p_m}$. Then for $x > 0$ it follows that

$$\frac{1}{\varepsilon^{\alpha_m+p_m} L_m(\varepsilon)}(F(\varepsilon x) - \sum_{k=1}^{m} A_k f_{\alpha_k+p_m+1}(\varepsilon x)) \to 0, \ \varepsilon \to 0^+ \text{ in } S'_{r+p_m}(\mathbf{R}).$$

The proof is completed by using the equality

$$S_r[f_{\alpha_k+1}](\varepsilon) = \frac{\varepsilon^{\alpha_k - r}\Gamma(r - \alpha_k)}{\Gamma(r+1)}, \ \varepsilon > 0, \ r > \alpha_k. \qquad \square$$

In a similar way, we can prove the assertion for the Laplace transform.

Let us remark that there is a lot of published papers dealing with Abelian type theorems for the integral transforms of generalized functions. We mention some of them: ([22], [23], [35], [65], [90], [96], [98], [125], [126], [190], [203]).

4.2   *Tauberian type theorems*

Recall the celebrated *Wiener–Tauberian theorem* [198]: *Suppose that* $f \in L^\infty(\mathbf{R})$, $k \in L^1(\mathbf{R})$ *and* $\mathcal{F}[k](y) \neq 0$, *for* $y \in \mathbf{R}$ ($\mathcal{F}[k]$ *is the Fourier transform of* k). *If*

$$\lim_{x \to \infty} \int_\mathbf{R} f(y)k(x - y)dy = a\int_\mathbf{R} k(y)dy, \ a \in \mathbf{R},$$

then for every $G \in L^1(\mathbf{R})$

$$\lim_{x \to \infty} \int_{\mathbf{R}} f(y)G(x-y)dy = a \int_{\mathbf{R}} G(y)dy.$$

This theorem has been much used in various branches of mathematics and so generalizations of it are, even now, important. Pitt's form of Wiener's theorem [140], [141] describes the behavior of the function f as $x \to \infty$, with some additional conditions on f. Many generalizations of these two basic results have been proved (c.f. [9]). For the Tauberian remainder theorems, we refer to Ganelius [62]. Recently, some interesting results on Tauberian type theorems have been published in [28].

The Wiener–Pitt type theorems can be proved in a quick and elegant way by a Banach-algebra approach (see [142]), or by the method of generalized functions as in [60]. One can make use of the direct connection between spectral synthesis and Tauberian theorems (see [7]).

Tauberian type results related to generalized asymptotic behavior have been elaborated only for some special integral transforms of distributions (see for example [135], [192]). It was natural to expect theorems of Wiener's or Pitt's form. In [105] Peetre proved a Wiener–Tauberian theorem when the kernel k belongs to a Banach space W, with a translation invariant norm and with the properties that the space of rapidly decreasing functions is a dense subspace of W and $W * W' \subset L^1$.

We shall discuss the Wiener–Tauberian theorems in a more general setting.

4.2.1 *Convolution type transforms in spaces of distributions*

We quote the next two theorems only with the ideas of the proof. For the complete proof cf. [129]. Similar theorems are Theorems 4.9 and 4.10, but in the space of ultradistributions, which are with the complete proof.

Theorem 4.7. *Let $f \in \mathcal{B}'(\mathbf{R})$ and $K \in \mathcal{D}_{L^1}(\mathbf{R})$ be such that $\mathcal{F}[K](\xi) \neq 0$, for $\xi \in \mathbf{R}$. If*

$$\lim_{x \to \infty} (f * K)(x) = A \int_{\mathbf{R}} K(t)dt, \quad A \in \mathbf{R},$$

then for every $\psi \in \mathcal{D}_{L^1}(\mathbf{R})$,

$$\lim_{x\to\infty} (f * \psi)(x) = A \int_{\mathbf{R}} \psi(t)dt.$$

Proof. We follow the proof of the Wiener–Tauberian Theorem given in ([31], pp. 234–235). Let M be the subspace of $\mathcal{D}_{L^1}(\mathbf{R})$ consisting of all finite linear combinations of translations of K. First, we show that M is dense in $\mathcal{D}_{L^1}(\mathbf{R})$.

We recall that by Theorem XXV, Chapter VI in [146], $f \in \mathcal{B}'(\mathbf{R})$ if and only if

$$f = \sum_{i=0}^{m} F_i^{(i)}, \qquad (4.22)$$

where F_i, $i = 0, 1, \ldots, m$, are bounded continuous functions on $\mathbf{R}$ and the derivatives are in the distributional sense.

Let $\psi \in \mathcal{D}_{L^1}(\mathbf{R})$ and $\varepsilon > 0$. There is $H \in M$ such that

$$\|\psi^{(k)} - H^{(k)}\|_{L^1} < \varepsilon, \ k = 0, \ldots, m.$$

By Lebesgue's theorem and by (4.22)

$$|(f * \psi)(x) - a \int_{\mathbf{R}} \psi(t)dt| \leq \left(\sum_{i=0}^{m} \|F_i\|_{L^\infty} + a + 1 \right)\varepsilon.$$

This implies the assertion. □

We denote by η a smooth function which is equal to 1 in a neighborhood of ∞, and to 0 in a neighborhood of $-\infty$.

Theorem 4.8. *For $f \in \mathcal{D}'(\mathbf{R})$ and $K \in C^\infty(\mathbf{R})$, we assume that*

(i) the set $\{f(\cdot + h)/(c); h \in \mathbf{R}\}$ is bounded in $\mathcal{D}'(\mathbf{R})$,

(ii) there exists $\delta > 0$ such that

$$\eta \check{K} \exp((\alpha + \delta)\cdot), \ (1 - \eta)\check{K} \exp(\beta\cdot) \in \mathcal{D}_{L^1}.$$

Then

(a) $\mathcal{F}[K \exp(-\alpha\cdot)](\xi) = \mathcal{F}[K](\xi - i\alpha)$, for $\xi \in \mathbf{R}$, exists,

(b) the convolution $f * K$ exists.

Moreover, if we assume that

(iii) $\mathcal{F}[K](\xi - i\alpha) \neq 0$, *for* $\xi \in \mathbf{R}$,

(iv) $\displaystyle\lim_{x\to\infty} \frac{(f * K)(x)}{L(\exp x)\exp(\alpha x)} = A \int_{\mathbf{R}} K(t)\exp(-\alpha t)dt$, *for* $A \in \mathbf{R}$,

then

(c) for every $\psi \in C^\infty(\mathbf{R})$ *for which* $\eta\check{\psi}\exp((\alpha+\delta)\cdot), (1-\eta)\check{\psi}\exp(\beta\cdot) \in \mathcal{D}_{L^1}$,

$$\lim_{x\to\infty} \frac{(f * \psi)(x)}{L(\exp x)\exp(\alpha x)} = A \int_{\mathbf{R}} \psi(t)\exp(-\alpha t)dt.$$

Proof. We shall use the functions L^*, c, e and e_δ defined in 4.1 by (4.4), (4.5) and (4.6), respectively in which we replace a by α and b by β.

First step: The proof of (a) and (b). From assumption (ii), we have $\check{K}(\alpha\cdot) \in \mathcal{D}_{L^1}(\mathbf{R})$ and this implies (a). To prove (b), we first prove that $f/e \in \mathcal{B}'(\mathbf{R})$.

Then the proof of assertion (b) follows from the equality

$$(f * K)(h) = \langle f(x+h), \check{K}(x)\rangle = \left\langle \frac{f(x+h)}{e(x+h)}, e(x+h)\check{K}(x)\right\rangle, \quad h \in \mathbf{R},$$

if we prove that for every $h \in \mathbf{R}$, $e(\cdot + h)\check{K} \in \mathcal{D}_{L^1}(\mathbf{R})$.

Second step: We prove that assumption (iv) (with (ii) and (i)) implies that

$$\left(\frac{f}{e} * (K\exp(-\alpha\cdot))\right)(h) = \left\langle \frac{f(x+h)}{e(x+h)} \cdot \check{K}(x)\exp(\alpha x)\right\rangle$$

$$\to A \int_{\mathbf{R}} K(t)\exp(-\alpha t)dt, \quad \text{as } h \to \infty. \tag{4.23}$$

Third step. We prove that $(f * \psi)(h)$, $h \in \mathbf{R}$, exists and that (c) holds.

Proposition 4.7. *Let* $f \in \mathcal{D}'(\mathbf{R})$ *be such that the set*

$$\{f(\cdot + h)/(L(e^h)e^{\alpha h}); \ h > 0\}$$

is bounded in $\mathcal{D}'(\mathbf{R})$ *and let* $\phi \in \mathcal{D}(\mathbf{R})$ *be such that the Fourier transform of* $\phi\exp(-\alpha\cdot)$ *is different from zero on* $\mathbf{R}$. *If*

$$\lim_{x\to\infty} \frac{(f * \phi)(x)}{L(\exp x)\exp(\alpha x)} = A \int_{\mathbf{R}} \phi(t)\exp(-\alpha t)dt, \ A \in \mathbf{R},$$

then for every $\psi \in \mathcal{D}(\mathbf{R})$

$$\lim_{x \to \infty} \frac{(f * \psi)(x)}{L(\exp x) \exp(\alpha x)} = A \int_{\mathbf{R}} \psi(t) \exp(-\alpha t) dt.$$

Proof. If in Theorem 4.8, we take $K \in \mathcal{D}(\mathbf{R})$, then assumption (ii) is satisfied for every $\alpha, \beta, \delta \in \mathbf{R}$, and assumptions (i), (iii) and (iv) of Theorem 4.8 imply the assertion in Proposition 4.7. $\qquad\square$

Remark. A direct consequence of Proposition 4.7 is: If

$$\lim_{h \to \infty} \langle f(x + h)/(e^{\alpha h} L(h)), \ \phi(x) \rangle$$

exists for a ϕ such that $\mathcal{F}[e^{-\alpha \phi \cdot}](x) \neq 0$, $x \in \mathbf{R}$ and the set $\{f(x + h)/(e^{\alpha h} L(h)); h > 0\}$ is bounded in $\mathcal{D}'(\mathbf{R})$, then f has the S-asymptotics related to $e^{\alpha h} L(e^h)$.

Proposition 4.8. *Let f and k be from $L^1_{loc}(\mathbf{R})$ and let $\alpha > \beta$ be such that:*

(i) $f/c \in L^\infty(\mathbf{R})$, where $c(x) = L(\exp x) \exp(\alpha x)$, if $x \geq 0$ and $c(x) = \exp(\beta x)$ if $x < 0$;

(ii) $\check{k} \exp((\alpha + \delta) \cdot) \in L^1((a, \infty))$ and $\check{k} \exp(\beta \cdot) \in L^1((-\infty, b))$ for some $a, b \in \mathbf{R}$, and some $\delta > 0$, where $\check{k}(x) = k(-x)$.

Then $(f * k)(x)$, $x \in \mathbf{R}$, exists and $\check{k} \exp(\alpha \cdot) \in L^1(\mathbf{R})$.

Moreover, if we assume:

(iii) $\mathcal{F}[k \exp(-\alpha \cdot)](y) \neq 0$, for $y \in \mathbf{R}$,

(iv) $$\lim_{x \to \infty} \frac{(f * k)(x)}{L(\exp x) \exp(\alpha x)} = A \int_{\mathbf{R}} k(t) \exp(-\alpha t) dt, \quad A \in \mathbf{R},$$

then

$$\lim_{x \to \infty} \frac{(f * \psi)(x)}{L(\exp x) \exp(\alpha x)} = A \int_{\mathbf{R}} \psi(t) \exp(-\alpha t) dt$$

for every $\psi \in L^1_{loc}(\mathbf{R})$ such that $\check{\psi} \exp((\alpha + \delta) \cdot) \in L^1((a, \infty))$ and $\check{\psi} \exp(\beta \cdot) \in L^1((-\infty, b))$.

In particular, if $\beta + 1 > 0$, then

$$\int_0^x f(\log t) dt \sim \frac{A}{\alpha + 1} x^{\alpha + 1} L(x), \quad x \to \infty.$$

Proof. Proposition 4.8 is not a direct consequence of Theorem 4.8, but the proof of it is a version similar to the proof of Theorem 4.8 (cf. [129]).

Remark. Theorem 1.7.5 in [9] gives conditions on f under which

$$f(t) \sim AL(\exp t)\exp(\alpha t), \ t \to \infty.$$

Proposition 4.8 contains both forms, Wiener's and Pitt's, of a Tauberian theorem (cf. [9], [29], [140]).

For simplicity, all results are given in one-dimensional case. They can also be obtained for a many-dimensional case with the asymptotic behavior in a cone.

Let us denote by W' the subspace of $\mathcal{D}'(\mathbf{R})$ consisting of those T for which $\{T(x + h); \ h \in \mathbf{R}\}$ is a bounded subset in $\mathcal{D}'(\mathbf{R})$. This space is the union of all the spaces introduced in [105]. Let θ be a smooth function on $\mathbf{R}$ such that $\operatorname{supp} \theta \subset [a, \infty)$ for some $a \in \mathbf{R}$ and $\theta(x) = 1$ for $x > b$ for some $b > a$. Let $f(x) \overset{g.s}{\sim} c(x)$, $x \to \infty$. Then by Theorem VI, Chapter VII in [146], we have, $\theta f/c \in \mathcal{S}'$; $\theta(x + h)f(x + h)/c(x + h) \to 1$, $h \to \infty$, in the sense of convergence in $\mathcal{S}'(\mathbf{R})$.

By Theorem 1 in [105], we directly obtain

Proposition 4.9. *Let $f \in \mathcal{D}'(\mathbf{R})$, $c \in C^\infty(\mathbf{R})$, $c(x) \neq 0$, $x \in \mathbf{R}$, and $\varphi \in \mathcal{S}(\mathbf{R})$ be such that the Fourier transform of φ, denoted by $\widehat{\varphi}$, is different from 0 on $\mathbf{R}$. If $\theta f/c \in \mathbf{W}'$ and $\langle f(x+h)/c(x+h), \varphi(x)\rangle \to A \cdot B$, $h \to \infty$, where $A \neq 0$, $B = \int\limits_{-\infty}^{+\infty} \varphi(x)dx = \widehat{\varphi}(0) \neq 0$, then $f(x) \overset{g.s}{\sim} A \cdot c(x)$, $x \to \infty$.*

We assume in the next proposition that $\tilde{c}(x), x > 0$, is constructed as $e(x)$ in Lemma 1.4.

Proposition 4.10. *Let f be a non-negative distribution on $\mathbf{R}$ such that $f/\tilde{c} \in W'$. Let $\varphi \in \mathcal{D}(\mathbf{R})$ and φ be different from 0 on $\mathbf{R}$ and such that*

$$\lim_{h \to \infty} \langle f(x + h)/c(h), \varphi(x)\rangle = C\langle 1, \varphi(x)\rangle, \ C \neq 0. \qquad (4.24)$$

Then, $f(x + h) \overset{s}{\sim} c(h) \cdot C$, $h \in \mathbf{R}_+$.

Proof. By Lemma 1.4, we have $c(h)/\tilde{c}(x+h) - 1 \to 0$, $h \to \infty$, uniformly on $\operatorname{supp} \varphi$. Because of that, for a sequence of positive numbers $\langle \varepsilon_n \rangle_{n \in \mathbf{N}}$

which tends monotonically to 0 there is an increasing sequence $\langle h_n \rangle_{n \in \mathbf{N}}$ such that

$$-\varepsilon_n \varphi(x) \leq (c(h)/\tilde{c}(x+h) - 1)\varphi(x) \leq \varepsilon_n \varphi(x), \quad x \in K, \ h > h_n.$$

This implies (for $h > h_n$)

$$-\varepsilon_n \left\langle \frac{f(x+h)}{c(h)}, \varphi(x) \right\rangle \leq \left\langle \frac{f(x+h)}{c(h)}, \left(\frac{c(h)}{\tilde{c}(x+h)} - 1 \right) \varphi(x) \right\rangle$$

$$\leq \varepsilon_n \left\langle \frac{f(x+h)}{c(h)}, \varphi(x) \right\rangle,$$

and because of (4.24)

$$\left\langle \frac{f(x+h)}{c(h)}, \left(\frac{c(h)}{\tilde{c}(x+h)} - 1 \right) \varphi(x) \right\rangle \to 0 \quad \text{as} \ h \to \infty.$$

Whence

$$\lim_{h \to \infty} \left\langle \frac{f(x+h)}{\tilde{c}(x+h)}, \varphi(x) \right\rangle = \lim_{h \to \infty} \left(\left\langle \frac{f(x+h)}{c(h)}, \varphi(x) \right\rangle \right.$$

$$\left. + \left\langle \frac{f(x+h)}{c(h)}, \left(\frac{c(h)}{\tilde{c}(x+h)} - 1 \right) \varphi(x) \right\rangle \right) = C \langle 1, \varphi(x) \rangle. \quad (4.25)$$

Now Proposition 4.9 implies that for suitable $A \in \mathbf{R}$, $f(x) \overset{g.s}{\sim} A\tilde{c}(x)$, $x \to \infty$. This implies that (4.25) holds for any test function from $\mathcal{S}(\mathbf{R})$ and the assertion is proved. □

Proposition 4.11. *Let f be a non-negative distribution, $f/\tilde{c} \in W'$, and $\varphi \in \mathcal{D}(\mathbf{R})$ such that φ is different from 0 on $\mathbf{R}$. Let*

$$\lim_{h \to \infty} \langle f(x+h)/c(h), e^{-\alpha x} \varphi(x) \rangle \to C \langle 1, \varphi(x) \rangle, \quad \alpha \neq 0.$$

Then $f(x+h) \overset{s}{\sim} c(h) C e^{\alpha x}$, $h \to \infty$.

Proof. We apply the preceding proof on $f(x)e^{-\alpha x}$, $x \in \mathbf{R}$. □

Let $c(h) = h^\beta L(h)$, $h > 0$, where $\beta \in \mathbf{R}$ and $L(h)$, $h > 0$, is a slowly varying function for which we assume to be monotonous and to belong to C^∞ on $(0, \infty)$. Proposition 1.6 with the Remark asserts. "If $f \in L^1_{loc}$ and for some $m_0 \in \mathbf{N}_0$ and $x_0 \in \mathbf{R}$, $f(x)x^{m_0}$ is non-decreasing for $x > x_0$, then the assumption $f(x+h) \overset{s}{\sim} c(h) \cdot 1$, $h \in \mathbf{R}_+$, implies $f(x) \sim c(x)$, $x \to \infty$". This assertion and Proposition 4.10 imply the following result.

Proposition 4.12. *Let $f \in L^1_{loc}(\mathbf{R})$ be such that for some $m_0 \in \mathbf{N}$ and $x_0 \in \mathbf{R}$ the function $f(x)x^{m_0}$ is non-negative and non-decreasing for $x >$*

x_0. *If for some $\varphi \in \mathcal{S}(\mathbf{R})$ the Fourier transform is different from 0 on $\mathbf{R}$, and*

$$\frac{1}{h^\beta L(h)} \int_{x_0}^{\infty} f(x+h)\varphi(x)dx \to C \int_{-\infty}^{\infty} \varphi(x)dx, \ h \to \infty,$$

then $f(x) \sim Cx^\beta L(x)$, $x \to \infty$.

Proof. We consider the function $x \to \theta(x)f(x)$, where $\theta \in C^\infty(\mathbf{R})$, θ is non-decreasing, $\theta(x) = 0$ for $x \leq x_0$, $\theta(x) = 1$ for $x \geq x_0 + 1$, and apply the previous assertions. $\qquad\square$

4.2.2 Convolution type transforms in other spaces of generalized functions

First, we consider in the space of ultra distributions $\mathcal{D}^*(\mathbf{R})$ (cf. **0.5**) in which (M.1), (M.2) and (M.3)' are satisfied. We use the function c_0 and its regularization c given by

$$c_0(x) = \begin{cases} L(e^x)e^{\alpha x}, & x \geq 0, \\ e^{\beta x}, & x < 0, \end{cases} \quad \alpha, \ \beta \in \mathbf{R}, \ \alpha > \beta. \tag{4.26}$$

and

$$c(x) = (c_0 * \omega)(x), \ x \in \mathbf{R}, \tag{4.27}$$

where $\omega \in \mathcal{D}^*$, $\operatorname{supp}\omega \subset [-1, 1]$, $\omega \geq 0$ and $\int_{\mathbf{R}} \omega(t)dt = 1$.

Let us denote by η a function from $\mathcal{E}^*$ with the properties

$$\eta(x) = 1, \ x > x_0 > 0, \ \eta(x) = 0, \ x < -x_0. \tag{4.28}$$

We shall use the following property of a slowly varying function (cf. [9]).

For every $\delta > 0$ there is $C_\delta > 0$ such that

$$\frac{1}{C_\delta} \min\left\{ \left(\frac{x}{y}\right)^\delta, \left(\frac{y}{x}\right)^\delta \right\} \leq \frac{L(x)}{L(y)} \leq C_\delta \max\left\{ \left(\frac{x}{y}\right)^\delta, \left(\frac{y}{x}\right)^\delta \right\}, \ x > 0, \ y > 0.$$
$$\tag{4.29}$$

Theorem 4.9. *Let $f \in \mathcal{B}'^*(\mathbf{R})$ and $K \in \mathcal{D}^*_{L^1}(\mathbf{R})$ such that $\mathcal{F}[K](\xi) \neq 0$, $\xi \in \mathbf{R}$. If*

$$\lim_{x\to\infty}(f * K)(x) = a \int_{\mathbf{R}} K(t)dt, \ a \in \mathbf{R},$$

then for every $\phi \in \mathcal{D}_{L^1}^*(\mathbf{R})$,

$$\lim_{x \to \infty} (f * \psi)(x) = a \int_{\mathbf{R}} \psi(t)dt.$$

Proof. First, we need the following version of Beurling's theorem ([31]) for bounded ultradistributions.

"*Let* $f \in \mathcal{B}'^*(\mathbf{R})$. *A point* ξ_0 *belongs to* supp $\hat{f}$ *if and only if there is a sequence of functions* $\langle \varphi_n \rangle_{n \in \mathbf{N}}$ *from* $\mathcal{S}^*(\mathbf{R})$ *such that*

$$f_n(x) = (f * \varphi_n)(x), \quad x \in \mathbf{R}, \ n \in \mathbf{N},$$

converges narrowly to $f_0(x) = e^{ix\xi_0}$, $x \in \mathbf{R}$, $n \to \infty$".

Recall from [31] that a sequence of continuous and bounded functions f_n on $\mathbf{R}$ *converges narrowly* to a continuous bounded function f_0 if and only if f_n converges to f_0 uniformly on bounded sets in $\mathbf{R}$ and

$$\|f_n\|_{L^\infty} \to \|f_0\|_{L^\infty}, \quad n \to \infty.$$

The proof of this assertion is the same as for bounded distributions since all the properties of Schwartz's test functions which were used in [31], pp. 230–231], have been proved in [79] and [123] for ultradifferentiable functions. The same holds for the next assertion, based on the previous one, which is analogous to Theorem on p. 232 in [31].

"*Let* $f \in \mathcal{B}'^*(\mathbf{R})$ *and* $K \in \mathcal{D}_{L^1}^*(\mathbf{R})$. *If* $K * f = 0$ *on* $\mathbf{R}$, *then* $\hat{K}(\xi) = 0$ *for* $\xi \in$ supp $\hat{f}$".

First, we shall prove that the set M which consists of finite linear combinations of translations of $K \in \mathcal{D}_{L^1}^*(\mathbf{R})$ is dense in $\mathcal{D}_{L^1}^*(\mathbf{R})$. By the property of dual pairing, M is dense in $\mathcal{D}_{L^1}^*(\mathbf{R})$ if and only if for every $S \in \mathcal{B}'^*(\mathbf{R})$, $S * \check{K} = 0 \iff S = 0$. For, if M were not dense, we would have an $S_0 \in \mathcal{B}'^*(\mathbf{R}), S_0 \neq 0$ such that $S_0 * \check{K} = 0$. Thus $\mathcal{F}[\check{K}](\xi) = 0$, $\xi \in$ supp $\mathcal{F}[S_0]$. Since we assume that $\mathcal{F}[\check{K}](\xi) = \mathcal{F}[K](-\xi)$ is never zero, we conclude that M is dense in $\mathcal{D}_{L^1}^*(\mathbf{R})$.

From that and previous statement, we obtain the proof of the quoted Wiener theorem.

Note (cf. [123]), $f \in \mathcal{B}'^*(\mathbf{R})$ if and only if it is of the form

$$f = \sum_{\alpha=0}^{\infty} D^\alpha F_\alpha, \quad F_\alpha \in L^\infty, \ \alpha \in \mathbf{N}_0,$$

where D is the derivative in $\mathcal{B}'^*(\mathbf{R})$ and F_α, $\alpha \in \mathbf{N}_0$, are such that for some $h > 0$ (in the (M_α)-case), respectively, for every $h > 0$ (in the $\{M_\alpha\}$-case)

$$\sum_{\alpha=0}^{\infty} \frac{M_\alpha}{h^\alpha} \|F_\alpha\|_{L^\infty} = K_h < \infty. \tag{4.30}$$

Let $\psi \in \mathcal{D}_{L^1}^*(\mathbf{R})$. Since M is dense in $\mathcal{D}_{L^1}^*(\mathbf{R})$, then: In the (M_p) case, for every $\varepsilon > 0$ and every $h > 0$, there is $H_h \in \mathbf{M}$ such that

$$\|H_h - \psi\|_{L^1, h} < \varepsilon. \tag{4.31}$$

In the $\{M_p\}$ case, we have that for every $\varepsilon > 0$ there is $h > 0$ and H_h such that (4.31) holds.

In the $\{M_p\}$ case, the assumption of the theorem and Lebesgue's theorem give that for $x > x_0(\varepsilon)$, where $x_0(\varepsilon)$ is large enough, we have

$$|(f * \psi)(x) - a\int_{\mathbf{R}} \psi(\xi)dt| \leq |((\psi - H) * f)(x) - a\int_{\mathbf{R}} (\psi(t) - H(t))dt|$$

$$+ |(H * f)(x) - a\int_{\mathbf{R}} H(t)dt|$$

$$\leq \left| \sum_{i=0}^{\infty} \int_{\mathbf{R}} (\psi - H)^{(\alpha)}(t) F_\alpha(x - t)dt \right|$$

$$+ a\int_{\mathbf{R}} |\psi(t) - H(t)|dt + |(H * f)(x) - a\int_{\mathbf{R}} H(t)dt|$$

$$\leq \sup_\alpha \frac{h^\alpha}{M_\alpha} \|(\psi - H)^{(\alpha)}\|_{L^1} \sum_{\alpha=0}^{\infty} \frac{M_\alpha}{h^\alpha} \|F_\alpha\|_{L^\infty} + a\varepsilon$$

$$+ |(H * f)(x) - a\int_{\mathbf{R}} H(t)dt| \leq \varepsilon K_h + a\varepsilon + \varepsilon.$$

The (M_p)-case can be proved similarly. The proof is completed. $\qquad \square$

Theorem 4.10. *Let $f \in \mathcal{D}'^*(\mathbf{R})$ and $K \in C^\infty(\mathbf{R})$. Assume:*

(i) $f/c \in \mathcal{B}'^*(\mathbf{R})$.

(ii) There exists $\delta > 0$ such that $\eta \check{K} e^{(\alpha+\delta)\cdot}$, $(1 - \eta)\check{K} e^{\beta\cdot} \in \mathcal{D}_{L^1}^*(\mathbf{R})$.

(iii) $\mathcal{F}[K](\xi - i\alpha) \neq 0$, $\xi \in \mathbf{R}$.

(iv) $\lim\limits_{x \to \infty} \dfrac{(f * K)(x)}{L(x^x)e^{\alpha x}} = a \int\limits_{\mathbf{R}} K(t)e^{-\alpha t}dt, \ a \in \mathbf{R}.$

Then, for every $\psi \in C^{\infty}(\mathbf{R})$ *for which*

$$\eta\check{\psi}e^{(\alpha+\delta)\cdot}, \ \ (1-\eta)\check{\psi}e^{\beta\cdot} \in \mathcal{D}_{L^1}^*(\mathbf{R}), \tag{4.32}$$

there holds

$$\lim\limits_{x \to \infty} \dfrac{(f * \psi)(x)}{L(e^x)e^{\alpha x}} = a \int\limits_{\mathbf{R}} \psi(t)e^{-\alpha t}dt.$$

Proof. We shall only prove the (M_p)-case because this proof can be simply transferred to the $\{M_p\}$-case.

The proof is organized as follows. We shall prove in part I estimations (4.33), (4.34) and (4.35) which will be used in part II to prove that $\mathcal{F}(Ke^{-\alpha\cdot})(\xi)$, $\xi \in \mathbf{R}^n$, and $f * K$ exist. In part III, we will prove the assertion of Theorem 4.10.

Part I. Note, from the assumption that $(1-\eta)\check{K}e^{\beta\cdot}$ and $\eta\check{K}e^{(\alpha+\delta)\cdot}$ belong to $\mathcal{D}_{L^1}^{(M_p)}(\mathbf{R})$ and from (M.2) it follows that for every $r > 0$

$$\sup\left\{ \dfrac{r^m}{M_m} \Big[\|e^{\beta x}((1-\eta(x)\check{K}(x))^{(m)}\|_{L^1} + \right. \tag{4.33}$$

$$\left. +\|e^{(\alpha+\delta)x}(\eta(x)\check{K}(x))^{(m)}\|_{L^1} \Big], \ m \in \mathbf{N}_0 \right\} < \infty.$$

Since $e^{(\alpha+\delta)x} \le e^{\beta x}$, for $x < 0$, we also have that for every $r > 0$

$$\sup\left\{ \dfrac{r^m}{M_m} \|e^{(\alpha+\delta)x}((1-\eta(x))\check{K}(x))^{(m)}\|_{L^1(-\infty,0)}, \ m \in \mathbf{N} \right\} < \infty. \tag{4.34}$$

We need the following estimate:

For every $r > 0$ there is $C > 0$ such that

$$\sup\limits_{k \in \mathbf{N}_0}\left\{ \dfrac{r^k}{M_k} \left| \left(\dfrac{c(x+h)}{L(e^h)e^{\alpha h}} - e^{\alpha x} \right)^{(k)} \right| \right\} \le \begin{cases} Ce^{\alpha x + \delta|x|}, & x + h > 0, \\ C^{\beta x}, & x + h < 0, \end{cases} \tag{4.35}$$

where we choose δ such that $0 < \delta < \alpha - \beta$.

Let $r > 0$, $k \in \mathbf{N}_0$ and $x + h > 1$. By (4.27) and (4.29) we have (with suitable constants)

$$\dfrac{r^k}{M_k} \left| \left(\dfrac{c(x+h)}{L(e^h)e^{\alpha h}} - e^{\alpha x} \right)^{(k)} \right|$$

$$\leq \frac{r^k}{M_k} \int\limits_{-1}^{1} \frac{L(e^{x+h-t})}{L(e^h)} e^{\alpha(x-t)} |\omega^{(k)}(t)| dt + \frac{|r\alpha|^k}{M_k} e^{\alpha x}$$

$$\leq C_1 e^{\alpha x + \delta|x|} \|\omega\|_{[-1,1],r} + C_2 e^{\alpha x} \leq C e^{\alpha x + \delta|x|}.$$

Similarly, for $x + h < -1$, we get that for a given $r > 0$ there is $C > 0$ such that

$$\sup_{k \in \mathbf{N}_0} \left\{ \frac{r^k}{M_k} \left| \left(\frac{c_0(x+h)}{L(e^h)e^{\alpha h}} - e^{\alpha x} \right)^{(k)} \right| \right\} \leq C e^{\beta x}.$$

Let $u = x + h \in [-1,1], r > 0$ and $k \in \mathbf{N}_0$. From (4.29), we have that

$$\frac{1}{L(e^{u-x})} \leq C e^{\delta|x|}, \quad x \in \mathbf{R},$$

where $C > 0$ and $\delta > 0$. Thus,

$$\frac{r^k}{M_k} \left| \left(\frac{c(x+h)}{L(e^h)e^{\alpha h}} - e^{\alpha x} \right)^{(k)} \right|$$

$$\leq \|\omega\|_{[-1,1],r} \sup_{t \in [-1,1]} \{c_0(x+h-t)\} \frac{1}{L(e^{u-x})e^{\alpha(u-x)}} + C_2 e^{\alpha x}$$

$$\leq C_3 e^{\alpha x + \delta|x|} + C_2 e^{\alpha x}.$$

These inequalities and the assumption $\delta \in (0, \alpha - \beta)$ imply (4.35).

Part II. Let $\psi \in \mathcal{E}^{(M_\alpha)}(\mathbf{R})$ be such that $\psi(x) = 1$ on $(-\infty, -1)$ and $\psi(x) = 0$ on $[0, \infty)$. We have

$$e^{\alpha x} \check{K}(x) = e^{\alpha x} \check{K}(x)(1 - \eta(x))\psi(x)$$
$$+ e^{\alpha x} \check{K}(x)(1 - \eta(x))(1 - \psi(x)) + e^{\alpha x} \check{K}(x)\eta(x), \quad x \in \mathbf{R}.$$
$$(4.36)$$

Since the multiplication in $\mathcal{D}_{L^1}^{(M_\alpha)}(\mathbf{R})$ is an inner operation, one can easily prove that all the members on the right-hand side of (4.36) are from $\mathcal{D}_{L^1}^{(M_\alpha)}(\mathbf{R})$, and so the same holds for $e^{\alpha \cdot} \check{K}$. This implies that $\mathcal{F}(Ke^{-\alpha \cdot})(\xi) = \mathcal{F}(K)(\xi - i\alpha), \xi \in \mathbf{R}^n$, exists.

Since

$$(f * K)(h) = \left\langle \frac{f(x+h)}{c(x+h)}, c(x+h)\check{K}(x) \right\rangle, \quad h \in \mathbf{R},$$

the existence of the convolution $f * K$ will be proved if we prove that for every $h \in \mathbf{R}$,

$$c(\cdot + h)\check{K} \in \mathcal{D}_{L^1}^{(M_\alpha)}(\mathbf{R}),$$

because by (i) $f(\cdot + h)/c(\cdot + h) \in \mathcal{B}'^{(M_p)}(\mathbf{R})$.

For a fixed $h \subset \mathbf{R}$ and a ψ as in (4.34), we have

$$c(x + h)\check{K}(x) = c(x + h)\check{K}(x)(1 - \eta(x))\psi(x)$$
$$+ c(x + h)\check{K}(x)(1 - \eta(x))(1 - \psi(x))$$
$$+ c(x + h)\check{K}(x)\eta(x), \quad x \in \mathbf{R}. \qquad (4.37)$$

By using (4.35), (4.33), (4.34) and that $e^{\alpha \cdot}\check{K} \in \mathcal{D}_{L^1}^{(M_\alpha)}(\mathbf{R})$, we prove that $c(\cdot + h)\check{K} \in \mathcal{D}_{L^1}^{(M_\alpha)}(\mathbf{R})$. We only prove that $c(\cdot + h)\check{K}(1 - \eta)\psi$ is from $\mathcal{D}_{L^1}^{(M_\alpha)}(\mathbf{R})$ for every $h \in \mathbf{R}$, because the proof that $c(\cdot + h)\check{K}\eta$ belongs to $\mathcal{D}_{L^1}^{(M_\alpha)}(\mathbf{R})$ is similar. Then, one can easily see that $c(\cdot + h)\check{K}(1 - \eta)(1 - \psi) \in \mathcal{D}_{L^1}^{(M_\alpha)}(\mathbf{R})$.

Since

$$c(x + h)\check{K}(x)(1 - \eta)(x))\psi(x) = \left[e^{\alpha h}L(e^h)\left(\frac{c(x + h)}{e^{\alpha h}L(e^h)} - e^{\alpha x} \right)\check{K}(x) \right.$$
$$\left. + e^{\alpha(x+h)}L(e^h)\check{K}(x) \right](1 - \eta(x))\psi(x), \quad x \in \mathbf{R},$$

we have to prove that

$$(1 - \eta)\psi\left(\frac{c(\cdot + h)}{e^{\alpha h}L(e^h)} - e^{\alpha \cdot} \right)\check{K} \in \mathcal{D}_{L^1}^{(M_p)}(\mathbf{R}).$$

For every $r > 0$, $x \in \mathbf{R}$, $k \in \mathbf{N}_0$, by using (M.2) and (4.35), we have

$$\frac{r^k}{K_k}\left\| \sum_{j=0}^{k} \binom{k}{j} \check{K}(x)(1 - \eta(x))\psi(x))^{(k-j)}\left(\frac{c(x + h)}{e^{\alpha h}L(e^h)} - e^{\alpha x} \right)^{(j)} \right\|_{L^1}$$

$$\leq \frac{AC}{2^k} \sum_{j=0}^{k} \binom{k}{j} \frac{(2rH)^{k-j}}{M_{k-j}}\{\|(\check{K}(x)(1 - \eta(x))\psi(x))^{(k-j)}e^{\beta x}\|_{L^1}\}$$

$$\leq C_1 \sup_{\alpha \in \mathbf{N}_0}\left\{ \frac{(2rH)^\alpha}{M_\alpha}\|(\check{K}(x)(1 - \eta(x))\psi(x))^{(\alpha)}e^{\beta x}\|_{L^1} \right\},$$

where C and C_1 are suitable constants.

To prove that the last supremum is bounded, we have to use the following estimates

$$\frac{(4rH)^\alpha}{M_\alpha} \frac{1}{2^\alpha} \sum_{j=0}^{\alpha} \binom{\alpha}{j} \| \check{K}(x)(1 - \eta(x))^{(j)} e^{\beta x} \psi^{(\alpha-j)} \|_{L^1}$$

$$\leq C_2 \sup_{j \in \mathbf{N}_0} \left\{ \frac{(4rH^2)^j}{M_j} \| \check{K}(x)(1 - \eta(x))^{(j)} e^{\beta x} \|_{L^1} \right\}$$

$$\times \sup_{\substack{\alpha \in \mathbf{N}_0 \\ j \leq \alpha}} \left\{ \frac{(4rH^2)^{k-j}}{M_{\alpha-j}} \| \psi^{(\alpha-j)} \|_{L^\infty(-1,0)} \right\} \leq C_3 \,,$$

where C_2 and C_3 are suitable constants. Thus, we have proved that the convolution $f * K$ exists.

Part III. We are going to prove that the assumptions of the theorem imply

$$\left(\frac{f}{c} * Ke^{-\alpha \cdot} \right)(h) = \left\langle \frac{f(x+h)}{c(x+h)}, \, \check{K}(x)e^{\alpha x} \right\rangle$$

$$\to a \int_{\mathbf{R}} K(t)e^{-\alpha t} dt, \; h \to \infty \,. \tag{4.38}$$

It is enough to prove that

$$\left\langle \frac{f(x+h)}{c(x+h)}, \left(\frac{c(x+h)}{e^{\alpha h}L(e^h)} - e^{\alpha x} \right) \check{K}(x) \right\rangle \to 0, \; \text{as} \; h \to 0 \,.$$

Since $\mathcal{B}'^*(\mathbf{R}) \ni f/c = \sum\limits_{i=0}^{\infty} D^i F_i$ in order that (4.38) holds, we have to prove that

$$S_h = \sum_{i=0}^{\infty} (-1)^i \int_{\mathbf{R}} F_i(x+h) \left(\left(\frac{c(x+h)}{L(e^h)e^{\alpha h}} - e^{\alpha x} \right) \check{K}(x) \right)^{(i)} dx \to 0, \; h \to \infty \,.$$

We have

$$S_h = \sum_{i=0}^{N} (-1)^i \int_{\mathbf{R}} F_i(x+h) \left(\left(\frac{c(x+h)}{L(e^h)e^{\alpha h}} - e^{\alpha x} \right) \check{K}(x) \right)^{(i)} dx$$

$$+ \sum_{i=N+1}^{\infty} (-1)^i \int_{\mathbf{R}} F_i(x+h) \left(\left(\frac{c(x+h)}{L(e^h)e^{\alpha h}} - e^{\alpha x} \right) \check{K}(x) \right)^{(i)} dx$$

$$= S_{h,N} + S_{h,\infty} \,.$$

Because the sum in $S_{h,N}$ is finite, the proof that $S_{h,N} \to 0$, $h \to \infty$, is the same as in the main assertion of [129]. By using (4.14), we obtain

$$S_{h,\infty} \leq \sum_{i=N+1}^{\infty} \frac{1}{2^i} \frac{M_i \|F_i\|_{L^\infty} (2r)^i}{r^i M_i} \left\| \left[\left(\frac{c(x+h)}{L(e^h)e^{\alpha h}} - e^{\alpha x} \right) \check{K}(x) \right]^{(i)} \right\|_{L^1}$$

$$\leq \frac{C}{2^{N+1}} \sum_{i=0}^{\infty} \frac{(2r)^i}{M_i} \left\| \left[\left(\frac{c(x+h)}{L(e^h)e^{\alpha h}} - e^{\alpha x} \right) \check{K}(x) \right]^{(i)} \right\|_{L^1}.$$

So, if we prove that the last series is bounded with respect to h for $h \geq h_0$, then the proof that $S_{h,\infty} \to 0$, $h \to \infty$, simply follows. Put

$$I_{m,h} = \left\| \left(\left(\frac{c(\cdot + h)}{e^{\alpha h} L(e^h)} - e^{\alpha \cdot} \right) \check{K} \right)^{(m)} \right\|_{L^1}, \ m \in \mathbf{N}_0, \ h \geq h_0.$$

We are going to prove that for every $r > 0$, there is a $C > 0$ such that

$$\sup_{m \in \mathbf{N}_0} \left\{ \frac{r^m}{M_m} I_{m,h} \right\} < C, \ h > h_0. \tag{4.39}$$

This implies that the quoted series is bounded.

Let $\eta(x), x_0 > 0$ be as in (4.28). We have

$$I_{m,h} = \int_{-\infty}^{-x_0} \left| \left[\left(\frac{c(x+h)}{e^{\alpha h} L(e^h)} - e^{\alpha x} \right) (1 - \eta(x)) \check{K}(x) \right]^{(m)} \right| dx$$

$$+ \int_{-x_0}^{x_0} \left| \left[\left(\frac{c(x+h)}{e^{\alpha h} L(e^h)} - e^{\alpha x} \right) \check{K}(x) \right]^{(m)} \right| dx$$

$$+ \int_{x_0}^{\infty} \left| \left[\left(\frac{c(x+h)}{e^{\alpha h} L(e^h)} - e^{\alpha x} \right) \eta(x) \check{K}(x) \right]^{(m)} \right| dx = I_1 + I_2 + I_3.$$

By the Leibnitz formula and by (4.35), (4.33), (4.34) and (4.29) there are constants C_1 and C which do not depend on m and p (but depend on r) such that for $\delta \in (0, \alpha - \beta)$

$$I_1 \leq \sum_{p=0}^{m} \binom{m}{p} \left[\int_{-\infty}^{-h} |(\frac{c(x+h)}{e^{\alpha h} L(e^h)} - e^{\alpha x})^{(p)} e^{-\beta x}| \|e^{\beta x}((1 - \eta(x))\check{K}(x))^{(m-p)}| dx \right.$$

$$+ \int_{-h}^{-x_0} \left| \left(\frac{c(x+h)}{e^{\alpha h} L(e^h)} - e^{\alpha x} \right)^{(p)} e^{-(\alpha x - \delta x)} \right| \left| e^{\alpha x - \delta x}((1 - \eta(x))\check{K}(x))^{(m-p)} \right| dx \right]$$

$$\leq \sum_{p=0}^{m} \binom{m}{p} C \left[\frac{M_p}{r^p} \frac{M_{m-p}}{r^{m-p}} \sup_{\substack{m,p \\ p \leq m}} \left\{ \frac{r^{m-p}}{M_{m-p}} \| e^{\beta x}(1 - \eta(x))(\check{K}(x))^{(m-p)} \|_{L^1} \right\} \right.$$

$$\left. + \frac{M_p}{r^p} \frac{M_{m-p}}{r^{m-p}} \sup_{\substack{m,p \\ p \leq m}} \left\{ \frac{r^{m-p}}{M_{m-p}} \left\| e^{\alpha x - \delta x}(1 - \eta(x))\check{K}(x) \|_{L^1_{(-\infty,0)}} \right\} \right]$$

$$\leq C_1 \sum_{p=0}^{m} \binom{m}{p} \frac{M_m}{r^m} = C_1 \frac{M_m}{(r/2)^m}.$$

This gives $\sup_{m \in \mathbf{N}_0} \left\{ \frac{(r/2)^m}{M_m} I_1 \right\} < C_1.$

In a similar way one can prove the corresponding estimates for I_2 and I_3, and the proof of (4.37) is completed.

Thus, we have proved (4.36).

If $\psi \in \mathcal{E}^{(M_\alpha)}(\mathbf{R})$ satisfies the assumption given in (4.32), then $\psi e^{\alpha \cdot} \in \mathcal{D}_{L^1}^{(M_\alpha)}(\mathbf{R})$, and we have

$$\left(\frac{f}{c} * \psi e^{-\alpha \cdot} \right)(h) \to a \int_{\mathbf{R}} \psi(x) e^{-\alpha x} dx, \quad h \to \infty.$$

As above, we can prove that $(f * \psi)(h)$, $h \in \mathbf{R}$, exists.

To prove that

$$\left\langle \frac{f(x+h)}{L(e^h) e^{\alpha h}}, \check{\psi}(x) \right\rangle \to a \int \psi(x) e^{-\alpha x} dx, \quad h \to \infty,$$

we have to prove that

$$\left\langle \frac{f(x+h)}{c(x+h)}, \left(\frac{c(x+h)}{e^{\alpha h} L(e^h)} - e^{\alpha x} \right) \psi(x) \right\rangle \to 0, \quad h \to \infty$$

but this has already been done (with K instead of ψ), and the proof of Theorem 4.10 is completed. □

Remark. It is an open problem whether the assumption that the set $\left\{ \frac{f(\cdot + h)}{c(h)}; \ h \in \mathbf{R} \right\}$ is bounded in $\mathcal{D}'^*(\mathbf{R})$ implies that $f/c \in \mathcal{B}'^*(\mathbf{R})$. Note that for distributions the corresponding assertion holds (see [146]).

Corollary 4.1. *Let* $f \in \mathcal{D}'^*(\mathbf{R})$ *be such that* $f/c \in \mathcal{B}'^*(\mathbf{R})$ *and let* $\phi \in \mathcal{D}^*(\mathbf{R})$ *be such that* $\mathcal{F}[\phi](\xi - i\alpha) \neq 0$, $\xi \in \mathbf{R}$. *If*

$$\lim_{x \to \infty} \frac{(f * \phi)(x)}{L(e^x)e^{\alpha x}} = a \int \phi(t)e^{-\alpha t}dt, \ a \in \mathbf{R},$$

then for every $\psi \in \mathcal{D}^*(\mathbf{R})$

$$\lim_{x \to \infty} \frac{(f * \psi)(x)}{L(e^x)e^{\alpha x}} = a \int \psi(t)e^{-\alpha t}dt.$$

The proof of corollary simply follows from the given Theorem 4.10 because ϕ in the corollary satisfies conditions assumed for K and functions from $\mathcal{D}^*(\mathbf{R})$ satisfy condition (4.32) of Theorem 4.10.

The next space of generalized functions in which we consider convolution transform will be the space of Fourier hyperfunctions $\mathcal{Q}(\mathbf{D}^n)(cf.0.5)$.

We quote only the results, which are published in [112], because the idea and the technique of the proofs are similar to those used for distributions and ultradistributions.

Proposition 4.13. *If* $f = [F] \in \mathcal{Q}(\mathbf{D}^n)$ *and* $\varphi \in P_*$, *then* $f * \varphi \in \mathcal{Q}(\mathbf{D}^n + iI^n)$ *for an appropriate interval* $I^n \subset \mathbf{R}^n, I^n \ni 0$ *(i.e.,* $f * \varphi$ *is a slowly increasing real analytic function).*

Proposition 4.14. *Let* $\varphi \in P_*$ *and* $\mathcal{F}(\varphi) = \psi$. *Assume that there exists* $\delta > 0$ *such that* $\psi^{-1}exp(-\delta\sqrt{.^2 + 1}) \in P_*$. *Let* M *be a subspace of* P_* *consisting of all finite linear combinations of* $\varphi(\cdot + x), x \in \mathbf{R}^n$. *Then* M *is dense in* P_*.

Theorem 4.11. *Let* $\varphi \in P_*, \psi = \mathcal{F}(\varphi), f = [F] \in \mathcal{Q}(\mathbf{D}^n), F \in \tilde{\mathcal{O}}((\mathbf{D}^n + iI)\#\mathbf{D}^n)(I = (-\alpha, \alpha)^n, \alpha > 0)$ *and let* c *be a positive and measurable function on* $\mathbf{R}^n$. *Assume:*

1. *(i)* $\lim_{x \to \infty} \dfrac{c(x)}{c(x + t)} = 1, t \in \mathbf{R}^n$.

   (ii) For every $\epsilon > 0$ *there exist positive constants* B *and* B_1 *such that*

   $$Be^{-\epsilon|t|} \leq \frac{c(x)}{c(x + t)} \leq B_1 e^{\epsilon|t|}, \ x, t \in \mathbf{R}^n.$$

2. *There exists* $\omega > 0$ *such that* $\psi^{-1}(z)e^{-\omega\sqrt{z^2+1}} \in P_*$.

3. *For every $\sigma \in \Lambda$, every compact set $K_\sigma \subset\subset I_\sigma = I \cap \Gamma_\sigma$ and for every $\eta > 0$, there exists $C > 0$ such that*

$$\left| \frac{F_\sigma(x + h + iy_\sigma)}{c(h)} \right| \leq C e^{\eta |x|}, \quad x \in \mathbf{R}^n, \; h \in \mathbf{R}^n_+, \; y_\sigma \in K_\sigma.$$

4. *There exists $A \in C$ such that $\lim\limits_{h \to \infty} (f * \varphi)(x + h)/c(h) = \langle A, \varphi \rangle$ in $\mathcal{Q}(\mathbf{D}^n)$.*

Then

$$\lim_{h \to \infty} (f * \lambda)(x + h)/c(h) = \langle A, \lambda \rangle \text{ in } \mathcal{Q}(\mathbf{D}^n), \text{ for every } \lambda \in \mathcal{P}_*. \quad (4.40)$$

Theorem 4.12. *Let $\varphi \in \mathcal{P}_*$, $\psi = \mathcal{F}(\varphi)$ and $f = [F] \in \mathcal{Q}$, $F \in \tilde{\mathbf{O}}((\mathbf{D}^n + iI)\#\mathbf{D}^n)$ $(I = (-\alpha, \alpha)^n$, $\alpha > 0)$. Let c and ψ satisfy assumptions 1 and 2 of Theorem 4.11.*

Assume also:

3. *For every $K_\sigma \subset\subset I_\sigma$, $\sigma \in \Lambda$, there exists $N > 0$ such that for every $y_\sigma \in K_\sigma$*

$$\left| \frac{F_\sigma(x + h + iy_\sigma)}{c(h)} \right| \leq N, \quad x \in \mathbf{R}^n, \; h \in \mathbf{R}^n_+.$$

4. *There exists $A \in \mathbf{C}$ such that $\lim\limits_{x \to \infty} (f * \varphi)(x)/c(x) = \langle A, \varphi \rangle$ in $\mathbf{C}$.*

Then

$$\lim_{x \to \infty} (f * \lambda)(x)/c(x) = \langle A, \lambda \rangle \quad \text{for every } \lambda \in \mathcal{P}_* \text{ in } \mathbf{C}. \quad (4.41)$$

Comments on Theorems 4.11 and 4.12

If $n = 1$, then assumption 1 (ii) follows from assumption 1 (i). This follows from Theorem 1.4.1 and Theorem 1.5.6 in [8] (with the change of variables $x = \log u$, $u > 0$). In fact, in this case we have that $c(x) = e^{\alpha x} L(e^x)$, $x > x_0$ (cf. [135]).

The function

$$c(x) = (x_1^2 + 1)^{p_1} L_1(e^{x_1}) \ldots (x_n^2 + 1)^{p_n} L_n(e^{x_n}), \quad x = (x_1, \ldots, x_n) \in \mathbf{R}^n,$$

where L_i, $i = 1, \ldots, n$ are slowly varying functions and $p = (p_1, \ldots, p_n) \in \mathbf{R}^n$, satisfies assumption 1 of Theorem 4.11 and 4.12.

2. Let $\varphi_\delta = \mathcal{F}^{-1}(\exp(-\delta\sqrt{\cdot^2 + 1}))$. Since $\exp(-\delta\sqrt{\cdot^2 + 1}) \in \mathcal{P}_*$, $\delta > 0$, it follows that $\varphi_\delta \in \mathcal{P}_*$. One can simply show that $\psi = \varphi_\delta$ satisfies assumption 2 of Theorem 4.11 and 4.12.

An important function which satisfies assumption 2 in Theorem 4.11 and 4.12 is the Fourier transformation of the function $\mathcal{K}$ introduced by Hörmander (cf. [70], Section 8.4):

$$\mathcal{K}(z) = \frac{1}{(2\pi)^n} \int_{\mathbf{R}^n} e^{i\langle z,\xi\rangle}/I(\xi)d\xi, \quad z \in \Omega = \{z \in \mathbf{C}^n; \ |\operatorname{Im} z| < 1\},$$

where $I(\xi) = \int_{|\omega|=1} e^{-i\langle \omega,\xi\rangle}d\omega$. Recall, $I(\xi) = I_0(\langle\xi,\xi\rangle^{1/2})$, $\xi \in \mathbf{R}^n$, where

$$I_0(\rho) = \frac{2\pi^{(n-1)/2}}{\Gamma((n-1)/2)} \int_{-1}^{1} (1-t^2)^{\frac{n-1}{2}-1} e^{-t\rho}dt, \quad \rho \in \mathbf{C}$$

is an entire function which satisfies the estimate

$$|I_0(\rho)| \le C(1+|\rho|)^{-(n-1)/2} e^{|\operatorname{Re}\rho|}, \quad \rho \in \mathbf{C} \tag{4.42}$$

(cf. Lemma 8.4.9 in [70]).

We shall prove that $\mathcal{K} \in \mathcal{P}_*$. By Lemma 8.4.10 in [70], $\mathcal{K}$ is analytic in every open connected set Ω satisfying $\Omega \subset \widetilde{\Omega} = \{z \in \mathbf{C}^n; \ \langle z,z\rangle \notin (-\infty,-1]\}$. One can simply prove that the strip $\Omega = \{z \in \mathbf{C}^n; \ |y_k| < 1/(2\sqrt{n}), \ k = 1,\ldots,n\}$ is a subset of $\widetilde{\Omega}$.

Let Γ be a closed cone such that if $z \in \Gamma \setminus \{0\}$, then $|x_k| > |y_k|$, $k = 1,\ldots,n$. If $z \in \Gamma \setminus \{0\}$, then $\langle z,z\rangle \notin (-\infty,0]$. By Lemma 8.4.10 in [70], there exists $c > 0$ such that $\mathcal{K}(z) = O(e^{-c|z|})$, $z \in \Gamma$, $|z| \to \infty$. Hence,

$$|\mathcal{K}(z)| \le C_{K,\varepsilon} e^{-(c-\varepsilon)|\operatorname{Re} z|}, \quad z \in \mathbf{R}^n + iK,$$

for every compact set $K \subset \{y \in \mathbf{R}^n; \ |y_k| < 1/(2\sqrt{n}), \ k = 1,\ldots,n\}$ and every $\varepsilon > 0$. Consequently $\mathcal{K} \in \mathcal{P}_*$.

We denote by ψ the Fourier transform of $\mathcal{K}$, i.e., $\psi = \mathcal{F}(\mathcal{K})$, and $\omega > 1$. Then $1/\psi(\zeta) = I(\zeta)$, $\zeta \in \mathbf{C}$ is an entire function. Let $\zeta = \xi + i\eta$, $|\eta| < 1$, then

$$|I(\zeta)| = \left| \int_{|\omega|=1} e^{-\langle \omega,\xi+i\eta\rangle}dxi \right| = I(\xi) = I_0(\xi), \ \xi \in \mathbf{C}.$$

Now, by (4.40)

$$|I(\zeta)| \le |I(\xi)| \le |I_0(|\zeta|)| \le C(1+|\zeta|)^{-(n-1)/2} e^{|\xi|+1}, \ \xi \in \mathbf{R}^n, |\eta| < 1.$$

This implies $I(\zeta)\exp(-\omega\sqrt{\zeta^2+1}) \in \mathcal{P}_*$.

3. The S-asymptotics of Fourier hyperfunctions is defined in the usual way:

Definition 4.3. Suppose that c is a positive function defined on $\mathbf{R}^n$ and $f \in \mathcal{Q}(\mathbf{D}^n)$. It is said that f has the S-asymptotics related to c with the limit $u \in \mathcal{Q}(\mathbf{D}^n)$ if

$$\lim_{x \to \infty} \langle f(t+x)/c(x), \varphi(t) \rangle = \langle u, \varphi \rangle \text{ for every } \varphi \in \mathcal{P}_* .$$

Theorem 4.12 asserts that if $f, \check{\varphi}$ and c satisfy suppositions of Theorem 4.12 and if

$$\lim_{x \to \infty} \left\langle \frac{f(\cdot + x)}{c(x)}, \check{\varphi} \right\rangle = \left\langle A, \check{\varphi} \right\rangle,$$

then f has the S-asymptotics related to c with the limit A.

Zharinov in [204] defined the space $\vec{\Phi}'$ which is isomorphic to $\mathcal{Q}(\mathbf{D}^n)$. But in the same paper he constructed the space $\vec{\Lambda}'(\Sigma) \subset \vec{\Phi}'$, where Σ is a domain in $\mathbf{R}^n$, $0 \in \Sigma$. He defined the quasi-asymptotics in $\vec{\Lambda}'(\Sigma)$.

Let Γ be a convex closed acute cone in $\mathbf{R}^n$. We denote by $\Sigma = \mathrm{int}\Gamma^*$, where Γ^* is the dual cone to Γ. We will follow definitions and results given in [204] and [205].

Let A and B be bounded domains in $\mathbf{R}^n$. Denote by $s_B(\xi) = \sup\{-y\xi; y \in B\}$ and by $\Lambda(A, B)$ the Banach space of functions holomorphic on $\mathbf{R}^n + iA$ and such that

$$\|\varphi\|^A_{-s_B} = \sup\{e^{-s_B(\xi)}|\varphi(\xi + i\eta)|; \zeta \in \mathbf{R}^n + iA\} < \infty$$

with the topology given by the norm $\|\cdot\|^A_{-s_B}$. It is easy to see that $\Lambda(A, B) \subset \Lambda(A', B')$, when $A' \subset A$ and $B \subset B'$. With the inclusion mapping $\rho_{AB,A'B'} : \Lambda(A, B) \to \Lambda(A', B')$, we can define

$$\vec{\Lambda}(\Sigma) = \mathrm{ind} \lim_{A \ni 0, B \subset \subset \Sigma} \Lambda(A, B); \quad \overleftarrow{\Lambda}(\Sigma) = \mathrm{proj} \lim_{B \subset \subset \Sigma, 0 \in A} \Lambda(B, A).$$

The space $\vec{\Lambda}(\Sigma)$ is a DFS space and its dual space $\vec{\Lambda}'(\Sigma)$ is an FS space. But $\overleftarrow{\Lambda}(\Sigma)$ is an FS space. Zharinov proved that $\vec{\Phi}'_\Gamma \subset \vec{\Lambda}'(\Sigma) \subset \vec{\Phi}'$, $0 \in \Sigma$, where $\vec{\Phi}'_\Gamma = \{g \in \vec{\Phi}'; supp\, g \subset \Gamma\}$.

Now, we can quote the definition of the quasi-asymptotics.

Definition 4.4. ([205]) Suppose that $g \in \vec{\Lambda}'(\Sigma)$ and that ρ is a positive and continuous function on $(0, \infty)$. If there exists

$$\lim_{t \to \infty} g(t\zeta)/\rho(t) = h(\zeta) \quad \text{in} \quad \vec{\Lambda}'(\Sigma), \; h \neq 0,$$

then it is said that g has the quasi-asymptotics related to ρ.

Since $\vec{\Lambda}'(\Sigma)$ is an FS space, the limit in Definition 4.4 is equivalent to

$$\lim_{t\to\infty} \langle g(t\xi)/\rho(t), \varphi(\xi) \rangle = \langle h, \varphi \rangle, \ h \neq 0$$

for every $\varphi \in \vec{\Lambda}(\Sigma)$.

In a similar way, as for the quasi-asymptotics of distributions, one can prove that ρ and h in Definition 4.4 have the following properties:

1) ρ has the form $\rho(t) = t^\alpha L(t)$, $\alpha \in \mathbf{R}$ and L is a slowly varying function

2) h is homogeneous of degree α.

We will present an Abelian type theorem for the Laplace transform of Fourier hyperfunctions (cf. [205]). But we have to define first the Laplace transform of elements belonging to $\vec{\Lambda}'(\Sigma)$.

For a fixed $z \in \mathbf{R}^n + iB$, where B is a bounded subset of Σ, $\xi \mapsto e^{iz\xi} \in \Lambda(A, B)$ for every bounded set A and

$$\|e^{iz\cdot}\|_{-s_B}^A = e^{s_A(x)}, \ z = x + iy.$$

Thus, for every fixed $z \in \mathbf{R}^n + i\Sigma$, $e^{iz\cdot} \in \vec{\Lambda}(\Sigma)$ (cf [204]).

Definition 4.5. The Laplace transform of $g \in \vec{\Lambda}'(\Sigma)$, $\mathcal{L}g$, is defined by

$$\mathcal{L}g(z) = \langle g(\xi), e^{iz\xi} \rangle, \ z \in \mathbf{R}^n + i\Sigma.$$

In [205] it has been proved that the Laplace transform defines an isomorphism $\vec{\Lambda}'(\Sigma)$ onto $\overleftarrow{\Lambda}(\Sigma)$. With this property and the mentioned properties of the family of functions $\{e^{iz\xi}; \ z \in \mathbf{R}^n + i\Sigma\}$ it is easy to prove the following Abelian type result.

Proposition 4.15. ([205]) *Suppose that* $g, h \in \vec{\Lambda}'(\Sigma)$ *and* $\rho(t) = t^\alpha L(t)$, $\alpha \in \mathbf{R}$. *If* $G = \mathcal{L}g$ *and* $H = \mathcal{L}h$, *then* $G, H \in \overleftarrow{\Lambda}(\Sigma)$. *If*

$$g(t\xi)/\rho(t) \to h(\xi), \ t \to \infty, \ \text{in} \ \vec{\Lambda}'(\Sigma),$$

then

$$G(z/t)/t^n\rho(t) \to H(z), \ t \to \infty, \ \text{in} \ \overleftarrow{\Lambda}(\Sigma).$$

In [205] one can find other properties of the quasi-asymptotics of the Fourier hyperfunctions.

4.2.3   Integral transforms of Mellin convolution type

Now, we discuss Tauberian theorems for integral transforms which are of the Mellin convolution type and whose kernels belong to suitable test function spaces. The results are based on the Wiener–Tauberian Theorem 4.8 and we apply then to the Laplace, Stieltjes, Weierstrass and Poisson transforms.

First, we recall some definitions and notions. By c and e, we denote the same functions as in 4.1, defined by (4.4), (4.5) and by q_δ as in Definition 4.2, respectively, with the change of a by α and b by β. We introduce function $c^*(x) = c(\log x)$ and $e^*(x) = e(\log x)$, $x \in \mathbf{R}_+$. Denote by $\mathcal{E}_{\alpha,\beta}(\mathbf{R}_+)$ the space from Definition 4.2 and by $\tilde{\mathcal{D}}_{L^1}(\mathbf{R}_+)$ the space of smooth functions ϕ defined on $\mathbf{R}_+$ for which all the seminorms

$$r_k(\phi) = \int\limits_{\mathbf{R}_+} |(Dx)^k \phi(x)| dx, \quad k \in \mathbf{N}_0,$$

are finite, where $(Dx)\phi(x) = \dfrac{d}{dx}(x\phi(x))$.

Note that r_0 is a norm. This sequence of seminorms defines the topological structure in $\tilde{\mathcal{D}}_{L^1}(\mathbf{R}_+)$.

Recall that the classical *Mellin transform* of $f \in L^1_{loc}(\mathbf{R}_+)$ is defined by

$$\mathcal{M}[f](s) = \int\limits_0^\infty f(x)x^{s-1} dx \tag{4.43}$$

for those $s \in \mathbf{C}$ for which this integral exists (see [202]).

Suppose that $k \in C^\infty(\mathbf{R}_+)$ and that the function: $t \mapsto (1/t)e^*(t)k(x/t)$, $t \in \mathbf{R}_+$, belongs to $\tilde{\mathcal{D}}_{L^1}(\mathbf{R}_+)$ for every $x \in \mathbf{R}_+$. Let $F \in \mathcal{D}'(\mathbf{R}_+)$ such that $F/e^* \in \tilde{\mathcal{D}}'_{L^1}(\mathbf{R}_+)$. Then, the *Mellin convolution* is defined by

$$(F \underset{M}{*} k)(x) = \left\langle F(t)/e^*(t), \frac{1}{t}e^*(t)k\left(\frac{x}{t}\right) \right\rangle$$

$$= \left\langle F(xt)/e^*(xt), \frac{1}{t}e^*(xt)k\left(\frac{1}{t}\right) \right\rangle, \quad x \in \mathbf{R}_+. \tag{4.44}$$

We refer to [202], Chapter 4, for this convolution. Note that several important integral transforms of distributions are of this form.

Define

$$\chi_1 : C^\infty(\mathbf{R}_+) \to C^\infty(\mathbf{R}), \quad \phi(x) \mapsto e^y \phi(e^y), \quad x = e^y, \ y \in \mathbf{R},$$

$$\chi_2 : C^\infty(\mathbf{R}) \to C^\infty(\mathbf{R}_+), \ \psi(y) \mapsto \frac{1}{x}\psi(\log x), \ y = \log x, \ x \in \mathbf{R}_+ \ .$$

Clearly, they are inverse to each other and thus, they are bijections.

Proposition 4.16. *The mapping χ_1 is a topological isomorphism of $\mathcal{D}(\mathbf{R}_+)$ onto $\mathcal{D}(\mathbf{R})$, and of $\tilde{\mathcal{D}}_{L^1}(\mathbf{R}_+)$ onto $\mathcal{D}_{L^1}(\mathbf{R})$. Its inverse is χ_2.*

Proof. We shall prove only that $\chi_1 : \tilde{\mathcal{D}}_{L^1}(\mathbf{R}_+) \to \mathcal{D}_{L^1}(\mathbf{R})$ is a topological isomorphism. The following formulas can be simply proved:

$$(Dx)^k \phi(x) = \sum_{j=0}^{k} a_{k,j} x^j \phi^{(j)}(x), \ x \in \mathbf{R}_+, \ k \in \mathbf{N}_0,$$

$$e^{(p+1)y}\phi^{(p)}(e^y) = \sum_{i=0}^{p} b_{p,i}(e^y\phi(e^y))^{(i)}, \ y \in \mathbf{R}, \ p \in \mathbf{N}_0, \ \phi \in C^\infty(\mathbf{R}_+),$$

where all the coefficients $a_{k,j}$ and $b_{p,i}$ are different from zero. This implies

$$\int_0^\infty |(Dx)^k\phi(x)|dx \le \sum_{j=0}^{k} c_{k,j} \int_{-\infty}^\infty |(e^y\phi(e^y))^{(j)}|dy, \ k \in \mathbf{N}_0,$$

and

$$\int_{-\infty}^\infty |(e^y\phi(e^y))^{(p)}|dy \le \sum_{i=0}^{p} d_{p,i} \int_0^\infty |(Dx)^i\phi(x)|dx, \ p \in \mathbf{N}_0,$$

where $c_{k,j}$ and $d_{p,i}$ are suitable positive constants. This implies that χ_1 is a topological isomorphism. $\square$

Proposition 4.13 and $\mathcal{D}(\mathbf{R}) \hookrightarrow \mathcal{D}_{L^1}(\mathbf{R})$ imply that $\mathcal{D}(\mathbf{R}_+) \hookrightarrow \tilde{\mathcal{D}}_{L^1}(\mathbf{R}_+)$, where $\hookrightarrow$ means that the left space is dense in the right one and that the inclusion mapping is continuous. Clearly, $\mathcal{S}_+$ is a subspace of $\tilde{\mathcal{D}}_{L^1}(\mathbf{R}_+)$, and since it contains $\mathcal{D}(\mathbf{R}_+)$, it follows $\mathcal{S}_+ \hookrightarrow \tilde{\mathcal{D}}_{L^1}(\mathbf{R}_+)$. This implies that $\tilde{\mathcal{D}}'_{L^1}(\mathbf{R}_+) \subset \mathcal{D}'(\mathbf{R}_+)$ and that all the elements of $\tilde{\mathcal{D}}'_{L^1}(\mathbf{R}_+)$ are tempered distributions with the support contained in $[0, \infty)$.

Proposition 4.17. *(i) The mapping $\chi_3 : \phi(x) \mapsto q_\delta(e^{-y})\phi(e^{-y}), x = e^{-y}, \ y \in \mathbf{R}$, is a topological isomorphism of $\mathcal{E}_{\alpha+\delta,\beta}$ onto $\mathcal{D}_{L^1}(\mathbf{R})$. (For q_δ cf. Definition 2).*

(ii) If $k \in \mathcal{E}_{\alpha+\delta,\beta}(\mathbf{R}_+)$ and $K(t) = k(e^t), \ t \in \mathbf{R}$, then $\eta \check{K} e^{(\alpha+\delta)}, (1 - \eta)\check{K} e^{\beta \cdot} \in \mathcal{D}_{L^1}(\mathbf{R})$, where $\check{K}$ denotes the function defined by $\check{K}(x) = K(-x), \ x \in \mathbf{R}$.

(iii) If $k \in \mathcal{E}_{\alpha+\delta,\beta}(\mathbf{R})$ and e is given by (4.5), then $\chi_3(k)e(\cdot + h) \in \mathcal{D}_{L^1}(\mathbf{R})$ for every $h \in \mathbf{R}$.

Proof. (i) One can prove by induction that for $m \in \mathbf{N}_0$ and $\phi \in \mathcal{E}_{\alpha+\delta,\beta}(\mathbf{R}_+)$,

$$(q_\delta(e^{-t})\phi(e^{-t}))^{(m)} = \sum_{k=0}^{m} a_{k,m} e^{(\alpha+\delta-k)t}\phi^{(k)}(e^{-t}), \ t > 1,$$

and

$$(q_\delta(e^{-t})\phi(e^{-t}))^{(m)} = \sum_{k=0}^{m} b_{k,m} e^{(\beta-k)t}\phi^{(k)}(e^{-t}), \ t < 0.$$

This implies that there is a constant $C > 0$ such that

$$\|(q_\delta(e^{-t})\phi(e^{-t}))^{(m)}\|_{L^1} \le C\sum_{i=0}^{m} \alpha_i(\phi),$$

where α_i are seminorms in $\mathcal{E}_{\alpha+\delta,\beta}(\mathbf{R}_+)$ (see Definition 4.2).

Thus the mapping $\mathcal{E}_{\alpha+\delta,\beta}(\mathbf{R}_+) \to \mathcal{D}_{L^1}(\mathbf{R})$ is continuous. Let us prove the continuity of $\mathcal{D}_{L^1}(\mathbf{R}) \to \mathcal{E}_{\alpha+\delta,\beta}(\mathbf{R}_+)$.

Let $\psi \in \mathcal{D}_{L^1}(\mathbf{R})$. Then, there exists $\phi \in \mathcal{E}_{\alpha+\delta,\beta}(\mathbf{R}_+)$ such that $q_\delta(e^{-y})\phi(e^{-y}) = \psi(y)$, $y \in \mathbf{R}$. Indeed, we have to prove that $\phi(x) = \dfrac{1}{q_\delta(x)}\psi(\log x)$, $x \in \mathbf{R}_+$, is in $\mathcal{E}_{\alpha+\delta,\beta}(\mathbf{R}_+)$. It is easy to prove that

$$q_\delta(x)x^{k-1}\phi^{(k)}(x) = \sum_{i=0}^{k} c_{i,k} x^{-1}\psi^{(i)}(\log x) \quad \text{for} \ \ x > 1/e.$$

This implies

$$\int\limits_{1}^{\infty} q_\delta(x)x^{k-1}|\phi^{(k)}(x)|dx \le \sum_{i=0}^{k}|c_{i,k}|\int\limits_{1}^{\infty}|\psi^{(i)}(\log x)|\frac{dx}{x}$$

$$\le \sum_{t=0}^{k}|c_{i,k}|\int\limits_{0}^{\infty}|\psi^{(i)}(y)|dy.$$

In the same way, we can prove that

$$\int\limits_{0}^{1} q_\delta(x)x^{k-1}|\phi^{(k)}(x)|dx < C\sum_{i=0}^{k}\|\psi^{(i)}\|_{L^1}.$$

This completes the proof of (i).

(ii) It follows by direct computation of $\mathcal{D}_{L^1}$-seminorms for given functions.

(iii) Let $i, j \in \mathbf{N}_0$, $h \in \mathbf{R}$ be fixed and $x_0 = \max(1 - h, h)$. Then $x + h > 1$ when $x > x_0$ and $x + h < 0$ when $x < -x_0$.

Let $K(t) = k(e^t)$, $t \in \mathbf{R}$. Using the estimate, for fixed h,

$$|e^{(i)}(x + h)\check{K}^{(j)}(x)| \leq \begin{cases} c_{i,j}\exp(\alpha x + \delta x)|\check{K}^{(j)}(x)|, & x + h > 1 \\ c_{i,j}\exp(\beta x)|\check{K}^{(j)}(x)|, & x + h < 0, \end{cases}$$

proved in the first step of the proof of Theorem 4.8, it follows that $e(\cdot + h)\check{K} \in \mathcal{D}_{L^1}(\mathbf{R})$. $\qquad\square$

Theorem 4.13. *Let* $F \in \mathcal{D}'(\mathbf{R}_+)$, $k \in C^\infty(\mathbf{R}_+)$ *and* $\alpha > \beta$ *such that*

(i)' the set $\left\{ \dfrac{F(u\cdot)}{c^*(u)};\ u \in \mathbf{R}_+ \right\}$ *is bounded in* $\mathcal{D}'(\mathbf{R}_+)$ *and*

(ii)' the kernel k *belongs to* $\mathcal{E}_{\alpha+\delta,\beta}(\mathbf{R}_+)$ *for some* $\delta > 0$.

Then,

a)' $\mathcal{M}[x^{-\alpha}k(x)](i\xi)$, $\xi \in \mathbf{R}$, *exists and*

b)' $(F \underset{M}{*} k)$, $k \in \mathbf{R}_+$, *exists, as well.*

Moreover, if we assume that

(iii)' $\mathcal{M}[x^{-\alpha}k(x)](i\xi) \neq 0$, $\xi \in \mathbf{R}$, *and*

(iv)' $\displaystyle\lim_{x\to\infty} (F \underset{M}{*} k)(x)/c^*(x)$

$$= \lim_{x\to\infty} \frac{1}{c^*(x)}\langle \frac{F(t)}{e^*(t)}, \frac{e^*(t)}{t}k(\frac{x}{t})\rangle = A \int\limits_{\mathbf{R}_+} t^{-\alpha-1}k(t)dt,\ A \in \mathbf{R},$$

then,

c)' the Mellin convolution $(F \underset{M}{*} g)(x)$, $x \in \mathbf{R}_+$, *exists for every* $g \in \mathcal{E}_{\alpha+\delta,\beta}(\mathbf{R}_+)$ *and*

$$\frac{(F \underset{M}{*} g)(x)}{c^*(x)} \to A \int\limits_{\mathbf{R}_+} t^{-\alpha-1}g(t)dt,\ x \to \infty.$$

Proof. We will show that the assumptions of theorem imply the assumptions of Theorem 4.8. Then, we will prove the assertions using the conclusions of Theorem 4.8.

Put $K(t) = k(e^t)$, $t \in \mathbf{R}$. By Proposition 4.8 (ii), Assumption (ii)' implies that

$$\eta \check{K} e^{(\alpha+\delta)\cdot}, \; (1-\eta)\check{K}e^{\beta\cdot} \in \mathcal{D}_{L^1}(\mathbf{R})\,.$$

Assumptions $\alpha > \beta$ and $k \in \mathcal{E}_{\alpha+\delta,\beta}(\mathbf{R}_+)$, imply that

$$\int\limits_0^\infty x^{-\alpha} k(x) x^{-i\xi-1} dx$$

is finite for every $\xi \in \mathbf{R}$. Thus, we proved a)'.

Let as prove the existence of $F \underset{M}{*} k$.

Since $k \in \mathcal{E}_{\alpha+\delta,\beta}(\mathbf{R}_+)$, it follows that for every $x \in \mathbf{R}_+$ the function: $t \mapsto (1/t)e^*(t)k(x/t)$, $t \in \mathbf{R}_+$ belongs to $\tilde{\mathcal{D}}_{L^1}(\mathbf{R}_+)$. Proposition 4.14 (iii) implies $e(\cdot + h)\check{K} \in \mathcal{D}_{L^1}(\mathbf{R})$ for every $h \in \mathbf{R}$.

Since

$$\chi_2[e(\cdot + h)\check{K}](t) = \frac{1}{t}\,e(\log t + \log x)k(e^{-\log t})$$

$$= \frac{1}{t}\,e^*(tx)k\left(\frac{1}{t}\right), \; t \in \mathbf{R}_+, \; x = e^h, \; h \in \mathbf{R}\,,$$

it follows that for every $x \in \mathbf{R}_+$, $t \mapsto (1/t)e^*(tx)k(1/t)$, $t \in \mathbf{R}_+$, belongs to $\tilde{\mathcal{D}}_{L^1}(\mathbf{R}_+)$.

Next, we will prove $F/e^* \in \tilde{\mathcal{D}}'_{L^1}(\mathbf{R}_+)$. Proposition 4.16 implies that

$$\phi \mapsto \langle F(x), \frac{1}{x}\,\phi(\log x)\rangle, \;\; \phi \in \mathcal{D}(\mathbf{R}),$$

defines a distribution $f \in \mathcal{D}'(\mathbf{R})$. Assumption (i)' and

$$\left\langle \frac{f(t+h)}{c(h)}, \phi(t) \right\rangle = \left\langle \frac{F(xp)}{c^*(p)}, \frac{1}{x}\,\phi(\log x) \right\rangle, \;\; h = \log p, \; p > 0\,,$$

imply that for every $\phi \in \mathcal{D}(\mathbf{R})$

$$\left\{ \left\langle \frac{f(t+h)}{c(h)}, \phi(t) \right\rangle; \; h \in \mathbf{R} \right\}$$

is a bounded set of continuous functions on $\mathbf{R}$. This implies that $f/e \in \mathcal{B}'(\mathbf{R})$. Let $\varphi \in \mathcal{D}(\mathbf{R}_+)$. We have

$$\left\langle \frac{F(x)}{e^*(x)}, \varphi(x) \right\rangle = \left\langle \frac{f(y)}{e(y)}, e^y\varphi(e^y) \right\rangle\,.$$

Proposition 4.13 implies that $F/e^* = (f/e) \circ \chi_1$ on $\mathcal{D}_{L^1}(\mathbf{R}_+)$. Since $\mathcal{D}(\mathbf{R})$ is dense in $\tilde{\mathcal{D}}_{L^1}(\mathbf{R})$, we have $F/e^* = (f/e) \circ \chi_1$ on $\tilde{\mathcal{D}}_{L^1}(\mathbf{R}_+)$. Hence, the Mellin convolution $F \underset{M}{*} k$ exists and

$$(f * K)(h) = (F \underset{M}{*} k)(p), \quad p = e^h, \ h \in \mathbf{R}. \tag{4.45}$$

Thus, we have proved b)'. Note that assumption (iii)' is equivalent to the assumption $F[K](\xi - i\alpha) \neq 0$, $\xi \in \mathbf{R}$. By (4.45), we have that (iv)' implies (iv) in Theorem 4.8. Now, Theorem 4.8 c) and (4.45), with g instead of K, imply assertion c)'. ☐

Corollary of Theorem 4.13. *If in Theorem 4.13, $F \in \mathcal{S}'_+$, $A \neq 0$ and $\alpha > \beta > -1$, then the assumptions (i)'–(iv)' imply that F has the quasi-asymptotics related to c^* with the limit $A\Gamma(\alpha+1)f_{\alpha+1}$.*

Proof. Formally, we have

$$\frac{(F \underset{M}{*} g)(x)}{c^*(x)} = \frac{1}{c^*(x)} \left\langle F(t), \frac{1}{t} g\left(\frac{x}{t}\right) \right\rangle = \frac{1}{c^*(x)} \langle F(tx), \ \psi(t) \rangle,$$

where $\psi(x) = \dfrac{1}{x} g\left(\dfrac{1}{x}\right)$, $x \in \mathbf{R}_+$.

Let $\delta > 0$. If we prove that for every $\psi \in \mathcal{S}_+$ there exists $g \in \mathcal{E}_{\alpha+\delta,\beta}(\mathbf{R}_+)$ such that $x^{-1}g(x^{-1}) = \psi(x)$, $x \in \mathbf{R}_+$, the above formal calculation holds (since $F \in \mathcal{S}'_+$) and the assertion of the corollary follows from c)' because it implies that for every $\psi \in \mathcal{S}_+$, $\displaystyle\lim_{x \to \infty} \frac{1}{c^*(x)} \langle F(tx), \ \psi(t) \rangle$ exists.

So, we have to prove that $g(x) = (1/x)\psi(1/x)$, $x \in \mathbf{R}$, belongs to $\mathcal{E}_{\alpha+\delta,\beta}(\mathbf{R}_+)$. Since for $k \in \mathbf{N}$ and $x \in \mathbf{R}_+$,

$$g^{(k)}(x) = \frac{1}{x^{k+1}} \sum_{i=0}^{k} a_{i,k} \frac{1}{x^i} \psi^{(i)}\left(\frac{1}{x}\right); \ a_{i,k} \in \mathbf{R}, \ k \in \mathbf{N}_0,$$

where $a_{i,k}$ are suitable numbers, we have

$$|q_\delta(x)x^{k-1}g^{(k)}(x)| \leq \sum_{i=0}^{k} |a_{i,k}| x^{-2-i-\alpha-\delta} \left|\psi^{(i)}\left(\frac{1}{x}\right)\right| < C, \ 0 < x < e^{-1},$$

and

$$|q_\delta(x)x^{k-1}g^{(k)}(x)| \leq \sum_{i=0}^{k} |a_{i,k}| x^{-\beta-i-2} \left|\psi^{(i)}\left(\frac{1}{x}\right)\right| \leq Cx^{-\beta-2}, \ x > 1.$$

This proves that $g \in \mathcal{E}_{\alpha+\delta,\beta}(\mathbf{R}_+)$, if $\alpha > \beta > -1$. ☐

Let $\{f_\alpha;\ \alpha \in \mathbf{R}\}$ be the family of distributions given in **0.4** and $F \in \mathcal{S}'_+$. Let $F^{(-\alpha)} = f_\alpha * F$, $\alpha \in \mathbf{R}$. If $\varphi \in \mathcal{S}_+$, and $m \in \mathbf{N}$, then

$$u^{-m}\langle F^{(-m)}(u\xi), \varphi(\xi)\rangle\rangle = \langle F(u\xi),\ \langle f_m(t), \varphi(\xi+t)\rangle\rangle, \qquad (4.46)$$

where the mapping $\varphi \mapsto \langle f_m(t),\ \varphi(\cdot + t)\rangle$ is an automorphism on $\mathcal{S}_+$ (cf. §3 in [192]).

The following theorem is formulated for future use.

Theorem 4.14. *Let $f \in \mathcal{S}'_+$, $m \in \mathbf{N}_0$ and $\alpha > \beta$. Assume:*

(i") The set $\left\{\dfrac{f(\cdot u)}{u^{-m}c^*(u)};\ u \in \mathbf{R}_+\right\}$ is bounded in $\mathcal{S}'_+$.

(ii") The kernel k_m belongs to $\mathcal{E}_{\alpha+\delta,\beta}(\mathbf{R})$ for some $\delta > 0$.

Then,

a") $\mathcal{M}[x^{-\alpha}k_m(x)](i\xi)$, $\xi \in \mathbf{R}$ exists and

b") $(f^{(-m)} \overset{*}{_M} k_m)(x)$, $x \in \mathbf{R}_+$, exists, as well.

Moreover, if

(iii") $\mathcal{M}[x^{-\alpha}k_m(x)](i\xi) \neq 0$, $\xi \in \mathbf{R}$, and

(iv") $\lim_{x\to\infty} (f^{(-m)} \overset{*}{_M} k_m)(x)/c^*(x) = A(m) \int_{\mathbf{R}_+} \dfrac{1}{x}k_m(\dfrac{1}{x})x^\alpha dx$,

where $A(m) \in \mathbf{R}$, then:

c") For every $g \in \mathcal{E}_{\alpha+\delta,\beta}$, the Mellin convolution $(f^{(-m)} \overset{*}{_M} g)(x)$, $x \in \mathbf{R}_+$, exists and

d") $\lim_{x\to\infty} \dfrac{(f^{(-m)} \overset{*}{_M} g)(x)}{c^*(x)} = A(m) \int_{\mathbf{R}_+} \dfrac{1}{t}g(\dfrac{1}{t})t^\alpha dt$.

In particular, if $\beta > -1$ and $A(m) \neq 0$, then $F \overset{q}{\sim} A(m)\Gamma(\alpha+1)f_{\alpha+1-m}$ related to $u^{-m}c^*(u)$.

Proof. If f satisfies (i"), then (4.46) implies that $\left\{\dfrac{f^{(-m)}(u\cdot)}{c^*(u)};\ u \in \mathbf{R}_+\right\}$ is a bounded set in $\mathcal{D}'(\mathbf{R}_+)$ as well. Thus, $f^{(-m)}$ satisfies condition (i') in Theorem 4.8. Applying Theorem 4.8 to $f^{(-m)}$ and k_m, we obtain a"), b"), c") and d") in Theorem 4.13. By the corollary of Theorem 4.13 it follows that

$$f^{(-m)} \overset{q}{\sim} A(m)\Gamma(\alpha+1)f_{\alpha+1} \quad \text{related to } c^*(u).$$

Theorem 2.1 implies

$$f \overset{q}{\sim} A_m \Gamma(\alpha + 1) f_{\alpha+1-m} \quad \text{related to} \quad u^{-m} c^*(u). \qquad \square$$

Theorem 4.7 and the change of variables $t = e^x$, $x \in \mathbf{R}$, imply the following theorem.

Theorem 4.15. *Let F, $k \in L^1_{loc}(\mathbf{R}_+)$ and $\alpha > \beta$, $\delta > 0$, be such that*

i) $F/c^* \in L^\infty(\mathbf{R}_+)$,

ii) $k(t)t^{-(\alpha+\delta+1)} \in L^1((0,1))$ and $k(t)t^{-(\beta+1)} \in L^1((1,\infty))$.

Then, there exist

$$\left(F \underset{M}{*} k\right)(x) = \int\limits_0^\infty F(t) k\left(\frac{x}{t}\right) \frac{dt}{t}, \quad x \in \mathbf{R}_+,$$

and $\mathcal{M}[x^{-\alpha} k(x)](i\xi)$, $\xi \in \mathbf{R}$. If

iii) $\mathcal{M}[x^{-\alpha} k(x)](i\xi) \neq 0$, $\xi \in \mathbf{R}$, and

iv) $\lim\limits_{x \to \infty} \left(F \underset{M}{*} k\right)(x)/c^*(x) = A \int\limits_0^\infty K(t) t^{-\alpha-1} dt$,

then

$$\lim_{x \to \infty} \left(F \underset{M}{*} \psi\right)(x)/c^*(x) = A \int\limits_0^\infty \psi(t) t^{-\alpha-1} dt, \quad A \in \mathbf{R},$$

for every $\psi \in L^1_{loc}(\mathbf{R}_+)$ such that $\psi(t)t^{-(\alpha+\delta+1)} \in L^1((0,1))$ and $\psi(t)t^{-\beta-1} \in L^1((1,\infty))$ hold. In particular, if $\beta > -1$, then

$$\int\limits_0^x F(t)dt \sim \frac{A}{\alpha + 1} x^{\alpha+1} L(x), \quad x \to \infty.$$

Remark. The last relation and Theorem 1.7.5 in [9], imply that $F(x) \sim A x^\alpha L(x)$, $x \to \infty$.

In the context of Theorem 4.17 see also [76] and Theorem 4.8.3 in [9].

4.2.4   *Special integral transforms*

We shall discuss some special cases of integral transforms.

As we mentioned in 4.1.2, Zemanian, introduced in [202] the spaces $L_{a,b}$ and $M_{a,b}$ in order to define the Laplace and Mellin integral transforms of generalized functions.

If the kernel $\check{K}$ belongs to $L_{a,b}$, then it satisfies condition (ii) of Theorem 4.8 for $a > \alpha > \beta > b$. This follows from

$$(\eta(x)\exp((\alpha + \delta)x)\check{K}(x))^{(m)}$$

$$= \sum_{i=0}^{m} \binom{m}{i}(\alpha + \delta)^{m-j}e^{-(a-\alpha-\delta)x}e^{ax}\check{K}^{(i)}(x), \quad x > \omega,$$

and

$$((1 - \eta(x))\exp(\beta x)\check{K}(x))^{m}$$

$$= \sum_{i=0}^{m} \binom{m}{i}\beta^{m-j}e^{(\beta - b)x}e^{bx}\check{K}^{(i)}(x), \quad x < -\omega,$$

where $0 < \delta < a - \alpha$, ω is a positive number large enough and $\eta \in C^{\infty}$, $\eta(x) = 1$ if $x \in (1, \infty)$ and $\eta(x) = 0$ if $x \in (-\infty, -1)$.

In such a way Theorem 4.8 can be applied to all transforms of convolution type whose kernels belong to $L_{a,b}$ with $a > \alpha > \beta > b$, $0 < \delta < a - \alpha$.

If $\psi \in M_{a,b}$ and $a > \alpha + 1 > \beta + 1 > b$, $0 < \delta < a - \alpha - 1$, then $\psi \in \mathcal{E}_{\alpha+\delta,\beta}(\mathbf{R}_{+})$. This follows from

$$|x^{k-1}q_{\delta}(x)\psi^{(k)}(x)| \leq x^{-(\alpha-a+\delta+2)}|x^{k+1-a}\psi^{(k)}(x)|$$

$$\leq C_1 x^{-(\alpha-a+\delta+2)}, \quad 0 < x < e^{-1},$$

$$|x^{k-1}q_{\delta}(x)\psi^{(k)}(x)| \leq x^{-(\beta-b+2)}|x^{k+1-b}\psi^{(k)}(x)|$$

$$\leq C_2 x^{-(\beta-b+2)}, \quad x > e^{-1}.$$

Hence, Theorem 4.14 can be applied to transforms which are of the Mellin convolution type with kernels in $M_{a,b}$, $a > \alpha + 1 > \beta + 1 > b$, $0 < \delta < a - \alpha - 1$.

Laplace transform. The most precise Tauberian theorem for the Laplace transform of distributions can be found in [192]. Tauberian theorems concerning positive measures are also considered in [149].

Recall, the Laplace transform of an $S \in \mathcal{S}'_{+}$, is defined by

$$\mathcal{L}[S](z) = \langle S(t), e^{izt}\rangle, \quad z = x + iy \in \mathbf{R} + i\mathbf{R}_{+}.$$

Theorem 4.16. *Let* $f \in \mathcal{S}'_+$, $m \in \mathbf{N}_0$, $\alpha > \beta > -1$. *Assume:*

a) *The set* $\left\{ \dfrac{f(u \cdot)}{u^{-m} c^*(u)}; \; u \in \mathbf{R}_+ \right\}$ *is bounded in* $\mathcal{S}'_+$.

b) $\qquad \lim\limits_{y \to 0^+} \dfrac{y}{y^m c^*(\frac{1}{y})} \langle f(t), e^{-yt} \rangle = A \displaystyle\int\limits_0^\infty e^{-t} t^\alpha \, dt, \; A \in \mathbf{R}.$

Then, $f(kx) \overset{q}{\sim} c^*(k) A \Gamma(\alpha + 1) f_{\alpha - m + 1}(x)$, $k \to \infty$.

Proof. If $y > 0$, then $e^{-y} \in \mathcal{S}_+$ and

$$\langle f(t), e^{-yt} \rangle = y^m \langle f^{(-m)}(t), e^{-yt} \rangle, \; y > 0.$$

Put $k_m(t) = t^{-1} \exp(-1/t)$, $t \in \mathbf{R}_+$. Then $k_m \in \mathcal{E}_{\alpha+\delta,\beta}(\mathbf{R}_+)$ for any $\alpha > \beta$, $\beta > -1$ and $\delta > 0$. Therefore,

$$\left(f^{(-m)} \underset{M}{*} k_m \right) \left(\frac{1}{y} \right) = y^{-m+1} \langle f(t), e^{-yt} \rangle$$

and

$$\mathcal{M}(t^{-\alpha} k_m(t))(i\xi) = \int\limits_0^\infty e^{-t} t^{\alpha + i\xi} \, dt = \Gamma(\alpha + 1 + i\xi) \neq 0, \; \xi \in \mathbf{R}, \; \alpha + 1 > 0.$$

Now, Theorem 4.16 follows from Theorem 4.14 (with $k = k_m$). $\qquad \square$

Remark. In [192] Theorem 2, Section 7, the authors proved the assertion of Theorem 4.16 under the assumptions b) and, instead of a), the following one:

There are $M > 0$, $\sigma > 0$ and $y_0 > 0$ such that

$$\left| \frac{y^{n-m+1}}{c^*(\frac{1}{y})} \langle f(t), e^{-yt} \rangle \right| \leq M n! \, n^\sigma, \; 0 < y \leq y_0, \; n \in \mathbf{N}. \qquad (4.47)$$

Since a) implies (4.47), (4.48) and b) is equivalent to

$$f(kx) \overset{q}{\sim} c^*(k) A \Gamma(\alpha + 1) f_{\alpha - m + 1}(x), \; k \to \infty,$$

it follows that conditions a) and b) are equivalent to (4.47) and b).

Stieltjes transform. We shall use the same notation as in 4.1. Reformulate (4.20) in another form

$$\frac{s^{r+m+1}}{(r+1)_m} \mathcal{S}_r[f](s) = \left\langle f^{(-m)}(t), \left(\frac{t}{s} + 1 \right)^{-r-m-1} \right\rangle, \; s \in \mathbf{R}_+. \qquad (4.48)$$

The next theorem follows from Theorem 4.14.

Theorem 4.17. *Let* $f \in \mathcal{S}'_+$, $m \in \mathbf{N}_0$, $r + m + 1 > \alpha + 1 > \beta + 1 > 0$ *and* $\delta > 0$. *Assume:*

a)
$$\left\{ \frac{f(u\cdot)}{u^{-m}c^*(u)}; \ u \in \mathbf{R}_+ \right\}$$

is bounded in $\mathcal{S}'_+$. *Then,* $\mathcal{S}_r[f](u)$, $u \in \mathbf{R}_+$, *exists. If moreover,*

b)
$$\lim_{u \to \infty} \frac{1}{u^{-(r+m)}c^*(u)} \mathcal{S}_r[f](u) = C_m, \ C_m \in \mathbf{R},$$

then for every $g \in \mathcal{E}_{\alpha+\delta,\beta}(\mathbf{R}_+)$,

$$\lim_{u \to \infty} \left\langle \frac{f^{(-m)}(\xi u)}{c^*(u)}, \frac{1}{\xi} g\left(\frac{1}{\xi}\right) \right\rangle = C_m \frac{\Gamma(r+1)}{\Gamma(r+m-\alpha)} \left\langle f_{\alpha+1}(\xi), \frac{1}{\xi} g\left(\frac{1}{\xi}\right) \right\rangle.$$
(4.49)

Hence,

$$f^{(-m)}(ux) \overset{q}{\sim} c^*(u) C_m \frac{\Gamma(r+1)}{\Gamma(r+m-\alpha)} f_{\alpha+1}(x), \ u \to \infty \tag{4.50}$$

and

$$f(ux) \overset{q}{\sim} u^{-m} c^*(u) C_m (\Gamma(r+1)/\Gamma(r+m-\alpha)) f_{\alpha+1-m}(x), \ u \to \infty. \tag{4.51}$$

Proof. Let $F = f^{(-m)}$ and

$$k_m(t) = \frac{1}{t}\left(\frac{1}{t} + 1\right)^{-(r+m+1)}, \ t \in \mathbf{R}_+.$$

Since $r + m + 1 > \alpha + 1 > \beta + 1 > 0$ and

$$q_\delta(x) x^{p-1} k_m^{(p)}(x) = \sum_{i=0}^{p} a_{i,p} q_\delta(x) x^{-i-2} \left(\frac{1}{x} + 1\right)^{-(r+m+i+1)}, \ x \in \mathbf{R}_+,$$

it follows that $k_m \in \mathcal{E}_{\alpha+\delta,\beta}(\mathbf{R}_+)$. Note that $t^{-1} k_m(t^{-t}) = (t+1)^{-(r+m+1)}$, $t \in \mathbf{R}_+$. This and (4.48) imply.

$$(f^{(-m)} \overset{*}{_M} k_m)(u) = \langle f^{(-m)}(\xi u), (\xi + 1)^{-(r+m+1)} \rangle$$

$$= \frac{1}{u}\langle f^{(-m)}(t), \left(\frac{t}{u} + 1\right)^{-(r+m+1)} \rangle$$

$$= \frac{u^{r+m}}{(r+1)_m} \mathcal{S}_r[f^{(-m)}](u), \ u \in \mathbf{R}_+. \tag{4.52}$$

We shall show that conditions of Theorem 4.14 hold for f and $k = k_m$. In fact, (i") is a) and (iii") follows from

$$\mathcal{M}[x^{-\alpha}k_m(x)](i\xi) = \frac{\Gamma(\alpha + 1 + i\xi)\Gamma(r + m - \alpha - i\xi)}{\Gamma(r + m + 1)} \neq 0, \ \xi \in \mathbf{R}.$$

Condition b) and (4.52) imply (iv") of Theorem 4.14

$$\lim_{u \to \infty} \frac{(f^{(-m)} \overset{*}{\underset{M}{}} k_m)(u)}{c^*(u)} = \lim_{u \to \infty} \frac{u^{r+m}}{(r+1)_m c^*(u)} \mathcal{S}_r[f](u)$$

$$= \frac{C_m}{(r+1)_m} = A_m \int_0^\infty \frac{t^\alpha}{(t+1)^{r+m+1}} dt = A_m \int_0^\infty \frac{1}{x} \, k_m\left(\frac{1}{k}\right) x^\alpha dx,$$

where $A_m = \dfrac{C_m \Gamma(r+1)}{\Gamma(r+m-\alpha)}$.

Thus, Theorem 4.14 implies (4.49) but (4.50) follows from $\mathcal{S}_+ \subset \mathcal{E}_{\alpha+\delta,\beta}(\mathbf{R}_+)$. As in the proof of Theorem 4.14, we have that the quasi-asymptotic behavior of $f^{(-m)}$ imply the appropriate quasi-asymptotic behavior of $(f^{(-m)})^{(m)} = f$ and this is given in (4.51). $\qquad\square$

Weierstrass transform. As earlier we will use the Kernel

$$k(s,t) = \frac{1}{(4\pi t)^{1/2}} \exp(-s^2/4t), \ \ s \in \mathbf{C}, \ t > 0.$$

The Weierstrass transform, $\mathcal{W}_t(f)$, $t > 0$, of $f \in \mathcal{K}'_1$ is defined by

$$\mathcal{W}_t[f](s) = \langle f(x), k(s - x, t)\rangle = (f * k(\cdot, t))(s), \ s \in \mathbf{C}, \ |\arg s| < \pi/4.$$

A Tauberian type theorem for this transform, in the case when $c(h) \equiv 1$, is given in [108]. The next one is more general.

Theorem 4.18. *Let $f \in \mathcal{D}'(\mathbf{R}), \alpha > \beta$, and the set $\left\{\dfrac{f(\cdot + h)}{c(h)}; \ h \in \mathbf{R}\right\}$ be bounded in $\mathcal{D}'(\mathbf{R})$. Then, $\mathcal{W}_t(f)(s)$ exists for $s \in \mathbf{C}$, $|\arg s| \le \pi/4$ and $t > 0$. Moreover, if we assume*

$$\lim_{h \to \infty} \frac{1}{c(h)} \mathcal{W}_{t_0}[f](h) = \frac{A(t_0)}{(4\pi t_0)^{1/2}} \int_{-\infty}^{\infty} \exp(-x^2/4t_0)\exp(-\alpha x)dx$$

$$= A(t_0)\exp(\alpha^2 t_0) = B(t_0),$$

then

$$\lim_{h \to \infty} \left\langle \frac{f(x+h)}{c(h)}, \psi(x) \right\rangle = B(t_0)\exp(-\alpha^2 t_0)\langle e^{\alpha x}, \psi(x)\rangle \qquad (4.53)$$

for every $\psi \in C^\infty(\mathbf{R})$ such that $\eta\psi\exp((\alpha+\delta)\cdot)$ and $(1-\eta)\check{\psi}\exp(\beta\cdot)$ belong to $\mathcal{D}_{L^1}(\mathbf{R})$.

The proof follows by showing that the conditions of Theorem 4.8 are satisfied. We omit the details. □

Remark. Of $A(t_0) \neq 0$, then (4.36) implies that f has the S-asymptotics related to c with the limit $B(t_0) \exp(-\alpha^2 t_0) \exp(\alpha x)$ in $\mathcal{D}'(\mathbf{R})$ and in $\mathcal{K}'_1(\mathbf{R})$.

Poisson transform. The Poisson transform of a distribution $f \in \mathcal{D}'(\mathbf{R})$ is defined by

$$\mathcal{P}[f](x,y) = \left(f(t) * \frac{y}{t^2 + y^2} \right)(x) = \left\langle f(t), \frac{y}{(x-t)^2 + y^2} \right\rangle, \ y > 0, \ x \in \mathbf{R}.$$

The equality

$$\int\limits_{-\infty}^{\infty} e^{-i\xi t} \frac{y}{t^2 + y^2} dt = e^{-|y\xi|}, \ y > 0, \ \xi \in \mathbf{R},$$

and Theorem 4.13 imply:

Theorem 4.19. *Let $f \in \mathcal{B}'(\mathbf{R})$. If for some $y > 0$*

$$\lim_{x \to \infty} \mathcal{P}[f](x,y) = a \int\limits_{-\infty}^{\infty} \frac{y}{t^2 + y^2} dt, \ a \in \mathbf{R},$$

then

$$\lim_{x \to \infty} \langle f(t+x), \psi(t) \rangle = \langle a, \psi(t) \rangle$$

for every $\psi \in \mathcal{D}_{L^1}(\mathbf{R})$.

Remark. If $a \neq 0$, then the last equality means that f has S-asymptotics in $\mathcal{B}'(\mathbf{R})$ related to $c = 1$.

4.2.5 *Localization of Tauberian type theorems*

In the Tauberian type theorems for generalized functions Vladimirov, Drozhzhinov, and Zavyalov ([192]) have used results on functions holomorphic in some tubular domains over cones. In [36] it is proved that such results can remain valid in the local variant, as well. Authors elaborated a technique which can be applied reconstruct asymptotic properties of a holomorphic function from the asymptotic behavior if its real part, a theorem on the non-compensation of singularities of holomorphic functions and a multidimensional theorem of Lindelöf type.

The basic idea lies in the following theorem.

Theorem 4.20. [36]. *Assume that C_1 and C_2 are open, strictly convex and acute cones in $\mathbf{R}^n$ with their vertices at the origin and such that*

$$C_1 \cap C_2 = C \neq 0, \quad ch(C_1 \cup C_2) = \mathbf{R}^n.$$

Let $f_1(z)$ and $f_2(z)$ be functions holomorphic in local tubular domains over the cones C_1 and C_2, respectively. Consider the sum of these two functions $\phi(z) = f_1(z) + f_2(z)$.

It is holomorphic in the local tubular domain over the cone C. Assume that for some $\alpha > 0$ there are constants M, N and r_0 such that

$$\left|\phi(z)\right| \leq \frac{M}{|z|^\alpha \varphi N} \ for \ z \in T^C, \ |z| < r_0,$$

where φ is the angle between the vector z and the boundary of the tubular domain T^C. Then there are constants M_1, N_1 such that

$$\left|f_1(z)\right| \leq \frac{M_1}{|z|^\alpha \varphi_1^{N_1}}, \ |z| < r_0,$$

where φ_1 is the angle between the vector z and the boundary of the tubular domain over the cone C_1. A similar estimate is valid for $f_2(z)$ with corresponding M_2, N_2 and φ_2.

In Theorem 4 of [36] is given a local variant of Theorem 4.16. First it was defined the space $\Lambda(\Gamma)$ of type $K\{M_p\}$ with a weight $M_p(t)$ (cf. [63], V.2) which is complete, perfect and nuclear countably normed space.

$\Lambda'(\Gamma)$ is the corresponding space of continuous linear functionals and Γ is a closed, convex acute, solid cone in $\mathbf{R}^n$ with vertex at the origin; $int\,\Gamma^* = C$. Let $C_{q,r} = \{y \in C; \ |y|^2 < q\Delta_C(y), \ |y| < r\}$ and let $H(T^{C_{q,r}})$ be the class of functions $g(z)$,

$$|g(x + iy)| \leq M(1 + |x|)^a / \Delta_{C_{q,r}}^b(y) \ \text{for} \ z \in T^{C_{q,r}}$$

for some a, b and M.

We recall also

Theorem 4.21. (cf. [36], Theorem 4). *Assume that $\mathcal{L}f \in H(T^{C_{q,r}})$ ($\mathcal{L}$ denote the Laplace transform) and let $\rho(k)$ be a regularly varying function of order α. Assume that the following conditions are fulfilled:*

a) *there is a domain* $\Omega \subset C$ *such that*

$$\lim_{k \to \infty} \frac{1}{k^n \rho(k)} \mathcal{L}f\left(\frac{iy}{k}\right) = \mathcal{L}g(iy), \quad y \in \Omega;$$

b) *there are* $y_0 \in prC$ *and constants* M, N *and* r_1 *such that*

$$\left|\mathcal{L}f(z)\right| \leq M \frac{\rho(1/|z|)}{|z|^n} \left|\frac{z}{y}\right|^N \quad \textit{if } z = x + i\tau y_0, \ \tau > 0, \ |z| < r_1.$$

Then the spectral function $f(t)$ *for* $\mathcal{L}f(z)$ *has quasi-asymptotics relative to* $\rho(k)$ *in* $\Lambda'(\Gamma)$. *In other words*

$$\lim_{k \to \infty} \frac{1}{\rho(k)} f(kt) = g(t) \quad \textit{in} \ \Lambda'(\Gamma).$$

For the asymptotics and integral transforms one can also consult the following papers: [117], [122] and [188].

5 Summability of Fourier series and integrals

In this part we discuss applications of the quasi-asymptotic behavior of distributions to the study of summability of one-dimensional Fourier series and integrals.

The study of the relationship between the local behavior of a periodic function and the convergence or summability of its Fourier series is an old and interesting problem. It has a long tradition [68], [206]. Since convergence fails in many interesting cases, one is led to use summability methods rather than ordinary convergence. In the case of periodic distributions, that is, periodic elements of the space $\mathcal{S}'(\mathbf{R})$, the local problems have classically focused around the notion of the value of a distribution at a point in the sense of Lojasiewicz [93], [94] (cf. **5.2**). A pioneer in this direction was G. Walter [195], [196] who study Cesàro and Abel summability of Fourier series under the presence of Lojasiewicz point values. Surprisingly, it is possible to characterize Lojasiewicz point values in terms of the summability of the Fourier series; a complete characterization in terms of asymmetric Cesàro limits of partial sums was obtained first by R. Estrada in [45]. The situation with Fourier integrals is similar to that of series, even in the classical cases one needs to use summability methods [170]. Recent studies

[49], [57], [58], [172], [173], [175], [176], [178], [181] and [182] have shown the deep connection between quasi-asymptotics and such problems, and it is exactly the approach that we will follow in this part of the book.

The first section **5.1** is of preliminary character. We start by discussing several summability procedures for series and integrals, and we then extend them to summability for distributional evaluations. We follow closely the expositions from [47], [56], [66] and [173].

The main section is **5.2** Let us state the problem to be considered. Fix the constants in the Fourier transform,

$$\hat{\phi}(t) = \int_{-\infty}^{\infty} e^{-itx}\phi(x)dx,$$

for $\phi \in \mathcal{S}(\mathbf{R})$, so that the Fourier inversion formula is

$$\phi(x) = \frac{1}{2\pi} \int_{-\infty}^{\infty} e^{ixt}\hat{\phi}(t)dt. \tag{5.1}$$

The first question that we should address is that of giving pointwise sense to (5.1) for very general tempered distributions, that is, if a distribution has a value at a point, we show that (5.1) holds pointwise for several summability methods. The second important problem that we want to consider is to find a summability method so that the pointwise Fourier inversion formula, interpreted with such a summability procedure, becomes a full characterization of Lojasiewicz point values. For this goals, we follow the results from [175], [176]; however, we present a simplification of the proofs based on the results from **2.10**. It is interesting to mention that the summability method which provides the right characterization is nothing but a structural characterization of a quasi-asymptotics, and it was actually the precedent to the structural theorems from **2.10**. In the remaining subsections, we discuss other related problems. Finally, we should mention that we do not discuss problems about summability order in our formulas; we refer to [181] for a complete study in that direction.

5.1   *The Cesàro behavior*

It is the intension of this section to introduce two methods of summability for distributional evaluations. We are only interested in the one-dimensional case; for the multidimensional case we refer to [173] and [182].

We start by presenting a very brief introduction to summability of divergent series and integrals. It will serve as a motivation to the study of more general notions applicable to Schwartz distribution. There is a very rich and extensive literature on this traditional subject; for instance, the reader is referred to the classical and beautiful book of Hardy [66]. See also [170], [206] for connections with Fourier series and integrals.

We will then discuss the Cesàro behavior of distributions and some basic properties of this concept, it will be the base to define limits of distributions and distributional evaluations in the Cesàro sense. We follow the approach from [47] and [56], where we refer for a more complete account. In **5.1.2**, we shall confine ourselves with the definition for integral Cesàro orders and comparison with respect to power functions. We point out that the Cesàro behavior of distributions can also be defined for fractional orders [181], [183] and [184], in addition, regularly varying functions may be included in the theory [183].

5.1.1 Cesàro, Riesz, and Abel summability of series and integral

We shall discuss the summability methods by Abel, Cesàro and Riesz means for series and integrals.

Let us start with *Cesàro summability* . In general we say that a numerical series $\sum_{n=0}^{\infty} c_n$, possibly divergent, is summable to a complex number γ in the average, or Cesàro, sense of order 1, if the averages of its partial sums converge to γ, that is,

$$\lim_{n \to \infty} \frac{s_0 + s_1 + s_2 + \cdots + s_n}{n+1} = \gamma, \tag{5.2}$$

where $s_n = \sum_{j=0}^{n} c_j$, in such a case one writes

$$\sum_{n=0}^{\infty} c_n = \gamma \quad (C, 1). \tag{5.3}$$

It is elementary to check that if the series is convergent, then it is summable by the $(C, 1)$ method, but the converse is naturally false. For example, one may take $\sum_{n=0}^{\infty} (-1)^n$, which is evidently divergent; but its average converges to $1/2$, hence $\sum_{n=0}^{\infty} (-1)^n = 1/2$ $(C, 1)$.

The Cesàro method of summability is important in the analysis of several series expansions of functions and generalized functions; in particular for Fourier series. In fact, it is a famous result of Féjer that the Fourier series of a continuous function, although not necessarily convergent, is $(C, 1)$

summable to the value of the function at any point [206]. Furthermore, Kolmogorov proved [206] that there are functions in the class $L^1[0, 2\pi]$ whose Fourier series diverge everywhere; therefore, even in the case of classical functions, it is imperative the use of summability methods for the pointwise analysis of trigonometric series. In **5.2.1**, we will generalize Féjer's classical result to include periodic distributions, for that we will use higher order Cesàro means.

We can extend the $(C, 1)$ to higher order average means. There are several approaches, and all of them are equivalent. Perhaps the simplest, but analytically inadequate, is that of Hölder means. We can define recursively the sequences, $s_n^k := (\sum_{j=0}^n s_j^{k-1})/(n + 1)$, with $s_n^0 := s_n = \sum_{j=0}^n c_j$. Then, we call s_n^k the *Hölder means* of order k of the series, and say that $\sum_{n=0}^\infty c_n = \gamma$ (H, k), if $s_n^k \to \gamma$ as $n \to \infty$. As we remarked before, Hölder means present serious difficulties associated with their analytical manipulation [66], we shall therefore avoid their use in the future.

Another approach to the extension of (5.3) is via higher order *Cesàro means*. Given a series $\sum_{n=0}^\infty c_n$ we define its Cesàro means of order β, $\beta > -1$, by

$$C_n^\beta = \frac{\Gamma(\beta + 1)}{n^\beta} \sum_{j=0}^n \binom{\beta + j}{\beta} c_{n-j}, \qquad (5.4)$$

then we say that the series is Cesàro summable of order β to γ, and write $\sum_{n=0}^\infty c_n = \gamma$ (C, β), if $C_n^\beta \to \gamma$ as $n \to \infty$. An interesting example is $\sum_{n=0}^\infty (-1)^n n^\alpha$, $\alpha > -1$, which is (C, β) summable whenever $\beta > \alpha$, oscillates finitely when $\beta = \alpha$, and oscillates infinitely for $\beta < \alpha$; we refer to [66] for a proof of this fact.

We shall also discuss the method of Marcel *Riesz* by *typical means* [66]. Actually, the Riesz method will be the most important for us in the subsequent sections. Let $\langle \lambda_n \rangle_{n=0}^\infty$ be an increasing sequence of non-negative numbers such that $\lambda_n \to \infty$ as $n \to \infty$. We say that a series is summable by the Riesz means, with respect to $\langle \lambda_n \rangle$, of order $\beta \geq 0$ if

$$\lim_{x \to \infty} \sum_{0 \leq \lambda_n < x} c_n \left(1 - \frac{\lambda_n}{x} \right)^\beta = \gamma; \qquad (5.5)$$

and then we write

$$\sum_{n=0}^\infty c_n = \gamma \qquad (\text{R}, \langle \lambda_n \rangle, \beta). \qquad (5.6)$$

These three methods of summability can be compared. If $\beta = m\mathbf{N}$, then the (H, m) and the (C, m) methods are equivalent [66]. While if $\beta \geq 0$ and $\lambda_n = n$, the (C, β) and the $(\mathrm{R}, \langle n \rangle, \beta)$ methods sum the same series to the same value, and so they are also equivalent [66]. Here the use of a continuous variable in (5.5) is absolutely necessary for the equivalence [66]. The Riesz method has an advantage over the other two methods, it is easily generalizable to integrals, even to distributions as we shall see later in **5.1.3**. Therefore, we advise the reader that whenever we talk about Cesàro summability, even if we write (C, β), the means should be thought as Riesz means.

Let now f be a locally integrable function supported in $[0, \infty)$. Let $\beta > 0$. We write

$$\lim_{x \to \infty} f(x) = \gamma \quad (\mathrm{C}, \beta), \tag{5.7}$$

if

$$\lim_{x \to \infty} \frac{\beta}{x} \int_0^x f(t) \left(1 - \frac{t}{x}\right)^{\beta - 1} dt = \gamma. \tag{5.8}$$

Note that (5.8) basically says that $f^{(-\beta)}(x)$, the β-primitive of f, is asymptotic to $\gamma x^\beta / \Gamma(\beta + 1)$ as $x \to \infty$. The last approach will allows us to consider Cesàro limits of distributions in the future. Suppose that f is a function of local bounded variation, then its distributional derivative is a Radon measure, a continuous linear functional over the space of continuous functions with compact support, say $f' = \mu$. Hence integration by parts in (5.8) shows that it is equivalent to

$$\lim_{x \to \infty} \int_0^x \left(1 - \frac{t}{x}\right)^\beta d\mu(t) = \gamma. \tag{5.9}$$

The latter can be taken as the definition of the relation

$$\int_0^\infty d\mu(x) = \gamma \quad (\mathrm{C}, \beta). \tag{5.10}$$

Observe that (5.6) holds if and only (5.10) holds for the Radon measure $\mu = \sum_{n=0}^\infty c_n \delta(\cdot - \lambda_n)$.

We end this discussion by considering *Abel summability* of series [66]. For a series $\sum_{n=0}^\infty c_n$, we consider its Abel means, that is, the power series $\sum_{n=0}^\infty c_n r^n$. We say that the series is Abel summable to γ, if $\sum_{n=0}^\infty c_n r^n$ is convergent for $|r| < 1$ and the power series approaches to the limit γ at the boundary point $r = 1$, i.e.,

$$\lim_{r \to 1_-} \sum_{n=0}^\infty c_n r^n = \gamma, \tag{5.11}$$

we write

$$\sum_{n=0}^{\infty} c_n = \gamma \quad \text{(A)}. \tag{5.12}$$

It will be more convenient for us to write $r = e^{-y}$, so that the power series becomes a Dirichlet series. So, we have a natural extension for measures supported in $[0, \infty)$ in terms of the Laplace transform. We say that $\int_0^{\infty} d\mu(x)$ is Abel summable to γ and write

$$\int_0^{\infty} d\mu(x) = \gamma \quad \text{(A)}, \tag{5.13}$$

if for any $y > 0$ the integral $\int_0^{\infty} e^{-yt} d\mu(t)$ exists as an improper integral, and

$$\lim_{y \to 0^+} \int_0^{\infty} e^{-yt} d\mu(t) = \gamma. \tag{5.14}$$

When the Radon measure is given by $\sum_{n=0}^{\infty} c_n \delta(x - \lambda_n)$, we write

$$\sum_{n=0}^{\infty} c_n = \gamma \quad (\text{A}, \langle \lambda_n \rangle), \tag{5.15}$$

if (5.14) holds, that is, if the Dirichlet series $\sum_{n=0}^{\infty} c_n e^{-y\lambda_n}$ is convergent for $y > 0$ and it tends to γ as $y \to 0^+$.

We finally comment some inclusion between the Cesàro and Abel method of summation, if (5.10) holds then (5.13) is satisfied, this is shown below (Corollary 5.1). In the case of power series this fact is the well known *Abel's theorem* [66]. Naturally, the converse is not true. The reader may wish to verify that the series whose coefficients are given by those of the power series $e^{\frac{1}{1-r}} = \sum_{n=0}^{\infty} c_n r^n$ is an explicit example of a series which is (A) summable but not (C, β) summable [66], no matter what value of β be taken. Furthermore, in [48], it is constructed a series which is Abel summable with coefficients $c_n = O(n^m)$, but it is not (C, β) summable for any β. The study of additional hypotheses to ensure the converse of Abel's theorem motivated the beginning of the Tauberian theory. For instance, Littlewood Tauberian condition $c_n = O(1/n)$ together with Abel summability imply the convergence of the series [66]. We will obtain a simple and quick proof of Littlewood's theorem in Theorem 5.17, as a direct consequence of the use of generalized asymptotic behavior.

5.1.2 The Cesàro behavior of distributions

Let us define the *Cesàro behavior of distributions* at infinity . It is studied by using the order symbols $O\left(x^{\alpha}\right)$ and $o\left(x^{\alpha}\right)$ in the Cesàro sense.

Definition 5.1. Let $f \in \mathcal{D}'(\mathbf{R})$, $m \in \mathbf{N}_0$, and $\alpha \in \mathbf{R} \setminus \{-1, -2, -3, \ldots\}$. We say that $f(x) = O\left(x^{\alpha}\right)$ as $x \to \infty$ in the Cesàro sense of order m (in the (C, m) sense) and write

$$f(x) = O\left(x^{\alpha}\right) \quad (C, m), \quad x \to \infty, \tag{5.16}$$

if each primitive F of order m, i.e., $F^{(m)} = f$, is an ordinary function for large arguments and satisfies the ordinary order relation

$$F(x) = p(x) + O\left(x^{\alpha+m}\right), \quad x \to \infty, \tag{5.17}$$

for some suitable polynomial p of degree at most $m - 1$. Similarly for the little o symbol. We say that f is asymptotic to Cx^{α} as $x \to \infty$ in the Cesàro sense of order m and write

$$f(x) \sim Cx_{+}^{\alpha} \quad (C, m), \quad x \to \infty, \tag{5.18}$$

if we have $f(x) - Cx_{+}^{\alpha} = o(x^{\alpha})$ $(C, m), x \to \infty$.

Notice that if $\alpha > -1$, then the polynomial p is irrelevant in (5.17). A similar definition applies when $x \to -\infty$. One may also consider the case when $\alpha = -1, -2, -3, \ldots$ [56] , but we shall not do so. Obviously, if f vanishes for large arguments, then $f(x) = o(x^{\alpha})$ (C, m), for any m and α. When we do not want to make reference to the order m in (5.16) or (5.18), we simply write (C), meaning (C, m) for some m.

For $\alpha = 0$, we obtain the notion of Cesàro limits at infinity.

Definition 5.2. Let $f \in \mathcal{D}'(\mathbf{R})$ and $m \in \mathbf{N}_0$. We say that f has a limit γ at infinity in the Cesàro sense of order m (in the (C, m) sense) and write

$$\lim_{x \to \infty} f(x) = \gamma \quad (C, m),$$

if we have that $f(x) = \gamma + o(1)$ (C, m), $x \to \infty$.

We may also consider Cesàro limits as $x \to -\infty$. We will be mainly concerned with limits and not asymptotics in the Cesàro; however, we want discuss the close relation between Cesàro asymptotics and the quasi-asymptotic behavior. For further properties, we refer to [56].

The next theorem shows that the Cesàro behavior, in the case $\alpha > -1$, is totally determined by the quasi-asymptotic properties of the distribution on intervals being bounded at the left.

Proposition 5.1. *Let $f \in \mathcal{D}'(\mathbf{R})$, $m \in \mathbf{N}_0$, and $\alpha > -1$. Let f_+ be any distribution supported on an interval of the form $[a, \infty)$, $a \in \mathbf{R}$, coinciding with f for large arguments, i.e., in some open interval with finite left end point. Then, we have the next equivalences.*

(i) The following two conditions are equivalent,

$$f(x) = O(x^\alpha) \quad (\mathrm{C}), \quad x \to \infty, \tag{5.19}$$

and f_+ belongs to $\mathcal{S}'(\mathbf{R})$ and is quasi-asymptotically bounded of degree α, i.e.,

$$f_+(\lambda x) = O(\lambda^\alpha) \quad \text{as } \lambda \to \infty \text{ in } \mathcal{S}'(\mathbf{R}). \tag{5.20}$$

(ii) The conditions,

$$f(x) \sim Cx_+^\alpha \quad (\mathrm{C}), \quad x \to \infty, \tag{5.21}$$

and $f_+ \in \mathcal{S}'(\mathbf{R})$ has the quasi-asymptotic behavior

$$f_+(\lambda x) \overset{q}{\sim} C\lambda^\alpha x_+^\alpha \quad \text{as } \lambda \to \infty \text{ in } \mathcal{S}'(\mathbf{R}), \tag{5.22}$$

are equivalent.

Proof. We can assume that $f = f_+$, and so the equivalence between (5.19) and (5.20) is precisely the structural theorem for quasi-asymptotic boundedness, see Section **2.12** On the other hand, the equivalence between (5.21) and (5.22) is precisely the content of the structural theorem for quasi-asymptotic behavior of degree $\alpha > -1$ (cf. Theorem 2.2 in **2.2** and Theorem 2.26 in **2.10**). $\square$

When $\alpha < -1$, we do not exactly obtain a characterization in terms of quasi-asymptotics because delta terms could appear in the expansion.

Proposition 5.2. *Let $f \in \mathcal{D}'(\mathbf{R})$, $m \in \mathbf{N}_0$, and $\alpha < -1$, $-\alpha \notin \mathbf{N}$. Let f_+ be any distribution supported on an interval of the form $[a, \infty)$, $a \in \mathbf{R}$, coinciding with f for large arguments. Then, we have the next equivalences.*

(i) The following two conditions are equivalent,

$$f(x) = O\left(x_+^\alpha\right) \quad (\mathrm{C}), \quad x \to \infty, \tag{5.23}$$

and there exist $n > -\alpha$ constants $a_0, \ldots, a_{n-1}$, in general depending on f_+, such that f_+ has the quasi-asymptotic expansion

$$f_+(\lambda x) = \sum_{j=0}^{n-1} a_j \frac{\delta^{(j)}(x)}{\lambda^{j+1}} + O(\lambda^\alpha) \quad \text{as } \lambda \to \infty \text{ in } \mathcal{S}'(\mathbf{R}). \quad (5.24)$$

(ii) The conditions,

$$f(x) \sim C x_+^\alpha \quad (C), \quad x \to \infty, \quad (5.25)$$

and the existence of constants $n > -\alpha$ constants $a_0, \ldots, a_{n-1}$, in general depending on f_+, such that

$$f_+(\lambda x) = C \lambda^\alpha x_+^\alpha + \sum_{j=0}^{n-1} a_j \frac{\delta^{(j)}(x)}{\lambda^{j+1}} + o(\lambda^\alpha) \quad \text{as } \lambda \to \infty \text{ in } \mathcal{S}'(\mathbf{R}),$$

$$(5.26)$$

are equivalent.

Proof. We can assume $f = f_+$. We only show (ii), the proof of (i) is similar to this case and is left to the reader. Assume (5.25), then there exist G_1, G_2, $m > -\alpha - 1$, and m constants $c_0, \ldots, c_{m-1}$ such that $f = G_1^{(m)} + G_2^{(m)}$, G_1 has compact support, G_2 is a locally integral functions with support on $[0, \infty)$, and

$$G_2(x) = \sum_{j=0}^{m-1} c_j \frac{x^j}{j!} + C \frac{\Gamma(\alpha+1)}{\Gamma(m+\alpha+1)} x^{m+\alpha} + o(x^{m+\alpha}), \quad x \to \infty.$$

Since G_1 has compact support, then $G_1(\lambda x) = O(\lambda^{-1})$, in $\mathcal{S}'(\mathbf{R})$, and so $G^{(m)}(\lambda x) = O(\lambda^{-m-1}) = o(\lambda^\alpha)$; then, since it does not contribute for (5.26), we can assume that $G_1 = 0$. On the other hand, the ordinary asymptotic expansion of G_2 implies

$$G_2(\lambda x) = \sum_{j=0}^{m-1} c_j \frac{(\lambda x)_+^j}{j!} + C \frac{\Gamma(\alpha+1)}{\Gamma(m+\alpha+1)} (\lambda x)_+^{m+\alpha} + o(\lambda^{m+\alpha})$$

in $\mathcal{S}'(\mathbf{R})$. Differentiating m-times the above asymptotic formula, and discarding the irrelevant constants, we obtain (5.26) with $a_j = c_{m-1-j}$. The converse follows from the structural theorem for quasi-asymptotics (cf. Theorem 2.2) applied to $f_+ - \sum_{j=0}^{n-1} a_j \delta^{(j)}$. $\square$

5.1.3 *Summability of distributional evaluations*

We now study two methods of summability for distributional evaluations, the two-sided Cesàro method, and Abel summability. Two more methods will be introduced in **5.2** (Definitions 5.6 and 5.8).

We start with summability in the Cesàro sense. First we assume that our distributions have support bounded at the left. Recall that H denotes the Heaviside function, i.e., the characteristic function of $(0, \infty)$.

Definition 5.3. Let $f \in \mathcal{D}'(\mathbf{R})$ have support bounded at the left. Let $\phi \in \mathcal{E}(\mathbf{R})$ and $m \in \mathbf{N}_0$. We say the evaluation $\langle f(x), \phi(x) \rangle$ has a value γ in the Cesàro sense of order m, and write

$$\langle f(x), \phi(x) \rangle = \gamma \quad (\mathrm{C}, m) \tag{5.27}$$

if the first order primitive $F = (\phi f)^{(-1)} = (\phi f) * H$, the first order primitive of ϕf with support bounded at the left, satisfies $\lim_{x \to \infty} F(x) = \gamma$ (C, m).

Example II.1. Let μ be a Radon measure with support on $[0, \infty)$. Then $\int_0^\infty d\mu(x) = \gamma$ (C, m) if and only if $\langle \mu(x), 1 \rangle = \gamma$ (C, m). In particular

$$\sum_{n=0}^\infty c_n = \gamma \quad (\mathrm{R}, \langle \lambda_n \rangle, m)$$

if and only if

$$\left\langle \sum_{n=0}^\infty c_n \delta(x - \lambda_n), 1 \right\rangle = \gamma \quad (\mathrm{C}, m).$$

If f has support bounded at the right then we say that $\langle f(x), \phi(x) \rangle$ (C) exists if and only if $\langle f(-x), \phi(-x) \rangle = \gamma$ (C) exists and we define $\langle f(x), \phi(x) \rangle = \gamma$ (C).

The distributional evaluations with respect to compactly supported distribution can always be computed in the (C) sense, actually with order $m = 0$.

Lemma 5.1. *Let $f \in \mathcal{E}'(\mathbf{R})$ and $\phi \in \mathcal{E}(\mathbf{R})$. Then $\langle f(x), \phi(x) \rangle$ $(\mathrm{C}, 0)$ always exists.*

Proof. We can assume that $\phi \equiv 1$. Consider $f^{(-1)}$, it is obviously constant for large arguments, we must show that it satisfies $f^{(-1)} = \langle f(x), 1 \rangle$

(a constant distribution) on certain interval (a, ∞). Decompose $f^{(-1)}(x) = g(x) + cH(x - a)$, where g has compact support and c and a are constants. Then $\langle f(x), 1 \rangle = \langle g'(x), 1 \rangle + \langle cH'(x - a), 1 \rangle = 0 + c\langle \delta(x - a), 1 \rangle = c$, from where the result follows. $\qquad \square$

We now define two-sided Cesàro evaluations.

Definition 5.4. Let $f \in \mathcal{D}'(\mathbf{R})$, $\phi \in \mathcal{E}(\mathbf{R})$, and $m \in \mathbf{N}_0$. We say the evaluation $\langle f(x), \phi(x) \rangle$ exists in the Cesàro sense of order m if there is a decomposition $f = f_- + f_+$, $\mathrm{supp} f_- \subseteq (-\infty, 0]$ and $\mathrm{supp} f_+ \subseteq [0, \infty)$, such that both evaluations $\langle f_\pm(x), \phi(x) \rangle = \gamma_\pm$ (C, m) exist. In this case we write

$$\langle f(x), \phi(x) \rangle = \gamma \ (\mathrm{C}, m), \tag{5.28}$$

where $\gamma = \gamma_- + \gamma_+$.

We must check the consistency of Definition 5.4. Let $f = f_1 + f_2 = g_1 + g_2$ be two decompositions such that f_2 and g_2 have supports bounded at the left, respectively, f_1 and g_1 have supports bounded at the right. Then $h = g_1 - f_1 = f_2 - g_2$ has compact support. If both $\langle f_j(x), \phi(x) \rangle = \gamma_j$ (C, m) exist, then, by Lemma 5.1, both $\langle g_j(x), \phi(x) \rangle = \beta_j$ (C, m) exist, and we have the two equalities $\beta_1 = \gamma_1 + \beta$ and $\beta_2 = \gamma_2 - \beta$, where $\beta = \langle f_j(x), \phi(x) \rangle$. Hence the number $\gamma = \gamma_1 + \gamma_2 = \beta_1 + \beta_2$ is independent on the choice of the decomposition.

Let us now define Abel summability for distributional evaluations.

Definition 5.5. Let $f \in \mathcal{D}'(\mathbf{R})$ and $\phi \in \mathcal{E}(\mathbf{R})$. We say the evaluation $\langle f(x), \phi(x) \rangle$ exists in the Abel sense if there is a decomposition $f = f_- + f_+$, $\mathrm{supp} f_- \subseteq (-\infty, 0]$ and $\mathrm{supp} f_+ \subseteq [0, \infty)$, such that both $e^{\mp yx} \phi(x) f_\pm \in \mathcal{S}'(\mathbf{R})$, for each $y > 0$, and

$$\lim_{y \to 0^+} \left(\langle \phi(x) f_-(x), e^{yx} \rangle + \langle \phi(x) f_+(x), e^{-yx} \rangle \right) = \gamma, \tag{5.29}$$

in this case we write $\langle f(x), \phi(x) \rangle = \gamma$ (A).

The notion of distributional evaluations in the Cesàro sense admits a characterization in terms of the quasi-asymptotic behavior.

Proposition 5.3. *Let $f \in \mathcal{D}'(\mathbf{R})$ and $\phi \in \mathcal{E}(\mathbf{R})$. Then $\langle f(x), \phi(x) \rangle = \gamma$ (C) if and only if there exist a decomposition $f = f_- + f_+$, where* supp$f_- \subseteq (-\infty, 0]$ and supp$f_+ \subseteq [0, \infty)$, and a constant β such that the following quasi-asymptotic behaviors hold

$$\phi(\lambda x) f_+(\lambda x) \overset{q}{\sim} \left(\frac{\gamma}{2} + \beta\right) \frac{\delta(x)}{\lambda} \quad \text{as } \lambda \to \infty \text{ in } \mathcal{S}'(\mathbf{R}) \tag{5.30}$$

and

$$\phi(\lambda x) f_-(\lambda x) \overset{q}{\sim} \left(\frac{\gamma}{2} - \beta\right) \frac{\delta(x)}{\lambda} \quad \text{as } \lambda \to \infty \text{ in } \mathcal{S}'(\mathbf{R}). \tag{5.31}$$

In particular, we obtain that $\phi f \in \mathcal{S}'(\mathbf{R})$ and it has the quasi-asymptotic behavior,

$$\phi(\lambda x) f(\lambda x) \overset{q}{\sim} \gamma \frac{\delta(x)}{\lambda} \quad \text{as } \lambda \to \infty \text{ in } \mathcal{S}'(\mathbf{R}). \tag{5.32}$$

Proof. We may assume that $\phi \equiv 1$. Put $f_-^{(-1)}$ equal to the primitive of $f_-(-x)$ with support on $[0, \infty)$. Because of the assumptions on the supports, note that (5.30) and (5.31) are equivalent to $\lim_{\lambda \to \infty} f_\pm^{(-1)}(\lambda x) = ((\gamma/2) \pm \beta) H(x)$ in $\mathcal{S}'(\mathbf{R})$. By Proposition 5.1, the latter are equivalent to $\lim_{\lambda \to \infty} f_\pm^{(-1)}(x) = (\gamma/2) \pm \beta$ (C), which are equivalent to $\langle f_\pm(x), 1 \rangle = (\gamma/2) \pm \beta$ (C). And so we obtain the equivalence with $\langle f(x), \phi(x) \rangle = \gamma$ (C). $\square$

We can use Proposition 5.3 to obtain Abel's theorem in the context of distributional evaluations. The converse is false [48].

Corollary 5.1. *Let $f \in \mathcal{D}'(\mathbf{R})$ and $\phi \in \mathcal{E}(\mathbf{R})$. Suppose that $\langle f(x), \phi(x) \rangle = \gamma$ (C), then $\langle f(x), \phi(x) \rangle = \gamma$ (A).*

Proof. Using Proposition 5.3, we obtain that, as $\lambda \to \infty$,

$$\left\langle \phi(x) f_-(x), e^{\frac{x}{\lambda}} \right\rangle + \left\langle \phi(x) f_+(x), e^{-\frac{x}{\lambda}} \right\rangle$$

$$= \lambda \left(\langle \phi(\lambda x) f_-(\lambda x), e^x \rangle + \langle \phi(\lambda x) f_+(\lambda x), e^{-x} \rangle \right)$$

$$= \left(\frac{\gamma}{2} - \beta\right) \langle \delta(x), e^x \rangle + \left(\frac{\gamma}{2} + \beta\right) \langle \delta(x), e^{-x} \rangle + o(1)$$

$$= \gamma + o(1). \qquad \square$$

5.2 Summability of the Fourier transform and distributional point values

We have arrived to the main section of **5**. We characterize the value of a tempered distribution at point in terms of the summability its Fourier transform. Let us adopt the following convention to denote the value of a distribution at a point. Let $f \in \mathcal{D}'(\mathbf{R})$ have the value γ at the point $x = x_0$ in the sense of Lojasiewicz [93, 94], that is, the following quasi-asymptotic behavior is satisfied

$$f(x_0 + \varepsilon x) = \gamma + o(1) \quad \text{as } \varepsilon \to 0 \text{ in } \mathcal{D}'(\mathbf{R}). \tag{5.33}$$

We will refer to Lojasiewicz point values as *distributional point values*, and will use the following notation for the existence of the distributional point value at $x = x_0$ with value γ,

$$f(x_0) = \gamma, \quad \text{distributionally.} \tag{5.34}$$

In order to characterize distributional point values. We will define an intermediate method of summability, weaker than Cesàro summability but stronger than Abel summability. We discuss applications of such a characterization to the study of convergence of Fourier series and integrals of distributions. Several Tauberian type results are also presented, in particular we obtain a quick simple distributional proof of the celebrated Littlewood's theorem [92, 177]. In the last part, we also formulate and solve the Hardy-Littlewood problem for Cesàro summability [68, 181] in the framework of distributions. The original sources for the results of this section are [172, 173, 176, 177, 181]. Finally, we mention references [181, 184], where more refined versions can be found, especially some results concerning to the order of summability that we do not treat here.

5.2.1 Characterization of distributional point values of tempered distributions

Let $f \in \mathcal{S}'(\mathbf{R})$ have distributional point γ at x_0. Then, the quasi-asymptotics (5.33) actually holds in the space $\mathcal{S}'(\mathbf{R})$(cf. Theorem 2.35 in **2.11.1**). Therefore, we can take Fourier transform in (5.33) and obtain the equivalent quasi-asymptotic expression

$$e^{i\lambda x_0 x} \hat{f}(\lambda x) \overset{q}{\sim} 2\pi\gamma \frac{\delta(x)}{\lambda} \quad \text{as } \lambda \to \infty \text{ in } \mathcal{S}'(\mathbf{R}). \tag{5.35}$$

Let us state this simple, but useful, observation as a lemma

Lemma 5.2. *Let* $f \in \mathcal{S}'(\mathbf{R})$. *Then,* $f(x_0) = \gamma$, *distributionally, if and only if the Fourier transform satisfies the quasi-asymptotic behavior (5.35).*

Therefore, on the Fourier side, distributional point values look like (5.35). Since our ultimate goal is to characterize distributional point values by certain type of summability of the Fourier transform, it is clear that our summability method should provide a characterization of the quasi-asymptotic behavior

$$g(\lambda x) \overset{q}{\sim} \gamma \frac{\delta(x)}{\lambda} \quad \text{as } \lambda \to \infty \text{ in } \mathcal{S}'(\mathbf{R}) \,. \tag{5.36}$$

A naive first approach to this problem might lead us to consider directly Cesàro summability. However, Proposition 5.3 tell us that it is not going to work: Cesàro summability is too strong to give a characterization. Let us be more precise on this matter. Observe that if $\langle g(x), 1 \rangle = \gamma$ (C), then Proposition 5.3 implies (5.36). However the converse is not true.

Example II.2. Consider the regular distributions $g(x) = (1/(x \log |x|))$ $H(|x| - 3)$. Note that for any $m \geq 0$,

$$\int_3^x \frac{1}{t \log t} \left(1 - \frac{t}{x} \right)^m dt = -\frac{1}{3 \log 3} + \frac{m}{x} \int_3^x \log(\log t) \left(1 - \frac{t}{x} \right)^{m-1} dt$$

$$\sim \log(\log x), \quad x \to \infty \,.$$

Then, the evaluation $\langle g(x), 1 \rangle$ does not exist in the Cesàro sense. However, $g(\lambda x) = o(\lambda^{-1})$ as $\lambda \to \infty$ in $\mathcal{S}'(\mathbf{R})$. In fact, if $\phi \in \mathcal{S}(\mathbf{R})$, then

$$\langle g(\lambda x), \phi(x) \rangle = \frac{1}{\lambda} \int_{\frac{3}{\lambda}}^{\infty} \frac{\phi(t) - \phi(-t)}{t \log(\lambda t)} dt = o\left(\frac{1}{\lambda} \right), \quad \lambda \to \infty \,.$$

Therefore, the Cesàro summability is not adequate for the characterization of distributional point values. If we now think carefully in (5.36), it is actually a quasi-asymptotic behavior of degree -1 with respect to the trivial slowly varying function, $L \equiv 1$, and we have already characterized the structure of such behaviors in **2.10.5**. Let us state the structural theorem for this particular case.

Proposition 5.4. *Let $g \in \mathcal{D}'(\mathbf{R})$, then,*

$$f(\lambda x) \overset{q}{\sim} \gamma \frac{\delta(x)}{\lambda} \quad \text{as } \lambda \to \infty \text{ in } \mathcal{D}'(\mathbf{R}) \,, \tag{5.37}$$

if and only if there exist $m \in \mathbf{N}$ and $(m + 1)$-primitive G_{m+1} of g, i.e., $G^{(m+1)} = f$, which is locally integrable for large positive and negative arguments, such that following limit holds for each $a > 0$

$$\lim_{x \to \infty} \left(a^{-m} G_{m+1}(ax) - (-1)^m G_{m+1}(-x) \right) = \gamma \,. \tag{5.38}$$

Furthermore, the above relation is satisfied if and only if there exists an asymptotically homogeneous function c of degree zero, i.e., $c(ax) = c(x) + o(1)$ as $x \to \infty$ for each $a > 0$, such that

$$G_{m+1}(x) = \frac{\gamma \mathrm{sgn} x}{2m!} x^m + c(|x|)\frac{x^m}{m!} + o(|x|^m), \quad |x| \to \infty, \tag{5.39}$$

in the ordinary sense. In such a case g is a tempered distribution and (5.37) holds in $\mathcal{S}'(\mathbf{R})$.

Proposition 5.4 implicitly suggests the method of summability: It should involve (5.38). Let us reformulate (5.38). Set $G = G^{(m)}$, then G is a first order primitive of g, and in fact (5.38) can be rewritten as

$$\lim_{x \to \infty} (G(ax) - G(-x) = \gamma \quad (C, m).$$

Hence, we have found the right summability method!

Definition 5.6. Let $g \in \mathcal{D}'(\mathbf{R})$, $\phi \in \mathcal{E}(\mathbf{R})$ and $m \in \mathbf{N}_0$. We say that the special value of $\langle g(x), \phi(x) \rangle$ exists in the Cesàro sense of order m (e.v. Cesàro sense), and write

$$\text{e.v.} \langle g(x), \phi(x) \rangle = \gamma \quad (C, m), \tag{5.40}$$

if for some primitive G of ϕg, i.e., $G' = \phi g$, and each $a > 0$, we have

$$\lim_{x \to \infty} (G(ax) - G(-x)) = \gamma \quad (C, m). \tag{5.41}$$

As a corollary we obtain.

Corollary 5.2. *Let $g \in \mathcal{D}'(\mathbf{R})$, $\phi \in \mathcal{E}(\mathbf{R})$. Then*

$$\text{e.v.} \langle g(x), \phi(x) \rangle = \gamma \quad (C) \tag{5.42}$$

if and only if

$$\phi(\lambda x)g(\lambda x) \overset{q}{\sim} \gamma \frac{\delta(x)}{\lambda} \quad \text{as } \lambda \to \infty \text{ in } \mathcal{S}'(\mathbf{R}). \tag{5.43}$$

In addition, we have that $\phi g \in \mathcal{S}'(\mathbf{R})$.

As expected, the Cesàro method is strictly stronger than the e.v Cesàro summability (see also Example II.2).

Proposition 5.5. *Let $g \in \mathcal{D}'(\mathbf{R})$, $\phi \in \mathcal{E}(\mathbf{R})$. Any evaluation summable (C, m) is also summable in e.v.(C, m), that is, $\langle g(x), \phi(x) \rangle = \gamma \quad (C, m)$, implies e.v. $\langle g(x), \phi(x) \rangle = \gamma \quad (C, m)$.*

Proof. Let G be a first order primitive of ϕg. Decompose it as $G = G_- + G_+$, with supp $G_- \subseteq (-\infty, 0]$ and supp $G_- \subseteq [0, \infty)$. Then, by Proposition 5.3,

$$\lim_{x \to \infty} \pm G_{\pm}(\pm x) = \frac{\gamma}{2} \pm \beta \quad (C, m),$$

for some β. Thus

$$\lim_{x \to \infty} (G(ax) - G(-x)) = \lim_{x \to \infty} (G_+(ax) - G_-(-x)) = \gamma \quad (C, m). \qquad \square$$

In summary, we succeeded characterizing distributional point values in terms of the summability of the Fourier inversion formula.

Theorem 5.1. *Let $f \in \mathcal{S}'(\mathbf{R})$. We have $f(x_0) = \gamma$, distributionally, if and only if there exists an $m \in \mathbf{N}_0$ such that*

$$\frac{1}{2\pi} \text{ e.v.} \left\langle \hat{f}(x), e^{ix_0 x} \right\rangle = \gamma \quad (C, m). \tag{5.44}$$

Proof. Combine Lemma 5.2 with Corollary 5.2. $\qquad \square$

Let $g = \mu$ be a Radon measure. It convenient in this case to write

$$\text{e.v.} \int_{-\infty}^{\infty} \phi(x) d\mu(x) = \gamma \quad (C, m) \tag{5.45}$$

for (5.40). When $m = 0$, we suppress $(C, 0)$ from the notation, and simply write

$$\text{e.v.} \int_{-\infty}^{\infty} \phi(x) d\mu(x) = \gamma.$$

In particular, if $\mu = \sum_{n=-\infty}^{\infty} c_n \delta(\cdot - n)$ and $\phi \equiv 1$, we use the notation

$$\text{e.v.} \sum_{n=-\infty}^{\infty} c_n = \gamma \quad (C, m), \tag{5.46}$$

omitting again $(C, 0)$ when $m = 0$.

Observe that if we use the family of summability kernels

$$\phi_a^m(x) = (1+x)^m (H(-x) - H(-1-x)) + \left(1 - \frac{x}{a}\right)^m (H(x) - H(x-a)), \tag{5.47}$$

where H is the Heaviside function, then (5.45) holds if and only if

$$\lim_{x \to \infty} \int_{-\infty}^{\infty} \phi_a^m \left(\frac{t}{x}\right) \phi(t) d\mu(t) = \gamma, \quad \text{for each } a > 0. \tag{5.48}$$

For series we obtain that (5.46) holds if and only if

$$\lim_{x \to \infty} \sum_{n=-\infty}^{\infty} \phi_a^m\left(\frac{n}{x}\right) c_n = \gamma, \quad \text{for each } a > 0. \tag{5.49}$$

Let us now discuss some immediate consequences of Theorem 5.1.

Corollary 5.3. *Let* $f \in \mathcal{S}'(\mathbf{R})$ *be such that* $\hat{f} = \mu$ *is a Radon measure. Then, we have* $f(x_0) = \gamma$, *distributionally, if and only if there exists an* $m \in \mathbf{N}_0$ *such that*

$$\frac{1}{2\pi} \text{e.v.} \int_{-\infty}^{\infty} e^{ix_0 x} d\mu(x) = \gamma \quad (C, m), \tag{5.50}$$

or which amounts to the same,

$$\lim_{x \to \infty} \frac{1}{2\pi} \int_{-\infty}^{\infty} e^{ix_0 t} \phi_a^m\left(\frac{t}{x}\right) d\mu(t) = \gamma, \quad \text{for each } a > 0. \tag{5.51}$$

The next corollary is a result of R. Estrada [45], the characterization of Fourier series having a distributional point value.

Corollary 5.4. *Let* $f \in \mathcal{S}'(\mathbf{R})$ *be a* 2π-*periodic distribution having Fourier series*

$$f(x) = \sum_{n=-\infty}^{\infty} c_n e^{inx}. \tag{5.52}$$

Then, we have $f(x_0) = \gamma$, *distributionally, if and only if there exists an* $m \in \mathbf{N}_0$ *such that*

$$\text{e.v.} \sum_{n=-\infty}^{\infty} c_n e^{inx_0} = \gamma \quad (C, m), \tag{5.53}$$

or which amounts to the same,

$$\lim_{x \to \infty} \sum_{n=-\infty}^{\infty} \phi_a^m\left(\frac{n}{x}\right) c_n e^{inx_0} = \gamma, \quad \text{for each } a > 0. \tag{5.54}$$

Proof. We have that $\hat{f}(x) = 2\pi \sum_{n=-\infty}^{\infty} c_n \delta(x - n)$, the rest follows from Corollary 5.3. $\square$

Let us state Corollary 5.3 when $\hat{f} \in L^1_{loc}(\mathbf{R})$. A particular case is obtained if $f \in L^p(\mathbf{R})$ with $1 \leq p \leq 2$, since $\hat{f} \in L^q(\mathbf{R})$ with $q = p/(p-1)$ [170].

Corollary 5.5. *Let* $f \in \mathcal{S}'(\mathbf{R})$ *be such that* $\hat{f} \in L^1_{loc}(\mathbf{R})$. *Then, we have* $f(x_0) = \gamma$, *distributionally, if and only if there exists an* $m \in \mathbf{N}_0$ *such that*

$$\frac{1}{2\pi} \text{e.v.} \int_{-\infty}^{\infty} e^{ix_0 x} \hat{f}(x) dx = \gamma \quad (C, m), \tag{5.55}$$

or which amounts to the same,

$$\lim_{x \to \infty} \frac{1}{2\pi} \int_{-\infty}^{\infty} \phi_a^m \left(\frac{t}{x} \right) e^{ix_0 t} \hat{f}(t) dt = \gamma, \quad \text{for each } a > 0. \tag{5.56}$$

It is important to observe that the characterization of distributional point values is given in terms of slightly asymmetric means and that the corresponding result for symmetric means does not hold. The result for separate integration over both the positive and negative parts of the spectrum does not hold either (we already discussed the latter in Example II.2). Let us provide two further examples.

Example II.3. If $f \in \mathcal{S}'(\mathbf{R})$ and $f(x_0) = \gamma$, distributionally, then by taking $a = 1$ we obtain that the symmetric means converge to γ, in the Cesàro sense, so that, in case $\hat{f}(t)e^{ix_0 t}$ is locally integrable,

$$\lim_{x \to \infty} \frac{1}{2\pi} \int_{-x}^{x} \hat{f}(t) e^{ix_0 t} dt = \gamma \quad (C, m), \tag{5.57}$$

for some m. However, (5.57) does not imply the existence of the distributional value $f(x_0)$. A simple example is provided by $f(x) = \delta'(x)$ at $x = 0$, since $\hat{f}(t) = it$, so that (5.57) exists and equals 0, but $f(0)$ does not exist.

Example II.4. If $f \in \mathcal{S}'(\mathbf{R})$ and the two Cesàro limits

$$\lim_{x \to +\infty} \frac{1}{2\pi} \int_0^x \hat{f}(t) e^{ix_0 t} dt = \gamma_+ \quad (C, k), \tag{5.58}$$

$$\lim_{x \to +\infty} \frac{1}{2\pi} \int_{-x}^0 \hat{f}(t) e^{ix_0 t} dt = \gamma_- \quad (C, k), \tag{5.59}$$

exist then the distributional value $f(x_0)$ exists and equals $\gamma = \gamma_+ + \gamma_-$. However, the existence of the distributional point value $f(x_0)$ does not imply the existence of both Cesàro limits. For instance, if

$$f(x) = \int_0^{\infty} \frac{\sin xt \, dt}{t \ln (t^2 + a^2)}, \tag{5.60}$$

for some $a > 1$, then f is continuous and $f(0) = 0$, but $\hat{f}(t) = -\pi i t^{-1}$ $(\log(t^2 + a^2))^{-1}$, and in that case both limits (5.58) and (5.59) give infinite results, $|\gamma_+| = |\gamma_-| = \infty$.

There is one case in which the distributional point values can be characterized by Cesàro summability of the Fourier inversion formula, not needing the asymmetric means, that is, when the distribution has support on a half-ray.

Theorem 5.2. Let $f \in \mathcal{S}'(\mathbf{R})$ have support bounded at the left . We have $f(x_0) = \gamma$, distributionally, if and only if there exists an $m \in \mathbf{N}_0$ such that

$$\frac{1}{2\pi} \left\langle \hat{f}(x), e^{ix_0 x} \right\rangle = \gamma \quad (C, m). \tag{5.61}$$

Proof. The converse follows from Proposition 5.5. Let now F be the primitive of $(1/2\pi)e^{ix_0 x}\hat{f}$ with support bounded at the left. Then, by Theorem 5.1, we have that

$$\lim_{x \to \infty} F(x) = \lim_{x \to \infty} (F(x) - F(-x)) = \gamma \quad (C, m). \qquad \square$$

Corollary 5.6. Let $f \in \mathcal{S}'(\mathbf{R})$ be such that $\hat{f} = \mu$ is a Radon measure supported on $[0, \infty)$. Then, we have $f(x_0) = \gamma$, distributionally, if and only if

$$\frac{1}{2\pi} \int_0^\infty e^{ix_0 x} d\mu(x) = \gamma \quad (C). \tag{5.62}$$

We also obtain a corresponding result for Riesz summability.

Corollary 5.7. Let $f = \sum_{n=0}^\infty c_n e^{i\lambda_n x}$ in $\mathcal{S}'(\mathbf{R})$, where $\lambda_n \nearrow \infty$. Then, we have $f(x_0) = \gamma$, distributionally, if and only if

$$\sum_{n=0}^\infty c_n e^{i\lambda_n x_0} = \gamma \quad (R, \langle \lambda_n \rangle). \tag{5.63}$$

These ideas can be applied to study some types of multiple series. It is not our scope to investigate problems in several variables (which are actually still open questions, see Open Problem 5.1 below), but the next theorem shows that some problems in summability of multiple series can be solved using this theory. The next result is an example of that.

Theorem 5.3. Let $f \in \mathcal{S}'(\mathbf{R})$ and be ρ be a real-valued function defined on $\mathbf{R}^d$ which only takes non-negative values at points $j \in \mathbf{Z}^d$. Suppose that

$$f(x) = \sum_{j \in \mathbf{Z}^d} c_j e^{i\rho(j)x} \text{ in } \mathcal{S}'(\mathbf{R}).$$

Enumerate the image $\rho(\mathbf{Z}^d)$ by an increasing sequence $\langle \lambda_n \rangle_{n=0}^{\infty}$ Then, $f(x_0) = \gamma$, distributionally, if and only if there exists an $m \in \mathbf{N}$ such that

$$\sum_{n=0}^{\infty} \left(\sum_{\rho(j)=\lambda_n} c_j e^{i\rho(j)x_0} \right) = \gamma \quad (\mathrm{R}, \langle \lambda_n \rangle, m), \tag{5.64}$$

or equivalently,

$$\lim_{\lambda \to \infty} \sum_{\rho(j) \leq \lambda} c_j e^{i\rho(j)x_0} \left(1 - \frac{\rho(j)}{\lambda} \right)^m = \gamma. \tag{5.65}$$

Proof. It follows immediately from Corollary 5.7, since

$$f(x) = \sum_{n=0}^{\infty} \left(\sum_{\rho(j)=\lambda_n} c_j e^{i\rho(j)x_0} \right) e^{i\lambda_n x}. \qquad \square$$

If in particular we take $\rho(y) = |y|^2$ (here $y \in \mathbf{R}^d$ and $|\cdot|$ is the standard euclidean norm) in Theorem 5.3, we obtain that $f(x_0) = \gamma$, distributionally, if and only if the multiple series is Bochner-Riesz summable by spherical means [25].

We end this subsection by pointing out an important open problem. While we have completely characterized the value of a one dimensional distribution at point by the summability of its Fourier transform in the e.v. Cesàro sense, the corresponding multidimensional problem is still an open question.

Open Problem 5.1. *For distributions in $\mathcal{S}'(\mathbf{R}^d)$, $d > 1$, find a characterization of distributional point values in terms of the summability of the multidimensional Fourier transform. The method of summability is part of the open question.*

5.2.2   Abel summability

We now analyze Abel summability of the Fourier inversion formula in the presence of distributional point values.

Let us first state an interesting theorem, which we may be regarded as a decomposition theorem for the quasi-asymptotic behavior (5.36).

Theorem 5.4. *Let* $g \in \mathcal{S}'(\mathbf{R})$. *Then*

$$g(\lambda x) = \gamma \frac{\delta(x)}{\lambda} + o\left(\frac{1}{\lambda}\right) \quad \text{as } \lambda \to \infty \text{ in } \mathcal{S}'(\mathbf{R}) \tag{5.66}$$

if and only if there exist a decomposition $g = g_- + g_+$, *where* supp $g_- \subseteq (-\infty, 0]$ *and* supp $g_+ \subseteq [0, \infty)$, *and an asymptotically homogeneous function* c *of degree zero, i.e.,* $c(ax) = c(x) + o(1)$ *as* $x \to \infty$ *for each* $a > 0$, *such that the following asymptotic relations hold*

$$g_+(\lambda x) = \left(\frac{\gamma}{2} + c(\lambda)\right)\frac{\delta(x)}{\lambda} + o\left(\frac{1}{\lambda}\right) \quad \text{as } \lambda \to \infty \text{ in } \mathcal{S}'(\mathbf{R}) \tag{5.67}$$

and

$$g_-(\lambda x) = \left(\frac{\gamma}{2} - c(\lambda)\right)\frac{\delta(x)}{\lambda} + o\left(\frac{1}{\lambda}\right) \quad \text{as } \lambda \to \infty \text{ in } \mathcal{S}'(\mathbf{R}). \tag{5.68}$$

Proof. Proposition 5.4 implies the existence of an $(m + 1)$-primitive of g, say G, such that

$$G(x) = \frac{\gamma \text{sgn} x}{2m!} x^m + c(|x|)\frac{x^m}{m!} + o(|x|^m), \quad |x| \to \infty. \tag{5.69}$$

Set $G_\pm(x) = G(x)H(\pm x)$, where H is the Heaviside function. We have that (cf. **2.10.2**)

$$c(\lambda x)H(x) = c(\lambda)H(x) + o(1) \quad \text{as } \lambda \to \infty \text{ in } \mathcal{S}'(\mathbf{R}),$$

which implies

$$G_\pm(\lambda x) = (\pm 1)^{m+1}\frac{\gamma}{2m!}(\lambda x)_\pm^m + (\pm 1)^m c(\lambda)\frac{(\lambda x)_\pm^m}{m!} + o(\lambda^m) \quad \text{as } \lambda \to \infty.$$

in $\mathcal{S}'(\mathbf{R})$. If we set $g_\pm = G_\pm^{(m+1)}$, differentiating $(m+1)$-times the last two asymptotic expressions we obtain (5.67) and (5.68). Conversely, setting $h_\pm(x) = g_\pm(x) \mp (c(x)H(x))'$, an application of the structural theorem for the quasi-asymptotic behavior of degree -1 with one-sided support to each $h_\pm$ implies that there exists m such that (5.69) is satisfied, and hence (5.66) follows. $\square$

Due to Corollary 5.2, Theorem 5.5 may also be stated in the following equivalent form.

Theorem 5.5. *Let $g \in \mathcal{D}'(\mathbf{R})$ and $\phi \in \mathcal{E}(\mathbf{R})$. Then e.v. $\langle f(x), \phi(x) \rangle = \gamma$ (C) if and only if there exist a decomposition $g = g_- + g_+$, where supp $g_- \subseteq (-\infty, 0]$ and supp $g_+ \subseteq [0, \infty)$, and an asymptotically homogeneous function c of degree zero, i.e., $c(ax) = c(x) + o(1)$ as $x \to \infty$ for each $a > 0$, such that the following asymptotic relations hold*

$$\phi(\lambda x) g_+(\lambda x) = \left(\frac{\gamma}{2} + c(\lambda) \right) \frac{\delta(x)}{\lambda} + o\left(\frac{1}{\lambda} \right) \quad \text{as } \lambda \to \infty \text{ in } \mathcal{S}'(\mathbf{R}) \quad (5.70)$$

and

$$\phi(\lambda x) g_-(\lambda x) = \left(\frac{\gamma}{2} - c(\lambda) \right) \frac{\delta(x)}{\lambda} + o\left(\frac{1}{\lambda} \right) \quad \text{as } \lambda \to \infty \text{ in } \mathcal{S}'(\mathbf{R}). \quad (5.71)$$

We can now obtain the next Abelian result.

Proposition 5.6. *Let $g \in \mathcal{D}'(\mathbf{R})$ and $\phi \in \mathcal{E}(\mathbf{R})$. Suppose that*

$$\text{e.v. } \langle g(x), \phi(x) \rangle = \gamma \quad (\mathrm{C}).$$

Then, $\langle g(x), \phi(x) \rangle = \gamma$ (A). Moreover, Let $g = g_- + g_+$ be a decomposition satisfying the support requirements of Theorem 5.5, then

$$\lim_{z \to 0} \left(\langle g_-(t), \phi(t) e^{i\bar{z}t} \rangle + \langle g_+(t), \phi(t) e^{izt} \rangle \right) = \gamma, \quad (5.72)$$

in any sector $\operatorname{Im} z \geq M \, |\operatorname{Re} z|$, with $M > 0$.

Proof. We may assume that $\phi \equiv 1$. We use (5.67) and (5.68). Write $z = (1/\lambda)(\tau + i)$, so $|\tau| \leq (1/M)$, hence, as $\lambda \to \infty$,

$$\langle g_-(t), e^{i\bar{z}t} \rangle + \langle g_+(t), e^{izt} \rangle$$

$$= \lambda \left(\left\langle g_-(-\lambda t), e^{-(i\tau+1)t} \right\rangle + \left\langle g_+(t), e^{(i\tau-1)t} \right\rangle \right)$$

$$= \left(\frac{\gamma}{2} - c(\lambda) \right) \left\langle \delta(t), e^{(i\tau+1)t} \right\rangle + \left(\frac{\gamma}{2} + c(\lambda) \right) \left\langle \delta(t), e^{(i\tau-1)t} \right\rangle + o(1)$$

$$= \gamma + o(1),$$

with uniform convergence since $\left\{ e^{(i\tau-1)t} H(t) \right\}_{M|\tau| \leq 1}$ is compact in $\mathcal{S}_+$. $\square$

So, we obtain the Fourier inversion formula in the Abel sense.

Corollary 5.8. *Let $g \in S'(\mathbf{R})$. Suppose $f(x_0) = \gamma$, distributionally. Then the Fourier inversion formula holds in the Abel sense, i.e.,*

$$\frac{1}{2\pi} \left\langle \hat{f}(x), e^{ix_0 x} \right\rangle = \gamma \quad (A). \tag{5.73}$$

Moreover, let $\hat{f} = \hat{f}_- + \hat{f}_+$, with $\mathrm{supp}\hat{f}_- \subseteq (-\infty, 0]$ and $\mathrm{supp}\hat{f}_+ \subseteq [0, \infty)$, then

$$\lim_{z \to x_0} \frac{1}{2\pi} \left(\left\langle \hat{f}_-(t), e^{i\bar{z}t} \right\rangle + \left\langle \hat{f}_+(t), e^{izt} \right\rangle \right) = \gamma, \tag{5.74}$$

in any sector $\mathrm{Im}\, z \geq M \,|\mathrm{Re}\, z - x_0|$, with $M > 0$.

In the case of Fourier series, we recover a result from [196].

Corollary 5.9. *Let $f \in S'(\mathbf{R})$ be a 2π-periodic distribution having Fourier series*

$$f(x) = \sum_{n=-\infty}^{\infty} c_n e^{inx}. \tag{5.75}$$

Suppose $f(x_0) = \gamma$, distributionally. Then

$$\lim_{z \to x_0} \left(c_0 + \sum_{n=1}^{\infty} \left(c_{-n} e^{-in\bar{z}} + c_n e^{inz} \right) \right) = \gamma \tag{5.76}$$

in any sector $\mathrm{Im}\, z \geq M \,|\mathrm{Re}\, z - x_0|$, with $M > 0$. In particular, if $a_n = c_{-n} + c_n$ and $b_n = c_n - c_{-n}$, we obtain that

$$\frac{a_0}{2} + \sum_{n=1}^{\infty} (a_n \cos nx_0 + b_n \sin nx_0) = \gamma \quad (A). \tag{5.77}$$

Proof. Relation (5.76) follows directly form Corollary 5.8. If we set $z = x_0 + iy$ in (5.76), we obtain

$$\lim_{y \to 0+} \left(\frac{a_0}{2} + \sum_{n=1}^{\infty} (a_n \cos nx_0 + b_n \sin nx_0) e^{-ny} \right) = \gamma$$

which gives (5.77). $\qquad\square$

Let now $f \in D'(\mathbf{R})$ have $f(x_0) = \gamma$, distributionally. We cannot longer talk about Abel summability of the Fourier inversion formula, since the Fourier transform is not available in $D'(\mathbf{R})$. Nevertheless, there is a substitute of Abel summability, if we interpret it as the boundary limit at $x = x_0$

of a harmonic representation. Recall that a harmonic function U, harmonic on Im $z > 0$, is called a harmonic representation [13, 5] of f if

$$\lim_{y \to 0^+} U(x + iy) = f(x), \quad \text{in } \mathcal{D}'(\mathbf{R}). \tag{5.78}$$

It is very well known that any distribution admits a harmonic representation [13].

Theorem 5.6. *Let* $f \in \mathcal{D}'(\mathbf{R})$. *Suppose that* U *is a harmonic representation of* f *in the upper half-plane* Im $z > 0$. *If* $f(x_0) = \gamma$, *distributionally, then*

$$\lim_{z \to x_0} U(z) = \gamma, \tag{5.79}$$

in any sector Im $z \geq M \left| \text{Re } z - x_0 \right|$, *with* $M > 0$.

Proof. We first see that we may assume $f \in \mathcal{S}'(\mathbf{R})$. Indeed we can decompose $f = f_1 + f_2$ where f_2 is zero in a neighborhood of x_0 and $f_1 \in \mathcal{S}'(\mathbf{R})$. Let U_1 and U_2 be harmonic representations of f_1 and f_2, respectively; then U_2 represents the zero distribution in a neighborhood of x_0. Then by applying the reflection principle to the real and imaginary parts of U_2 ([5] , [169]), we have that U admits a harmonic extension to a (complex) neighborhood of x_0, and so it is real analytic, therefore, $U(z) - U_1(z) = U_2(z) = O\left(|z - x_0|\right)$ as $z \to x_0$. Additionally, $f_1(x_0) = \gamma$, distributionally, thus, we can assume that $f = f_1$. The same argument with the reflection principle shows that (5.79) is independent of the choice of U. Therefore, we can assume that U is the Fourier-Laplace representation [13] of f, that is, let $\hat{f} = \hat{f}_+ + \hat{f}_-$ be a decomposition such that supp $\hat{f}_- \subseteq (-\infty, 0]$ and supp $\hat{f}_+ \subseteq [0, \infty)$, we can assume that

$$U(z) = \frac{1}{2\pi} \left(\left\langle \hat{f}_-(t), e^{i\bar{z}t} \right\rangle + \left\langle \hat{f}_+(t), e^{izt} \right\rangle \right), \quad Im \, z > 0.$$

But in this case, Corollary 5.8 yields (5.79). ☐

Naturally, the converse of Theorem 5.6 is not true.

5.2.3 *Convergence of Fourier Series*

We now analyze sufficient conditions under which the existence of the distributional point value implies the convergence of the Fourier series at the point. Note that, in particular, any result of this type gives a Tauberian condition for Cesàro summability of series. The next theorem is our first result in this direction. We denote by l^p, $1 \le p < \infty$, the set of those sequences $\langle c_n \rangle_{n=-\infty}^{\infty}$ such that $\sum_{n=-\infty}^{\infty} |c_n|^p < \infty$.

Theorem 5.7. *Let* $f(x) = \sum_{n=-\infty}^{\infty} c_n e^{inx}$ *in* $S'(\mathbf{R})$. *Suppose that* $\langle c_n \rangle \in l^p$, $1 \le p < \infty$ *and*

$$r_{N,p} = \sum_{|n| \ge N} |c_n|^p = O\left(\frac{1}{N^{p-1}}\right), \quad N \to \infty. \tag{5.80}$$

Then, $f(x_0) = \gamma$, *distributionally, implies*

$$\text{e.v.} \sum_{n=-\infty}^{\infty} c_n e^{inx_0} = \gamma, \tag{5.81}$$

or which amounts to the same

$$\lim_{x \to \infty} \sum_{-x \le n \le ax} c_n e^{inx_0} = \gamma, \tag{5.82}$$

for each $a > 0$.

Proof. If $p = 1$, it is trivial. Let us assume $1 < p < \infty$, and let us find q so that $\dfrac{1}{p} + \dfrac{1}{q} = 1$. If $f(x_0) = \gamma$, we have

$$\lim_{\varepsilon \to 0^+} \sum_{n=-\infty}^{\infty} c_n e^{ix_0 n} \psi(\varepsilon n) = \gamma \psi(0),$$

for each $\psi \in S(\mathbf{R})$. Choose $\phi \in \mathcal{D}(\mathbf{R})$ such that $0 \le \phi \le 1$, and $\phi(x) = 1$ for $x \in [-1, a]$. Hence

$$\sum_{n=-\infty}^{\infty} c_n e^{inx_0} \phi(n\varepsilon) = \sum_{-\frac{1}{\varepsilon} \le n \le \frac{a}{\varepsilon}} c_n e^{ix_0 n} + \sum_{\frac{a}{\varepsilon} < n} c_n e^{ix_0 n} \phi(n\varepsilon)$$

$$+ \sum_{\frac{1}{\varepsilon} < n} c_{-n} e^{-ix_0 n} \phi(-n\varepsilon) + o(1),$$

as $\varepsilon \to 0^+$. Therefore,

$$\limsup_{\varepsilon \to 0^+} \left| \sum_{-\frac{1}{\varepsilon} \le n \le \frac{a}{\varepsilon}} c_n e^{inx_0} - \gamma \right| \le \limsup_{\varepsilon \to 0^+} \left(\sum_{\frac{a}{\varepsilon} < n} |c_n| \, |\phi(\varepsilon n)| + \sum_{\frac{1}{\varepsilon} < n} |c_{-n}| \, |\phi(-\varepsilon n)| \right).$$

But,

$$\sum_{\frac{a}{\varepsilon}<n} |c_n|\,|\phi(\varepsilon n)| \leq \left\{\sum_{\frac{a}{\varepsilon}<n} |c_n|^p\right\}^{\frac{1}{p}} \left\{\sum_{\frac{a}{\varepsilon}<n} |\phi(\varepsilon n)|^q\right\}^{\frac{1}{q}},$$

By (5.80), we can find $M > 0$ such that

$$\sum_{\frac{a}{\varepsilon}<n} |c_n|\,|\phi(\varepsilon n)| \leq M a^{-\frac{1}{q}} \left\{\varepsilon \sum_{\frac{a}{\varepsilon}<n} |\phi(\varepsilon n)|^q\right\}^{\frac{1}{q}}.$$

Then,

$$\limsup_{\varepsilon\to 0+} \lim \sum_{\frac{a}{\varepsilon}<n} |c_n|\,|\phi(\varepsilon n)| \leq M a^{-\frac{1}{q}} \left\{\int_a^\infty |\phi(x)|^q\,dx\right\}^{\frac{1}{q}}.$$

Similarly, $\exists M' > 0$ such that

$$\limsup_{\varepsilon\to 0+} \lim \sum_{\frac{1}{\varepsilon}<n} |c_{-n}|\,|\phi(-\varepsilon n)| \leq M' a^{-\frac{1}{q}} \left\{\int_{-\infty}^{-1} |\phi(x)|^q\,dx\right\}^{\frac{1}{q}}.$$

Now, we are free to choose ϕ such that the right sides of the last two inequalities are both less than $\eta/2$. Therefore,

$$\limsup_{\varepsilon\to 0+} \lim \left| \sum_{-\frac{1}{\varepsilon}\leq n\leq\frac{a}{\varepsilon}} c_n e^{inx_0} - \gamma \right| < \eta.$$

Since this can be done for each $\eta > 0$, we conclude that

$$\lim_{x\to\infty} \sum_{-x\leq n\leq ax} c_n e^{inx_0} = \gamma,$$

as required. □

As an example of the use of Theorem 5.7, let us obtain a classical Tauberian result of Hardy for Cesàro summability of series [66].

Corollary 5.10. *Suppose that $\sum_{n=0}^\infty c_n = \gamma$ (C). The Tauberian condition $nc_n = O(1)$ implies the convergence of the series to γ.*

Proof. We associate to the sequence a Fourier series, $f(x) = \sum_{n=0}^\infty c_n e^{inx}$. The (C) summability to γ implies $f(0) = \gamma$, distributionally. Now, Hardy's Tauberian hypothesis obviously implies (5.80), so Theorem 5.7 gives the convergence. □

We can generalize Theorem 5.7 to other norms.

Theorem 5.8. *Let $f(x) = \sum_{n=-\infty}^{\infty} c_n e^{inx}$ in $\mathcal{S}'(\mathbf{R})$. Suppose that*

$$\sum_{|n| \geq N} |c_n|^p |n|^{-pr} = O\left(\frac{1}{N^{rp+p-1}}\right) \tag{5.83}$$

for some r and p with $1 < p < \infty$. If $f(x_0) = \gamma$, distributionally, then

$$\text{e.v.} \sum_{n=-\infty}^{\infty} c_n e^{inx_0} = \gamma, \tag{5.84}$$

or which amounts to the same

$$\lim_{x \to \infty} \sum_{-x \leq n \leq ax} c_n e^{inx_0} = \gamma, \tag{5.85}$$

for each $a > 0$.

Proof. Use the inequality

$$\sum_{\frac{a}{\varepsilon} < n} |c_n| |\phi(\varepsilon n)| \leq \left\{ \sum_{\frac{a}{\varepsilon} < n} |c_n|^p n^{-rp} \right\}^{\frac{1}{p}} \left\{ \sum_{\frac{a}{\varepsilon} < n} n^{rq} |\phi(\varepsilon n)|^q \right\}^{\frac{1}{q}}$$

and follow a similar argument as the one in the proof of Theorem 5.7. $\square$

If we take $r = (1/p) - 1$ in the last theorem, we obtain the following Hardy-Littlewood Tauberian condition for (C) summability [67].

Corollary 5.11. *If*

$$\sum_{n=0}^{\infty} c_n = \gamma \quad (C, k), \tag{5.86}$$

for some $k \in \mathbf{N}$, then the Tauberian condition $(p \geq 1)$

$$\sum_{n=0}^{\infty} n^{p-1} |c_n|^p < \infty \tag{5.87}$$

implies that $\sum_{n=0}^{\infty} c_n$ is convergent to γ.

Next, we would like to make some comments about the results we just discussed. If $\langle c_n \rangle \in l^p$ for $1 \leq p \leq 2$, then $f(x) = \sum c_n e^{inx}$ belongs to $L^q[0, 2\pi]$, but the converse is not true. Similarly, if $f \in L^p[0, 2\pi]$, $1 \leq p \leq 2$, then $\langle c_n \rangle$, belongs to l^q, but the converse is not necessarily true. Hence, the results for $\langle c_n \rangle \in l^p$ with $1 \leq p \leq 2$ are about functions. However, for $p > 2$, these results are about *distributions*, in general. For example, as follows from [206], if $\langle c_n \rangle \in l^p \setminus l^2$ for some $p > 2$ then for almost all choices of signs $\rho_n = \pm$ the distribution $\sum_{n=0}^{\infty} \rho_n c_n e^{inx}$ is not locally integrable; or if $\langle c_n \rangle \in l^p \setminus l^2$ is lacunary then $\sum_{n=0}^{\infty} c_n e^{inx}$ is never a regular distribution.

We conclude this subsection discussing a type of Tauberian result in summability of Fourier series of distributions, where the conclusion is not the convergence of the series but the (C,m) summability for a specific m. As it has been mentioned before, any result of this type gives a result in the theory of Cesàro summability of series. Let us suppose that f is a periodic distribution of period 2π, and $f(x) = \sum_{n=-\infty}^{\infty} c_n e^{inx}$. We want to find sufficient conditions under which the existence of $f(x_0)$, distributionally, implies that

$$\lim_{x \to \infty} \sum_{-x \leq n \leq ax} c_n e^{inx_0} = f(x_0) \quad (C, m), \tag{5.88}$$

for an specific positive integer m. A partial answer to this question is given in Theorem 5.9.

Theorem 5.9. *Let* $f \in \mathcal{S}'(\mathbf{R})$ *such that* $f(x) = \sum_{n=-\infty}^{\infty} c_n e^{inx}$. *Suppose that* $f(x_0) = \gamma$, *distributionally. If for a fixed* a

$$\sum_{-x \leq n \leq ax} c_n e^{inx_0} = O(1) \quad (C, m), \tag{5.89}$$

then

$$\lim_{x \to \infty} \sum_{-x \leq n \leq ax} c_n e^{inx_0} = \gamma \quad (C, m+1). \tag{5.90}$$

Proof. For $x > 0$, set

$$g_a(x) = \sum_{-x \leq n \leq ax} c_n e^{inx_0},$$

and put $g_a(x) = 0$ for $x \leq 0$. Condition (5.89) means that there is an m-primitive G of g_a, such that $\mathrm{supp}\, G \subseteq [0, \infty)$ and

$$G(x) = O(x^m), \quad x \to \infty.$$

In addition, since $f(x_0) = \gamma$, we have that

$$G(\lambda x) \overset{q}{\sim} \frac{\gamma \lambda^m x_+^m}{m!} \quad \text{as } \lambda \to \infty \text{ in } \mathcal{D}'(\mathbf{R}),$$

i.e., for each $\phi \in \mathcal{D}(\mathbf{R})$

$$\int_0^\infty G(\lambda x)\phi(x)\, dx = \frac{\gamma \lambda^m}{m!} \int_0^\infty x^m \phi(x)\, dx + o\left(\lambda^m\right).$$

Pick $\phi \in \mathcal{D}(\mathbf{R})$ such that $\phi(x) = 1$ for $x \in [-1, 1]$ and supp $\phi \subseteq [-1, 2]$. Evaluating G at ϕ, we obtain

$$\frac{1}{\lambda} \int_0^\lambda G(x)\, dx + \frac{1}{\lambda} \int_\lambda^{2\lambda} G(x)\phi\left(\frac{x}{\lambda}\right) dx$$

$$= \frac{\gamma \lambda^m}{(m+1)!} + \frac{\gamma \lambda^m}{m!} \int_1^2 x^m \phi(x)\, dx + o(\lambda^m), \quad \lambda \to \infty,$$

which implies

$$\left| \frac{(m+1)!}{\lambda^{m+1}} \int_0^\lambda G(x)\, dx - \gamma \right|$$

$$\leq o(1) + \gamma(m+1) \int_1^2 x^m \phi(x)\, dx + \frac{(m+1)!}{\lambda^{m+1}} \int_\lambda^{2\lambda} |G(x)|\, \phi\left(\frac{x}{\lambda}\right) dx$$

$$= o(1) + \{\gamma(m+1) + (m+1)!\, O(1)\} \int_1^2 x^m \phi(x)\, dx, \quad \lambda \to \infty,$$

since we can choose ϕ such that $\int_1^2 x^m \phi(x)\, dx$ is as small as we want, we conclude that

$$\lim_{\lambda \to \infty} \frac{(m+1)!}{\lambda^{m+1}} \int_0^\lambda G(x)\, dx = \gamma,$$

and the result follows. $\qquad\square$

We obtain the following interesting corollary of Theorem 5.9, known as convexity theorem [66].

Corollary 5.12. *Let $\langle c_n \rangle_{n \in \mathbf{N}_0}$ be a sequence of complex numbers. Suppose that*

$$\sum_{n=0}^\infty c_n = \gamma \quad (\mathrm{C}, k), \tag{5.91}$$

for some $k \in \mathbf{N}$. If the m-Cesàro means are bounded then

$$\sum_{n=0}^\infty c_n = \gamma \quad (\mathrm{C}, m+1). \tag{5.92}$$

5.2.4   Series with gaps

In this subsection we apply the ideas of the last subsection to series with gaps. In particular, we shall find examples of continuous functions whose distributional derivatives do not have distributional point values at any point.

Theorem 5.10. *Let* $f(x) = \sum_{n=0}^{\infty} c_n e^{inx}$, *in* $\mathcal{S}'(\mathbf{R})$. *In addition, suppose that* $\langle c_n \rangle_{n=0}^{\infty}$ *is lacunary, in the sense of Hadamard, i.e.,* $c_n = 0$ *except for a sequence* $n_k \in \mathbf{N}_0$ *with* $n_{k+1} > \alpha n_k$ *for some* $\alpha > 1$. *Then* $f(x_0) = \gamma$, *distributionally, if and only if*

$$\sum_{n=0}^{\infty} c_n e^{inx_0} = \gamma. \tag{5.93}$$

In particular, $c_{n_k} = o(1)$, $k \to \infty$.

Proof. Let $\phi \in \mathcal{D}(\mathbf{R})$ such that $0 \leq \phi \leq 1$, $\phi(x) = 1$ for $x \in [0,1]$ and supp $\phi \subseteq [-1, \alpha]$. Set $b_n = c_n e^{inx_0}$, for each $n \in \mathbf{N}$. We have that

$$M(\lambda) = \sum_{n_k \leq \lambda} b_{n_k} + \sum_{\lambda < n_k < \alpha\lambda} b_{n_k} \phi\left(\frac{n_k}{\lambda}\right) - \gamma = o(1), \ \lambda \to \infty.$$

Note that given $\lambda > 0$ there exists at most one k_λ such that $\lambda < k_\lambda < \alpha\lambda$. Therefore if $\lambda = n_m$, we obtain

$$M(n_m) = o(1), \ m \to \infty,$$

which is the same as

$$\sum_{k=0}^{m} b_{n_k} - \gamma = o(1), \ m \to \infty.$$

This completes the proof. $\square$

Moreover, with a little modification of the last argument, we obtain the following result.

Theorem 5.11. *Let* $f(x) = \sum_{n=-\infty}^{\infty} c_n e^{inx}$, *in* $\mathcal{S}'(\mathbf{R})$. *Suppose that* $\langle c_n \rangle_{n\in\mathbf{Z}}$ *is lacunary in both directions; then,* $f(x_0) = \gamma$, *distributionally, if and only if*

$$\lim_{x \to \infty} \sum_{-x \leq n \leq ax} c_n e^{inx_0} = \gamma,$$

for each $a > 0$.

We obtain several interesting corollaries from the last two theorems. The second part of the following corollary is a result of Kolmogorov.

Corollary 5.13. *If $f \in L^1[0, 2\pi]$ and $\langle c_n \rangle_{n \in \mathbf{Z}}$ lacunary, then the Fourier series of f converges to $f(x_0)$ at every point where $f(x_0)$ exists distributionally in the sense of Lojasiewicz. In particular, it converges almost everywhere.*

Proof. Indeed, the first part follows directly from Theorem 5.11, while the second statement is true because f has distributional point values at every point of the Lebesgue set of f. $\square$

Corollary 5.14. *Let $f(x) = \sum_{n=-\infty}^{\infty} c_n e^{inx}$, in $\mathcal{S}'(\mathbf{R})$. If $\langle c_n \rangle_{n \in \mathbf{Z}}$ is lacunary, but $c_n \neq o(1)$, then the distributional value $f(x_0)$ does not exist at any point x_0.*

The next corollary allows us to find examples of continuous functions whose distributional derivatives do not have point values anywhere.

Corollary 5.15. *Let $\langle c_n \rangle_{n \in \mathbf{Z}}$ be a lacunary sequence such that $c_n \neq o(1)$ but $c_n = O(1)$, $|n| \to \infty$. Then*

$$g(x) = \sum_{n=-\infty}^{\infty} \frac{c_n}{n} e^{inx}, \qquad (5.94)$$

is continuous but $g'(x)$ does not have distributional point values at any point; in particular, g is nowhere differentiable.

That g' does not have point values in the sense of Lojasiewicz at any point is stronger than the fact that g is nowhere differentiable. For example, consider $g(x) = x \sin x^{-1}$; $g'(0)$ does not exist in the usual sense, but g' has the value 0 at $x = 0$ distributionally [93].

A good illustration of Corollary 5.14 is obtained when we consider the two *Weierstrass functions*

$$f_\alpha(x) = \sum_{n=0}^{\infty} b^{-n\alpha} \cos(b^n x),$$

and

$$g_\alpha(x) = \sum_{n=0}^{\infty} b^{-n\alpha} \sin(b^n x),$$

where $b > 1$ is an integer and α is a positive number less or equal to 1. Observe that f_α and g_α are continuous. Weierstrass showed that for α small enough they are nowhere differentiable. The extension to $0 < \alpha \leq 1$ was first proved by Hardy. Using Corollary 5.15, we obtain a stronger result for it, namely, f_α' and g_α' do not have distributional point values at any point.

5.2.5   *Convergence of Fourier integrals*

We now extend the results from **5.2.3** to Fourier integrals.

Theorem 5.12. *Let $f \in \mathcal{S}'(\mathbf{R})$. Assume that $\hat{f} \in L^p$, $1 \leq p < \infty$, and*

$$r_{p,x} = \int_{|t| \geq x}^{\infty} \left| \hat{f}(t) \right|^p dt = O\left(\frac{1}{x^{p-1}} \right), \quad x \to \infty. \tag{5.95}$$

Then, $f(x_0) = \gamma$, distributionally, if and only if

$$\frac{1}{2\pi} \text{e.v.} \int_{-\infty}^{\infty} \hat{f}(t) e^{ix_0 t} dt = \gamma. \tag{5.96}$$

Proof. We only consider the case $1 < p < \infty$. Assume that $f(x_0) = \gamma$, distributionally. Fix $a > 0$. Taking Fourier transform in

$$f\left(x_0 + \frac{x}{\lambda} \right) = \gamma + o(1), \quad \lambda \to \infty, \tag{5.97}$$

we obtain

$$e^{ix_0 \lambda x} \hat{f}(\lambda x) = \frac{2\pi \gamma \delta(x)}{\lambda} + o\left(\frac{1}{\lambda} \right), \quad \lambda \to \infty, \text{ in } \mathcal{S}'(\mathbf{R}). \tag{5.98}$$

Set $g(x) = e^{ix_0 x} \hat{f}(x)$. Take $\phi \in \mathcal{D}(\mathbf{R})$, such that $\phi(x) = 1$ for $x \in [-1, a]$ and $0 \leq \phi \leq 1$. Take q such that $\frac{1}{p} + \frac{1}{q} = 1$. Thus, we have

$$\int_{-\lambda}^{\lambda a} g(t)\, dt - 2\pi\gamma = -\int_{-\infty}^{-\lambda} g(t)\phi\left(\frac{t}{\lambda} \right) dt - \int_{a\lambda}^{\infty} g(t)\phi\left(\frac{t}{\lambda} \right) dt + o(1),$$

as $\lambda \to \infty$. We show that

$$\lim_{\lambda \to \infty} \int_{a\lambda}^{\infty} g(t)\phi\left(\frac{t}{\lambda} \right) dt = 0, \tag{5.99}$$

and

$$\lim_{\lambda \to \infty} \int_{-\infty}^{-\lambda} g(t)\phi\left(\frac{t}{\lambda}\right) dt = 0. \tag{5.100}$$

We have

$$\left| \int_{a\lambda}^{\infty} g(t)\phi\left(\frac{t}{\lambda}\right) dt \right| \leq O\left(\lambda^{-\frac{1}{q}}\right) \left\{ \lambda \int_{a}^{\infty} |\phi(t)|^q \, dt \right\}^{\frac{1}{q}}$$

$$= O(1)\left\{ \int_{a}^{\infty} |\phi(t)|^q \, dt \right\}^{\frac{1}{q}}.$$

Since $\left\{ \int_{a}^{\infty} |\phi(t)|^q \, dt \right\}^{\frac{1}{q}}$ can be made arbitrarily small, we conclude (5.99). Similarly, (5.100) follows. □

Likewise, one can show.

Theorem 5.13. *Let $f \in \mathcal{S}'(\mathbf{R})$. Suppose that $\hat{f}$ is locally integrable and*

$$\int_{|t| \geq x} \left| \hat{f}(t) \right|^p |t|^{-rp} \, dt = O\left(\frac{1}{x^{pr+p-1}}\right), \quad x \to \infty.$$

for some $1 < p < \infty$ and $r \in \mathbf{R}$. Then, $f(x_0) = \gamma$, distributionally, if and only if

$$\frac{1}{2\pi} \text{e.v.} \int_{-\infty}^{\infty} \hat{f}(t)e^{ix_0 t} dt = \gamma.$$

If we take $r = (1/p) - 1$ in Theorem 5.13, we obtain the next corollary.

Corollary 5.16. *If $f \in \mathcal{S}'(\mathbf{R})$, $\hat{f}$ is locally integrable, and*

$$\int_{-\infty}^{\infty} |t|^{p-1} \left| \hat{f}(t) \right|^p \, dt < \infty, \tag{5.101}$$

then, $f(x_0) = \gamma$, distributionally, if and only if

$$\frac{1}{2\pi} \text{e.v.} \int_{-\infty}^{\infty} \hat{f}(t)e^{ix_0 t} dt = \gamma.$$

5.2.6   *Tauberian theorem for distributional point values*

We now turn our attention to Tauberian theorems for Abel summability. Our aim is to obtain a Tauberian converse to the following result of Constantinescu [26]:

Suppose that $f \in \mathcal{D}'(\mathbf{R})$ is the boundary value of a function F analytic in the upper half-plane, that is, $f(x) = F(x + i0)$; if $f(x_0) = \gamma$ distributionally, then $F(x_0 + iy) \to \gamma$ as $y \to 0^+$.

Notice that the above result is a particular case of Theorem 5.6, which we already remarked that can be viewed as Abel summability for nontempered distributions. Observe also that the converse is false, as we have pointed out many times before [48].

Our Tauberian condition will be *distributional boundedness* at a point.

Definition 5.7. Let $f \in \mathcal{D}'(\mathbf{R})$. We say that f is distributionally bounded at $x = x_0$ if it is quasi-asymptotic bounded of degree 0 at x_0 with respect to the trivial slowly varying function, that is,

$$f(x_0 + \varepsilon x) = O(1) \quad \text{as } \varepsilon \to 0^+ \text{ in } \mathcal{D}'(\mathbf{R}) . \tag{5.102}$$

Example II.5. The function $f(x) = |x|^i$ is bounded in the ordinary sense, and thus it defines a unique regular distribution which is distributionally bounded at $x = x_0$. It easy to see that $f(0)$ does not exist distributionally. In general the evaluation $\langle f(\varepsilon x), \phi(x) \rangle$ does not tend to a limit as $\varepsilon \to 0$ if $\phi \in \mathcal{D}(\mathbf{R})$.

Observe the Definition 5.7 is meaningful if f is just defined in a neighborhood of $x = x_0$, since the quasi-asymptotics at 0 are local properties.

We shall introduce the equivalent approach of Campos-Ferreira to distributional point values and distributional boundedness [21]. It is in somehow connected with the structure of these two quasi-asymptotic concepts. Let us introduce the operator μ_a which is defined on complex valued locally integrable functions defined in $\mathbf{R}$ as

$$\mu_a \{ f(t); x \} = \frac{1}{x-a} \int_a^x f(t) \, dt, \quad x \neq a, \tag{5.103}$$

while the operator ∂_a is the inverse of μ_a,

$$\partial_a (g) = ((x - a) g (x))' , \qquad (5.104)$$

and it is defined on distributions. Suppose first that $f_0 = f$ is *real*. Then if it is bounded near $x = a$, we can define

$$\overline{f_0} (a) = \limsup_{x \to a} f (x) , \qquad \underline{f_0} (a) = \liminf_{x \to a} f (x) . \qquad (5.105)$$

Then $f_1 = \mu_a (f)$ will be likewise bounded near $x = a$ and actually

$$\underline{f_0} (a) \le \underline{f_1} (a) \le \overline{f_1} (a) \le \overline{f_0} (a) \qquad (5.106)$$

and, in particular, if $f (a) = f_0 (a)$ exists, then $f_1 (a)$ also exists and $f_1 (a) = f_0 (a)$. The next lemma is not difficult to show, we leave the verification as an exercise to the reader (see also [21]).

Lemma 5.3. *A distribution $f \in \mathcal{D}' (\mathbf{R})$ is distributionally bounded at $x = x_0$ if and only if there exist $n \in \mathbf{N}$ and $f_n \in \mathcal{D}' (\mathbf{R})$, continuous and bounded in a pointed neighborhood $(x_0 - \varepsilon, x_0) \cup (x_0, x_0 + \varepsilon)$ of x_0, such that $f = \partial_{x_0}^n f_n$.*

If f_0 is distributionally bounded at $x = x_0$, then there exists a *unique* distributionally bounded distribution near $x = x_0$, f_1, with $f_0 = \partial_{x_0} f_1$. Therefore, ∂_{x_0} and μ_{x_0} are isomorphisms of the space of distributionally bounded distributions near $x = x_0$. Given f_0 we can form a sequence of distributionally bounded distributions $\{f_n\}_{n=-\infty}^{\infty}$ with $f_n = \partial_{x_0} f_{n+1}$ for each $n \in \mathbf{Z}$.

We have an analogous result for distributional point values, again, we leave the proof of the following lemma as an exercise for the reader (see also [21]).

Lemma 5.4. *A distribution $f \in \mathcal{D}' (\mathbf{R})$ satisfies $f(x_0) = \gamma$, distributionally, if and only if there exist $n \in \mathbf{N}$ and $f_n \in \mathcal{D}' (\mathbf{R})$, continuous near x_0, such that $f = \partial_{x_0}^n f_n$ and $f_n(x_0) = \gamma$. We call n the order of the point value.*

Observe also that if $f = \partial_{x_0}^n f_n$, and f_n is bounded near $x = x_0$, then $f (x_0)$ exists distributionally, and equals γ, if and only if $f_n (x_0) = \gamma$, distributionally.

We start with a Tauberian result for bounded analytic functions.

Theorem 5.14. *Let F be analytic and bounded in a rectangular region of the form $(a, b) \times (0, R)$. Suppose $f(x) = \lim_{y \to 0^+} F(x + iy)$ in the space $\mathcal{D}'(a, b)$. Let $x_0 \in (a, b)$ such that*

$$\lim_{y \to 0^+} F(x_0 + iy) = \gamma. \tag{5.107}$$

Then

$$f(x_0) = \gamma, \quad \text{distributionally.} \tag{5.108}$$

In fact, (5.108) is a point value of the first order, that is,

$$\lim_{x \to x_0} \frac{1}{x - x_0} \int_{x_0}^x f(t) \, dt = \gamma. \tag{5.109}$$

Proof. We shall first show that it is enough to prove the result if the rectangular region is the upper half-plane $\mathbf{H} = \{z \in \mathbf{C} : \operatorname{Im} z > 0\}$. Indeed, let C be a smooth simple closed curve contained in $(a, b) \times [0, R)$ such that $\mathsf{C} \cap (a, b) = [x_0 - \eta, x_0 + \eta]$, and which is symmetric with respect to the line $Re\ z = x_0$. Let φ be a conformal bijection from $\mathbf{H}$ to the region enclosed by C such that the image of the line $\operatorname{Re} z = x_0$ is contained in $\operatorname{Re} z = x_0$, so that, in particular, $\varphi(x_0) = x_0$. Then (5.107) holds if and only if $F \circ \varphi(x_0 + iy) \to \gamma$ as $y \to 0^+$, while (5.108) and (5.109) hold if and only if the corresponding equations hold for a distribution given locally as $f \circ \varphi$ near $x = x_0$.

In the sequel we will use the weak* topology of L^∞. This is the topology on L^∞ determined by the family of seminorms $f \to p_\varphi(f) = |\langle f, \varphi \rangle|$, $f \in L^\infty$, where $\varphi \in L^1(\mathbf{R})$.

Therefore we shall assume that $a = -\infty$, and $b = R = \infty$. In this case, f belongs to the Hardy space H^∞, the closed subspace of $L^\infty(\mathbf{R})$ consisting of the boundary values of bounded analytic functions on $\mathbf{H}$ ([82]); moreover, one easily verifies that H^∞ is a weak* closed subspace of L^∞, this fact will be used below. Let $f_\varepsilon(x) = f(x_0 + \varepsilon x)$. Then the set $\{f_\varepsilon : \varepsilon \neq 0\}$ is weak* bounded (as a subset of the dual space $(L^1(\mathbf{R}))' = L^\infty(\mathbf{R})$) and, consequently, a relatively weak* compact set. Suppose that $\{\varepsilon_n\}_{n=0}^\infty$ is a sequence of non-zero numbers with $\varepsilon_n \to 0$ such that the sequence $\{f_{\varepsilon_n}\}_{n=0}^\infty$ is weak* convergent to $g \in L^\infty(\mathbf{R})$. It will be shown that $g \equiv \gamma$. Since $g \in H^\infty$, we can write it as $g(x) = G(x + i0)$ where G is a bounded analytic function in $\mathbf{H}$, then the weak* convergence of f_{ε_n} to g implies that $F(x_0 + \varepsilon_n z)$ converges to $G(z)$ uniformly on compacts of $\mathbf{H}$, and thus $G(iy) = \gamma$ for all $y > 0$. It follows that $G \equiv \gamma$, and so $g \equiv \gamma$. Since any sequence $\{f_{\varepsilon_n}\}_{n=0}^{\infty}$ with $\varepsilon_n \to 0$ has a weak* convergent subsequence, and since that subsequence converges to the constant function γ, we conclude that $f_\varepsilon \to \gamma$ in the weak* topology of $L^\infty(\mathbf{R})$.

That $f(x_0) = \gamma$, distributionally, is now clear, because $\mathcal{D}(\mathbf{R}) \subset L^1(\mathbf{R})$.

On the other hand, (5.109) follows by taking $x = x_0 + \varepsilon$ and $\phi(t) = \chi_{[0,1]}(t)$, the characteristic function of the unit interval, in the limit $\lim_{\varepsilon \to 0} \langle f_\varepsilon(t), \phi(t) \rangle = \gamma \int_{-\infty}^{\infty} \phi(t)\, dt$, which in view of the previous argument holds now for $\phi \in L^1(\mathbf{R})$. $\qquad\square$

We can now prove our Tauberian theorem.

Theorem 5.15. *Let F be analytic in a rectangular region of the form $(a,b) \times (0,R)$. Suppose $f(x) = \lim_{y \to 0^+} F(x + iy)$ in the space $\mathcal{D}'(a,b)$. Let $x_0 \in (a,b)$ such that $\lim_{y \to 0^+} F(x_0 + iy) = \gamma$. If f is distributionally bounded at $x = x_0$ then $f(x_0) = \gamma$, distributionally.*

Proof. There exists $n \in \mathbf{N}$ and a function f_n bounded in a neighborhood of x_0 such that $f = \partial_{x_0}^n f_n$; notice that $f(x_0) = \gamma$, distributionally, if and only if $f_n(x_0) = \gamma$, distributionally. But $f_n(x) = F_n(x + i0)$ in $\mathcal{D}'(a,b)$, where F_n is analytic in $(a,b) \times (0,R)$; here F_n is the only angularly bounded solution of $F(z) = \partial_{x_0}^n F_n(z)$ (derivatives with respect to z). Since f_n is bounded near $x = x_0$, F_n is also bounded in a rectangular region of the form $(a_1, b_1) \times (0, R_1)$, where $x_0 \in (a_1, b_1)$. Clearly $\lim_{y \to 0^+} F_n(x_0 + iy) = \gamma$, so the Theorem 5.14 yields $f_n(x_0) = \gamma$, distributionally, as required. $\qquad\square$

Observe that in general the result (5.109) does not follow if f is not bounded but just distributionally bounded near x_0.

The condition (5.107) may seem weaker than the angular convergence of $F(z)$ to γ as $z \to x_0$, however, if F is angularly bounded, which is the case if f is distributionally bounded at $x = x_0$, then angular convergence and radial convergence are equivalent. In fact [27] both conditions are equivalent to the existence of an arc $\kappa : [0,1] \longrightarrow (a,b) \times [0,R)$ such that $\kappa([0,1)) \subset \{z \in \mathbf{C} : \mathrm{Im}\, , z \geq m |\mathrm{Re}\, z - x_0|\}$ for some $m > 0$ and such that $\kappa(1) = x_0$, for which

$$\lim_{t \to 1^-} F(\kappa(t)) = \gamma. \tag{5.110}$$

Therefore, we may use a conformal map to obtain the following general form of the Theorem 5.15.

Theorem 5.16. *Let* C *be a smooth part of the boundary* $\partial\Omega$ *of a region* Ω *of the complex plane. Let* F *be analytic in* Ω, *and suppose that* $f \in \mathcal{D}'$ (C) *is the distributional boundary limit of* F. *Let* $\xi_0 \in$ C *and suppose that* κ *is an arc in* Ω *that ends at* ξ_0 *and that approaches* C *angularly. If* $\lim_{t\to 1^-} F(\kappa(t)) = \gamma$ *and* f *is distributionally bounded at* $\xi = \xi_0$, *then* $f(\xi_0) = \gamma$, *distributionally.*

Theorem 5.15 may also be used to obtain Littlewood type Tauberian results for distributions. The first corollary is also contained in the general theorem of Vladimirov, Drozhzhinov, and Zavyalov [192], but it is convenient to state it here for future applications. We will use Corollary 5.17 to produce in **5.2.7** a simple proof of the celebrated Hardy-Littlewood Tauberian theorem for the converse of Abel's theorem.

Corollary 5.17. *Let* g *be a tempered distribution supported on* $[0, \infty)$. *Suppose that*

$$\lim_{y\to 0^+} \langle g(x), e^{-yx} \rangle = \gamma. \tag{5.111}$$

Then, the Tauberian condition

$$g(\lambda x) = O\left(\frac{1}{\lambda}\right) \quad as \ \lambda \to \infty \ in \ \mathcal{D}'(\mathbf{R}) \tag{5.112}$$

implies that g *has the quasi-asymptotic behavior*

$$g(\lambda x) \overset{q}{\sim} \gamma \frac{\delta(x)}{\lambda} \quad as \ \lambda \to \infty \ in \ \mathcal{S}'(\mathbf{R}). \tag{5.113}$$

Proof. Let f be such that $\hat{f} = g$. Then (5.111) translates into $F(iy) \to \gamma$ as $y \to 0^+$, where $F(z) = \langle g(x), e^{izx} \rangle$ (hence $f(x) = F(x + i0)$) and (5.112) corresponds to the statement f distributionally bounded at $x = 0$, by Theorem 5.15, we have that $f(0) = \gamma$, distributionally. Thus, Fourier inverse transform yields (5.113). $\qquad\square$

Corollary 5.18. *Let* g *be a tempered distribution supported on* $[0, \infty)$ *and* $\phi \in \mathcal{E}(\mathbf{R})$. *Suppose that* $\langle g(x), \phi(x) \rangle = \gamma$ (A). *Then, the Tauberian condition*

$$\phi(\lambda x)g(\lambda x) = O\left(\frac{1}{\lambda}\right) \quad as \ \lambda \to \infty \ in \ \mathcal{D}'(\mathbf{R})$$

implies that $\langle g(x), \phi(x) \rangle = \gamma$ (C).

Proof. Corollary 5.17 gives that

$$\phi(\lambda x)g(\lambda x) \overset{q}{\sim} \gamma \frac{\delta(x)}{\lambda} \quad \text{as } \lambda \to \infty \text{ in } \mathcal{S}'(\mathbf{R}),$$

which by Proposition 5.3 implies that $\langle g(x), \phi(x) \rangle = \gamma \ (C)$. $\qquad \square$

5.2.7 *Application: Littlewood's Tauberian theorem*

We now discuss a non-trivial application of Theorem 5.15. Our application is a proof of a famous Tauberian theorem of Hardy and Littlewood. This distributional proof was originally found in [177]. In fact, we give a proof for the version proposed by Littlewood but first proved by Ananda Rau [2]. We begin with a lemma whose proof can be tracked down to the proof of the original first Tauber's theorem ([66] , [168]).

Lemma 5.5. *Let* $\langle b_n \rangle_{n=0}^{\infty}$ *be a sequence of complex numbers. Suppose that* $\langle \lambda_n \rangle_{n=0}^{\infty}$ *is an increasing sequence of non-negative real numbers such that* $\lambda_n \to \infty$ *as* $n \to \infty$. *If*

$$b_n = O\left(\frac{\lambda_n - \lambda_{n-1}}{\lambda_n}\right), \tag{5.114}$$

then,

$$\sum_{n=0}^{\infty} b_n e^{-\lambda_n y} - \sum_{\lambda_n < \frac{1}{y}} b_n = O(1), \quad \text{as } y \to 0^+. \tag{5.115}$$

Proof. Choose M such that $|b_n| \leq M\lambda_n^{-1}(\lambda_n - \lambda_{n-1})$, for every n. Then,

$$\left| \sum_{n=0}^{\infty} b_n e^{-\lambda_n y} - \sum_{\lambda_n < \frac{1}{y}} b_n \right| \leq \sum_{\lambda_n < \frac{1}{y}} |b_n| \left(1 - e^{-\lambda_n y}\right) + \sum_{\frac{1}{y} \leq \lambda_n} |b_n| e^{-\lambda_n y}$$

$$\leq My \sum_{\lambda_n < \frac{1}{y}} (\lambda_n - \lambda_{n-1}) + My \sum_{\frac{1}{y} \leq \lambda_n} (\lambda_n - \lambda_{n-1}) e^{-\lambda_n y}$$

$$= O(1) + My \int_{\frac{1}{y}}^{\infty} e^{-yt} \, dt = O(1), \quad y \to 0^+,$$

as required. $\qquad \square$

Recall (cf. **5.1.1**) that a series $\sum_{n=0}^{\infty} c_n$ is $(A, \langle \lambda_n \rangle)$ summable to γ if $\sum_{n=0}^{\infty} c_n e^{-\lambda_n y} \to \gamma$ as $y \to 0^+$. When $\lambda_n = n$ we obtain the notion of Abel summability. Then we have the ensuing Hardy-Littlewood Tauberian theorem.

Theorem 5.17. *Suppose that* $\langle \lambda_n \rangle_{n=0}^{\infty}$ *is an increasing sequence of nonnegative real numbers such that* $\lambda_n \to \infty$, *as* $n \to \infty$. *If*

$$\sum_{n=0}^{\infty} c_n = \gamma \quad (A, \lambda_n), \tag{5.116}$$

and $c_n = O\left(\lambda_n^{-1}(\lambda_n - \lambda_{n-1}) \right)$, *then* $\sum_{n=0}^{\infty} c_n = \gamma$.

Proof. The plan of the proof is to associate to the series the tempered distribution $g(x) = \sum_{n=0}^{\infty} c_n \delta(x - \lambda_n)$ and show that $g(\lambda x) \overset{q}{\sim} \gamma \delta(\lambda x)$, as $\lambda \to \infty$ in $\mathcal{S}'(\mathbf{R})$, based on this conclusion, we will deduce the convergence of the series.

Let us first verify that f defines a tempered distribution; indeed from Lemma 5.5 and the assumption (5.116), we have that $G(x) = \sum_{\lambda_n < x} c_n$ is a bounded function, hence g is a tempered distribution; furthermore, $G(\lambda x) = O(1)$ in $\mathcal{S}'(\mathbf{R})$, and therefore by differentiating $G(\lambda x)$ with respect to x, we obtain that $\lambda g(\lambda x)$ is bounded in $\mathcal{S}'(\mathbf{R})$. Therefore, by Corollary 5.17, we obtain that

$$g(\lambda x) = \gamma \frac{\delta(x)}{\lambda} + o\left(\frac{1}{\lambda} \right), \quad \text{in } \mathcal{S}'(\mathbf{R}). \tag{5.117}$$

So, we have that

$$\lim_{\lambda \to \infty} \sum_{n=0}^{\infty} c_n \phi\left(\frac{\lambda_n}{\lambda} \right) = \gamma, \quad \text{for each } \phi \in \mathcal{D}(\mathbf{R}). \tag{5.118}$$

To conclude the proof, we take in (5.118) suitable test functions. Let $\eta > 0$ and let us choose the test function $\phi \in \mathcal{D}(\mathbf{R})$ such that $0 \leq \phi \leq 1$, $\phi(x) = 1$ for $x \in [0,1]$, $\operatorname{supp} \phi \subseteq [-1,2]$, ϕ is decreasing on the interval $(1,2)$, and such that $\int_1^2 \phi(x) dx < \eta$. Then

$$\limsup_{N \to \infty} \left| \sum_{n=0}^{N} c_n - \gamma \right| \leq \limsup_{N \to \infty} \sum_{\lambda_N < \lambda_n \leq 2\lambda_N} \frac{\lambda_n - \lambda_{n-1}}{\lambda_N} \phi\left(\frac{\lambda_n}{\lambda_N} \right)$$

$$\leq \int_1^2 \phi(x) \, dx < \eta.$$

Since η was arbitrary, we conclude that $\displaystyle\sum_{n=0}^{\infty} c_n = \gamma.$ $\qquad\qquad\qquad$ $\square$

5.2.8 *Solution to the Hardy-Littlewood* (C) *summability problem for distributions*

As an application of Theorem 5.1, we now formulate and solve the so-called Hardy-Littlewood (C) summability problem in the context of tempered distributions. This classical problem aims to characterize trigonometric series, in sine-cosine form, which are (C) summable to some value at a point $x = x_0$, that is,

$$\frac{a_0}{2} + \sum_{n=1}^{\infty}(a_n \cos nx_0 + b_n \sin nx_0) = \gamma \quad (\mathrm{C}, m),$$

for some γ and $m \in \mathbf{N}_0$. One also imposes the restrictions $a_n = O(n^k)$ and $b_n = O(n^k)$, for some k; thus, the trigonometric series represents a tempered distribution! The problem for trigonometric series was first formulated by Hardy and Littlewood in [68]; a complete treatment with historical remarks is found in [206] (see also [56])

In order to formulate the problem for tempered distributions, we need the following summability notion for distributional evaluations.

Definition 5.8. Let $g \in \mathcal{D}'(\mathbf{R})$, $\phi \in \mathcal{E}(\mathbf{R})$, and $m \in \mathbf{N}_0$. We say that the principal value evaluation p.v. $\langle g(x), \phi(x)\rangle$ exists and is equal to γ in the Cesàro sense of order m, and write

$$\text{p.v. } \langle g(x), \phi(x)\rangle = \gamma \quad (\mathrm{C}, m), \tag{5.119}$$

if some first order primitive G of ϕg, i.e., $G' = \phi g$, satisfies

$$\lim_{x \to \infty} (G(x) - G(-x)) = \gamma \quad (\mathrm{C}, m). \tag{5.120}$$

Note that e.v. $\langle g(x), \phi(x)\rangle = \gamma$ (C, m) implies p.v. $\langle g(x), \phi(x)\rangle = \gamma$ (C, m), as the reader can easily verify. On the other hand the converse is not true; take for example p.v. $\langle x, 1\rangle = 0$ (C, 0), but clearly the evaluation e.v. $\langle x, 1\rangle$ (C) does not exist.

When $g = \mu$ is a Radon measure, we write

$$\text{p.v.} \int_{-\infty}^{\infty} \phi(x)d\mu(x) = \gamma \quad (C, m) \tag{5.121}$$

for (5.119). Observe that (5.121) explicitly means that

$$\lim_{x \to \infty} \int_{-x}^{x} \phi(t) \left(1 - \frac{t}{|x|} \right)^m d\mu(t) = \gamma. \tag{5.122}$$

If $\mu = \sum_{n=-\infty}^{\infty} c_n \delta(\cdot - n)$ and $\phi \equiv 1$, then we write (5.121) as

$$\text{p.v.} \sum_{n=-\infty}^{\infty} c_n = \gamma \quad (C, m), \tag{5.123}$$

which is equivalent to have

$$c_0 + \sum_{n=1}^{\infty} (c_n + c_{-n}) = \gamma \quad (C, m). \tag{5.124}$$

Example II.6. Consider the trigonometric series $\sum_{n=-\infty}^{\infty} c_n e^{inx_0}$ then

$$\text{p.v.} \sum_{n=-\infty}^{\infty} c_n e^{inx_0} = \gamma \quad (C, m)$$

if and only if

$$\frac{a_0}{2} + \sum_{n=1}^{\infty} (a_n \cos nx_0 + b_n \sin nx_0) = \gamma \quad (C, m),$$

with $a_n = c_n + c_{-n}$ and $b_n = c_n - c_{-n}$.

We can now formulate our problem: *we want to characterize tempered distributions f such that*

$$\frac{1}{2\pi} \text{p.v.} \left\langle \hat{f}(x), e^{ix_0 x} \right\rangle = \gamma \quad (C). \tag{5.125}$$

We study some properties of the principal value evaluations in the (C) sense. They admit a quasi-asymptotic characterization, but unlike e.v. Cesàro evaluations, the existence of p.v. $\langle g(x), \phi(x) \rangle = \gamma$ (C) does not imply that $\phi g \in \mathcal{S}'(\mathbf{R})$. We first need the following lemmas.

Lemma 5.6. *Let $g \in \mathcal{D}'(\mathbf{R})$ be an even distribution. There exists $h \in \mathcal{D}'(\mathbf{R})$ such that supp $h \subseteq [0, \infty)$ and $g(x) = h(x) + h(-x)$.*

Proof. Decompose $g = g_- + g_+$, where supp $g_- \subseteq (-\infty, 0]$ and supp $g_+ \subseteq [0, \infty)$. The parity of g implies that $g_+(x) - g_-(-x)$ is concentrated at the origin, and so there exist constants such that

$$g_-(x) = g_+(-x) + \sum_{j=0}^{n} a_j \delta^{(j)}(x), \tag{5.126}$$

Since, $g(x) - g_+(-x) - g_+(x) = \sum_{j=0}^{n} a_j \delta^{(j)}(x)$ is even, it follows that $a_j = 0$ whenever j is odd. So, $n = 2k$, and hence $h = g_+ + (1/2) \sum_{j=0}^{k} a_{2j} \delta^{(2j)}$ satisfies the requirements. $\qquad\square$

Lemma 5.7. *Let $g \in \mathcal{D}'(\mathbf{R})$ be an even distribution. Then*

$$g(\lambda x) = \gamma \frac{\delta(x)}{\lambda} + o\left(\frac{1}{\lambda}\right) \quad \text{as } \lambda \to \infty \text{ in } \mathcal{D}'(\mathbf{R}) \tag{5.127}$$

if and only if any $h \in \mathcal{S}'(\mathbf{R})$ such that supp $h \subseteq [0, \infty)$, and $g(x) = h(x) + h(-x)$, satisfies

$$h(\lambda x) = \frac{\gamma \delta(x)}{2\lambda} + o\left(\frac{1}{\lambda}\right) \quad \text{as } \lambda \to \infty \text{ in } \mathcal{S}'(\mathbf{R}). \tag{5.128}$$

Proof. The converse is clear. On the other hand take h as in Lemma 5.6. Proposition 5.4 implies the existence of m such that

$$h^{(-2m)}(x) = \frac{\gamma x^{2m-1}}{2(2m-1)!} + c(x) \frac{x^{2m-1}}{(2m-1)!} + o(x^{2m-1})$$

and

$$h^{(-2m)}(x) = \frac{\gamma x^{2m-1}}{2(2m-1)!} - c(|x|) \frac{x^{2m-1}}{(2m-1)!} + o(x^{2m-1}),$$

$x \to \infty$, but comparison between the last two expressions gives that $c(x) = o(1)$, and hence

$$h^{(-2m)}(x) \sim \frac{\gamma x^{2m-1}}{2(2m-1)!}, \quad x \to \infty,$$

which implies (5.128). $\qquad\square$

Proposition 5.7. *Let $g \in \mathcal{D}'(\mathbf{R})$ and $\phi \in \mathcal{E}(\mathbf{R})$. Then,*

$$\text{p.v.}\,\langle g(x), \phi(x)\rangle = \gamma \quad (C) \tag{5.129}$$

if and only if

$$\phi(-\lambda x)g(-\lambda x) + \phi(\lambda x)g(\lambda x) = 2\gamma\frac{\delta(x)}{\lambda} + o\left(\frac{1}{\lambda}\right) \quad \text{as } \lambda \to \infty \text{ in } \mathcal{S}'(\mathbf{R}); \tag{5.130}$$

if and only if for any decomposition $g = g_- + g_+$, where supp $g_- \subseteq (-\infty, 0]$ and supp $g_+ \subseteq [0, \infty)$,

$$\phi(-\lambda x)g_-(-\lambda x) + \phi(\lambda x)g_+(\lambda x) = \gamma\frac{\delta(x)}{\lambda} + o\left(\frac{1}{\lambda}\right) \quad \text{as } \lambda \to \infty \text{ in } \mathcal{S}'(\mathbf{R}). \tag{5.131}$$

In particular, we obtain that $\phi(-x)g(-x) + \phi(x)g(x) \in \mathcal{S}'(\mathbf{R})$.

Proof. Assume that $\phi \equiv 1$. We have that $g(-x) + g(x)$ is an even distribution, then, by Lemma 5.6, we can find h with supp $h \subseteq [0, \infty)$ such that $g(-x) + g(x) = h(x) + h(-x)$. It is easy to see that (5.129) is equivalent to $\lim_{x\to\infty} h^{(-1)}(x) = \gamma$ (C) which holds if and only if (5.128) (with γ replaced by $\gamma/2$), and by Lemma 5.7, it is equivalent to (5.130). The equivalence with (5.131) follows by taking $h(x) = g_-(-x) + g_+(x)$. $\square$

The right notion to characterize (5.125) is that of distributional symmetric point values.

Definition 5.9. Let $f \in \mathcal{D}'(\mathbf{R})$. We say that f has a (distributional) symmetric point value at $x = x_0$ if the following quasi-asymptotic limit holds

$$\lim_{\varepsilon\to 0}(f(x_0 - \varepsilon x) + f(x_0 + \varepsilon x)) = \gamma, \quad \text{in } \mathcal{D}'(\mathbf{R}). \tag{5.132}$$

In this case we write $f_{\text{sym}}(x_0) = \gamma$, distributionally.

Observe that if we define $\chi^f_{x_0}(x) = f(x_0 - x) + f(x_0 + x)$, the symmetric part of f about the point $x = x_0$, then (5.132) holds if and only if $\chi^f_{x_0}(0) = \gamma$, distributionally. In addition, note that if $\chi^f_{x_0} \in \mathcal{S}'(\mathbf{R})$, then (5.132) actually holds in the space $\mathcal{S}'(\mathbf{R})$ (cf. Theorem 2.35). We have set the ground to solve our problem. The following theorem is the solution to the Hardy-Littlewood (C) problem for tempered distributions.

Theorem 5.18. *Let $f \in \mathcal{S}'(\mathbf{R})$. Then*

$$\frac{1}{2\pi}\text{p.v.}\left\langle \hat{f}(x), e^{ix_0x} \right\rangle = \gamma \quad (C) \tag{5.133}$$

if and only if $f_{\text{sym}}(x_0) = \gamma$, distributionally.

Proof. By definition $f_{\text{sym}}(x_0) = \gamma$, distributionally, if and only if,

$$f(x_0 - \varepsilon x) + f(x_0 + \varepsilon x) = \gamma + o(1) \quad \text{as } \varepsilon \to 0^+ \text{ in } \mathcal{S}'(\mathbf{R}),$$

which, by taking Fourier transform, is equivalent to

$$e^{-i\lambda x_0 x}\hat{f}(-\lambda x) + e^{i\lambda x_0 x}\hat{f}(\lambda x) = 2\pi\gamma\frac{\delta(x)}{\lambda} + o\left(\frac{1}{\lambda}\right) \quad \text{as } \lambda \to \infty \text{ in } \mathcal{S}'(\mathbf{R}),$$

and, by Proposition 5.7, the latter is equivalent to (5.133). $\square$

We immediately obtain, by Theorem 5.18 and Example II.6, the following result of Hardy and Littlewood. Naturally, the language in the original statement differs from ours, at that time distribution theory and quasi-asymptotics did not even exist!

Corollary 5.19. *Let $f \in \mathcal{S}'(\mathbf{R})$ be a 2π periodic distribution having Fourier series, in sine-cosine form,*

$$\frac{a_0}{2} + \sum_{n=1}^{\infty}(a_n \cos nx + b_n \sin nx).$$

Then,

$$\frac{a_0}{2} + \sum_{n=1}^{\infty}(a_n \cos nx_0 + b_n \sin nx_0) = \gamma \quad (C)$$

if and only if $f_{\text{sym}}(x_0) = \gamma$, distributionally.

We end this section by showing two Abelian results.

Theorem 5.19. *Let $g \in \mathcal{D}'(\mathbf{R})$ and $\phi \in \mathcal{E}(\mathbf{R})$. If*

$$\text{p.v.} \langle g(x), \phi(x) \rangle = \gamma \quad (C),$$

then,

$$\langle g(x), \phi(x) \rangle = \gamma \quad (A).$$

Proof. Take g_- and g_+ as in Proposition 5.7, then, by (5.131), as $\lambda \to \infty$,

$$\left\langle \phi(x)g_-(x), e^{\frac{x}{\lambda}} \right\rangle + \left\langle \phi(x)g_+(x), e^{-\frac{x}{\lambda}} \right\rangle$$

$$= \lambda \left\langle \phi(-\lambda x)g_-(-\lambda x) + \phi(\lambda x)g_+(\lambda x), e^{-x} \right\rangle$$

$$= \gamma \left\langle \delta(x), e^{-x} \right\rangle + o(1)$$

$$= \gamma + o(1). \qquad \square$$

For symmetric point values, we get a radial version of Theorem 5.6.

Theorem 5.20. *Let* $f \in \mathcal{D}'(\mathbf{R})$. *Let* U *be a harmonic representation of* f *on the upper half-plane* $\operatorname{Im} z > 0$. *If* $f_{\mathrm{sym}}(x_0) = \gamma$, *distributionally, then*

$$\lim_{y \to 0^+} U(x_0 + iy) = \gamma. \qquad (5.134)$$

Proof. As in the proof of Theorem 5.6, we may assume that f is a tempered distribution. If $\hat{f} = \hat{f}_+ + \hat{f}_-$ is a decomposition such that $\operatorname{supp}\hat{f}_- \subseteq (-\infty, 0]$ and $\operatorname{supp}\hat{f}_+ \subseteq [0, \infty)$, we can assume that

$$U(z) = \frac{1}{2\pi} \left(\left\langle \hat{f}_-(t), e^{i\bar{z}t} \right\rangle + \left\langle \hat{f}_+(t), e^{izt} \right\rangle \right), \quad \operatorname{Im} z > 0.$$

But in this case, Theorem 5.19 yields (5.134). $\qquad \square$

Bibliography

[1] M. Aguirre-Téllez, The expansion of $\delta^{(k-1)}(m^2 + P)$, Integral Transform Spec. Funct. 8 (1999), 139–148.

[2] K. Ananda Rau, On the converse of Abel's theorem, J. London Math. Soc. 3 (1928), 349–355.

[3] P. Antosik, J. Mikusiński, R. Sikorski, Theory of Distributions, The Sequential Approach, Polish Scientific Publishers, Warszawa, 1973.

[4] J. M. Ash, P. Erdős, L. A. Rubel, Very slowly varying functions, Aequationes Math. 10 (1974), 1–9.

[5] E. J. Beltrami, M.R. Wohlers , Distributions and the Boundary Values of Analytic Functions, Academic Press, New York, 1966.

[6] J. J. Benedetto, Analytic representation of generalized functions, Math. Z. 97 (1967), 303–319.

[7] J. J. Benedetto, Spectral synthesis, Academic Press, New York, 1975.

[8] S. Berceanu, A. Gheorghe, On the asymptotics of distributions with support in a cone, J. Math. Phys. 26 (1985), 2335–2341.

[9] N. H. Bingham, C. M. Goldie, J. L. Teugels, Regular Variation, Cambridge Univ. Press, Cambridge, 1989.

[10] N. N. Bogolubov, V. S. Vladimirov, A. N. Tavkelidze, On automodel asymptotics in quantum field theory, Proc. Steklov Inst. Math. 135 (1978), 27–54.

[11] R. Bojanić, J. Karamata, On a class of functions of regular asymptotic behavior, Math. Res. Centre, U.S. Army, Madison, Wis., Tech. Summary Rep. No. 436 (1963).

[12] N. Bourbaki, Topologie Générale, Chapitres 1, 2, Hermann, Paris, 1951.

[13] H. Bremermann, Distributions, Complex Variables and Fourier Transforms, Addison-Wesley, Reading, Massachusetts, 1965.

[14] N. G. de Bruijn, Asymptotic Methods in Analysis, Dover Publications, Inc., New York, 1981.

[15] Yu. A. Brychkov, Asymptotic Expansions of Generalized Functions I, Theoret. Mat. Fiz. 5 (1970), 98–109 (In Russian).

[16] Yu. A. Brychkov, On Asymptotic Expansions of Generalized Functions, Mat. Zametki 12 (1972), 131–138 (In Russian).

[17] Yu. A. Brychkov, Asymptotic expansions of generalized functions II, Theoret. Mat. Fiz. 15 (1973), 375-381 (In Russian).

[18] Yu. A. Brychkov, Asymptotic expansions of generalized functions III, Theoret. Mat. Fiz. 23 (1975), 191-198 (In Russian).

[19] Yu. A. Brychkov, A. P. Prudnikov, Integral Transforms of Generalized Functions, Gordon and Breach Sci. Publ., New York, 1989.

[20] Yu. A. Brychkov, Yu. M. Shirokov, Asymptotic behavior of the Fourier transforms, Teoret. Mat. Fiz. 4 (1970), 301-309 (In Russian).

[21] J. Campos Ferreira, Introduction to the Theory of Distributions, Longman, Essex, 1997.

[22] R. D. Carmichael, Abelian Theorems for the Stieltjes Transform of functions, Bull. Calcutta Math. Soc. 68 (1976), 49-52.

[23] R. D. Carmichael, E. O. Milton, Abelian theorems for the distributional Stieltjes transform, J. Math. Anal. Appl. 72 (1979), 195-205.

[24] R. D. Carmichael, R. S. Pathak, Asymptotic analysis of the H-function transform, Glasnik Math. 25 (1990), 103-127.

[25] K. Chandrasekharan, S. Minakshisundaram, Typical Means, Oxford University Press, 1952.

[26] F. Constantinescu, Boundary values of analytic functions, Comm. Math. Phys. 7 (1968), 225-233.

[27] J. B. Conway, *Functions of One Complex Variable II*, Springer Verlag, New York, 1995.

[28] J. Čižek, J. Jelinek, A Tauberian theorem for distributions, Comment. Math. Univ. Carolinae 37 (1996), 479-488.

[29] H. Delange, Sur les théorèmes inverses des procédés de somation des séries divergentes, I, II, Ann. Sci. Ecole Norm Sup. 67 (1950), 99-160, and 199-242.

[30] G. Doetsch, Handbuch der Laplace Transformation, I, Birkhäuser, Basel, 1950.

[31] W. F. Donoghue, Distributions and Fourier transforms, Academic Press, New York, 1969.

[32] Yu. N. Drozhzhinov, B. I. Zavyalov, Quasi-asymptotics of generalized functions and Tauberian theorems in the complex domain, Mat. Sb. 102 (1977), 372-390 (In Russian). English transl. Math. USSR Sb. 36 (1977).

[33] Yu. N. Drozhzhinov, B. I. Zavyalov, Tauberian theorems for generalized functions with supports in cones, Mat. Sb. 108 (1979), 78-90 (In Russian).

[34] Yu. N. Drozhzhinov, B. I. Zavyalov, Asymptotic properties of some classes of generalized functions, Izv. Akad. Nauk. SSR, Ser. Mat. 49 (1985), 81-140 (In Russian).

[35] Yu. N. Drozhzhinov, B. I. Zavyalov, Multidimensional Abelian and Tauberian comparing theorems, Mat. Sb. 180 (1989), 1234-1258 (In Russian).

[36] Yu. N. Drozhzhinov, B. I. Zavyalov, Local Tauberian theorems in spaces of distributions related to cones, and their applications, Izvestiya RAN. Ser. Mat. 61:6 (1997), 59-102.

[37] Yu. N. Drozhzhinov, B. I. Zavyalov, A Wiener-type Tauberian theorem for tempered distributions, Mat. Sb. 189 (1998), 91-130.

[38] Yu. N. Drozhzhinov, B. I. Zavyalov, Tauberian-type theorems for a generalized multiplicative convolution, Izv. Ross. Akad. Nauk Ser. Mat. 64 (2000), 37–94 (In Russian). Translation in Izv. Math. 64 (2000), 35-92.

[39] Y. N. Drozhzhinov, B. I. Zavyalov, Multidimensional Tauberian theorems for Banach-space valued generalized functions, Sb. Math. 194 (2003), 1599–1646.

[40] Yu. N. Drozhzhinov, B. I. Zavyalov, Asymptotically homogeneous generalized functions and boundary properties of functions holomorphic in tubular cones, Izv. Math. 70 (2006), 1117–1164.

[41] Yu. N. Drozhzhinov, B. I. Zavyalov, Applications of Tauberian theorems in some problems of mathematical physics, Teoret. Mat. Fiz. 157 (2008), 373–390 (In Russian). Translation in Theoret. and Math. Phys. 157 (2008), 1678–1693.

[42] Yu. N. Drozhzhinov, B. I. Zavyalov, Generalized functions asymptotically homogeneous with respect to special transformation groups, Sb. Math. 200 (2009), 803-844.

[43] A. Duran, Laguerre Expansions of Tempered Distributions and Generalized Functions, J. Math. Anal. Appl. 150 (1990), 166–180.

[44] A. Erdely, Asymptotic Expansions, Dover Publ., New York, 1956.

[45] R. Estrada, Characterization of the Fourier series of distributions having a value at a point, Proc. Amer. Math. Soc. 124 (1996), 1205–1212.

[46] R. Estrada, Regularization of distributions, Internat. J. Math. & Math. Sci. 21 (1998), 625–636.

[47] R. Estrada, The Cesàro behaviour of distributions, Proc. R. Soc. Lond. A 454 (1998), 2425–2443.

[48] R. Estrada, Boundary values of analytic functions without distributional point values, Tamkang J. Math. 35 (2004), 53–60.

[49] R. Estrada, A Distributional Version of the Ferenc-Lukács Theorem, Sarajevo J. Math. 1 (2005), 75–92.

[50] R. Estrada, S. A. Fulling, Distributional asymptotic expansion of spectral functions and of the associated Green kernels, Electronic J. Diff. Eqns. 6 (1999), 1–37.

[51] R. Estrada, R. P. Kanwal, Moment sequences for a certain class of distributions, Compl. Variabl. 9 (1987), 31–39.

[52] R. Estrada, R. P. Kanwal, A distributional theory of asymptotic expansions, Proc. Roy. Soc. London Ser. A 428 (1990), 399–430.

[53] R. Estrada, R. P. Kanwal, The asymptotic expansion of certain multidimensional generalized functions, J. Math. Anal. Appl. 163 (1992), 264–283.

[54] R. Estrada, R. P. Kanwal, Taylor expansions for distributions, Math. Methods Appl. Sci. 16 (1993), 297–304.

[55] R. Estrada. R. P. Kanwal, Asymptotic separation of variables, J. Math. Anal. Appl. 178 (1993), 130–142.

[56] R. Estrada, R. P. Kanwal, A Distributional Approach to Asymptotics. Theory and Applications, Second Edition, Birkhäuser, Basel, 2002.

[57] R. Estrada, J. Vindas, Determination of jumps of distributions by differentiated means, Acta Math. Hungar. 124 (2009), 215–241.

[58] R. Estrada, J. Vindas, On the Point Behavior of Fourier Series and Conjugate Series, Z. Anal. Anwend. 29 (2010), 487–504.

[59] R. Estrada, J. Vindas, A General Integral, preprint, 2011.

[60] W. Feller, An Introduction to Probability Theory and its Applications, I, Wiley, New York, 1968.

[61] A. Friedman, Generalized functions and partial differential equations, Prentice-Hall, New Jercey, London, 1963.

[62] T. H. Ganelius, Tauberian remainder theorems, Lect. Not. Math. 232, Springer, Berlin, 1971.

[63] I. M. Gelfand, G. E. Shilov, Generalized Functions 1, Academic Press, New York, 1964.

[64] J. L. Geluk, L. De Haan, Regular variation Extensions and Tauberian Theorems, CWI Tracts 40, Amsterdam, 1987.

[65] H. J. Glaske, D. Müller, Abelsche Sätze für die Laplace-Transformation von Distribution, Z. Anal. Anw. 10 (1991), 232–238.

[66] G. H. Hardy, Divergent Series, Clarendon Press, Oxford, 1949.

[67] G. H. Hardy, J. E. Littlewood, Some theorems concerning Dirichlet's series, Messenger of Mathematics 43 (1914), 134–147.

[68] G. H. Hardy, J. E. Littlewood, Solution of the Cesàro summability problem for power series and Fourier series, Math. Z. 19 (1923), 67–96.

[69] L. G. Hernández, R. Estrada, Solution of ordinary differential equations by series of delta functions, J. Math. Anal. Appl. 191 (1995), 40–55.

[70] L. Hörmander, Linear partial differential operators I, II, Springer, Berlin, 1963.

[71] I. Imai, Applied Hyperfunction Theory, Kluwer, Dordrecht, 1992.

[72] D. S. Jones, Generalized Functions, McGraw-Hill, New York, Toronto, Ont.-London, 1966.

[73] A. Kaneko, Representation of hyperfunctions by measures and some of its applications, J. Fac. Sci. Univ. Tokyo, Sec. 1A, 19 (1972), 321–352.

[74] A. Kaneko, On the Structure of Fourier Hyperfunctions, Proc. Japan Acad. 48 (1972), 651–653.

[75] A. Kaneko, Introduction to Hyperfunctions, Kluwer, Dordrecht, 1988.

[76] J. Karamata, Sur un mode de croissance regulière des functions, Mathematica (Cluj) 3 (1930), 33–48.

[77] J. Karamata, Sur un mode de croissance régulière. Théorèmes fondamentaux, Bull. Soc. Math. France 61 (1933), 55–62.

[78] T. Kawai, On the theory of Fourier hyperfuctions and its applications to partial differential equations with constant coefficients, J. Fac. Sci. Univ. Tokyo, 1A, 17 (1970), 467–517.

[79] H. Komatsu, Ultradistributions I, J. Fac. Sci. Univ. Tokyo Sect. 1A Mat. 20 (1973), 25–105.

[80] H. Komatsu, Laplace transform of hyperfunctions-A new fundation of the Heaviside calculus, J. Fac. Sci. Univ. Tokyo IA 34 (1987), 805–820.

[81] H. Komatsu, Microlocal analysis in Gevrey classes and in complex domains, Lect. Not. Math. 1726, 426–493, Springer-Verlag, Berlin, 1989.

[82] P. Koosis, Introduction to H_p spaces, London Mathematical Society

Lecture Note Series, 40. Cambridge University Press, Cambridge-New York, 1980.

[83] J. Korevaar, T. van Aardenne-Ehrenfest, N. G. de Bruijn, A note on slowly oscillating functions, Nieuw. Arch. Wisk. 23 (1949), 77–88.

[84] J. Korevaar, Distribution proof of Wiener Tauberian theorem, Proc. Amer. Math. Soc. 16 (1965), 353–355.

[85] D. Kovačević, S. Pilipović, On Homogeneous Generalized Functions, Publ. Math. Debrecen 41 (1992), 227–287.

[86] D. Kovačević, S. Pilipović, Structural properties of the space of tempered ultradistributions, in: Proc. Conf. Complex Analysis and Generalized Functions (Varna, 1991), 169–184, 1993.

[87] E. Landau, Sur les valueurs moyennes de certains functions arithmétiques, Bull. Acad. Roy. Belgique (1911), 443–472.

[88] J. Lavoine, Sur des théoèmes abéliens et taubériens de la transformation de Laplace, Ann. Inst. Henri Poincare Nouv. S. 4 (1966), 49–65.

[89] J. Lavoine, O. P. Misra, Théorèmes Abéliens pour la transformation de Stieltjes des distributions, C. R. Acad. Sci. Paris 279 (1974), 99–102.

[90] J. Lavoine, O. P. Misra, Abelian theorems for the distributional Stieltjes transformation, Math. Proc. Cambridge Philos. Soc. 86 (1979), 287–293.

[91] M. J. Lighthill, Introduction to Fourier Analysis and Generalized Functions, Cambridge Univ. Press, London, 1958.

[92] J. E. Littlewood, The converse of Abel's theorem on power series, Proc. London Math. Soc. 9 (1911), 434–448.

[93] S. Lojasiewicz, Sur la valeur et la limit d'une distribution dans un point, Studia Math. 16 (1957), 1–36.

[94] S. Lojasiewicz, Sur la fixation des variables dans une distribution, Studia Math. 17 (1958), 1–64.

[95] M. Mangad, Asymptotic expansions of Fourier transforms and discrete polyharmonic Green's functions, Pacific J. Math. 20 (1967), 85–98.

[96] V. Marić, M. Skendžić, A. Takači, On Stieltjes transform of distributions behaving as regularly varying functions, Acta. Sci. Math. Szeged 50 (1986), 405–410.

[97] T. Matsuzawa, A Calculus Approach to Hyperfunctions, II, Transactions of the American Mathematical Society 313 (1989), 619–654.

[98] O. P. Misra, Some Abelian theorems for distributional Stieltjes transform, J. Math. Anal. Appl. 39 (1972), 590–599.

[99] M. Nedeljkov, S. Pilipović, Generalized Solutions to a Semilinear Hyperbolic System with a non-Lipshitz Nonlinearity, Monatsh. Math. 125 (1998), 255–261.

[100] D. Nikolić-Despotović, S. Pilipović, Tauberian theorem for the distributional Stieltjes transform, Internat. J. Math. Math. Sci. 9 (1986), 521–524.

[101] D. Nikolić-Despotović, S. Pilipović, The quasiasymptotic expansion of tempered distributions and the Stieltjes transform, Univ. u Novom Sadu Zb. Rad. Prirod.-Mat. Fak. Ser. Mat. 18 (1988), 31–44.

[102] D. Nikolić-Despotović, S. Pilipović, The quasiasymptotic expansion at the origin. Abelian-type results for the Laplace and Stieltjes transforms, Math.

Nachr. 174 (1995), 231–238.

[103] N. Ortner, P. Wagner, Applications of weighted $\mathcal{D}'_{L^P}$-spaces to the convolution of distributions, Bull. Polish Acad. Sci. Math. 37 (1989), 579–595.

[104] R. S. Pathak, S. K. Misra, Asymptotic behavior of a class of integral transforms in complex domains, Proc. Edinburgh Math. Soc. 32 (1989), 231–247.

[105] J. Peetre, On the value of a distribution at a point, Portugal. Math. 27 (1968), 149–159.

[106] S. Pilipović, On the S-asymptotics of tempered distributions and $\mathcal{K}'_1$ distributions, Part I, Univ. u Novom Sadu Zb. Rad. Prirod.-Mat. Fak. Ser. Mat. 15 (1985), 47–58.

[107] S. Pilipović, On the S-asymptotics of tempered distributions and $\mathcal{K}'_1$ distributions, Part II, Univ. u Novom Sadu Zb. Rad. Prirod.-Mat. Fak. Ser. Mat. 15 (1985), 59–67.

[108] S. Pilipović, Asymptotic behavior of the distributional Weierstrass transform, Appl. Anal. 25 (1987), 171–179.

[109] S. Pilipović, Remarks on supports of distributions, Glasnik Mat. 22 (1987), 375–380.

[110] S. Pilipović, On the quasiasymptotic behavior of the Stieltjes transform of distributions, Publ. Inst. Math. Beograd 40 (1987), 143–152.

[111] S. Pilipović, Some properties of the quasi-asymptotics of Schwartz distributions, Part I, Quasiasymptotic at $\pm\infty$, Publ. Inst. Math. Beograd 43 (1988), 125–130.

[112] S. Pilipović, Some properties of the quasiasymptotic of Schwartz distributions, Part II-Qiasiasymptotic at 0, Publ. Inst. Math. Beograd 43 (57) (1988), 131–135.

[113] S. Pilipović, On the quasiasymptotic of Schwartz distributions, Math. Nachr. 137 (1988), 19–25.

[114] S. Pilipović, On the S-asymptotics of tempered distributions and $\mathcal{K}'_1$-distributions, Part III, Univ. u Novom Sadu, Zb. Rad. Prirod.-Mat. Fak. Ser. Mat. 18 (1988), 179–189.

[115] S. Pilipović, On the S-asymptotics of tempered and $\mathcal{K}'_1$ distributions, Part IV. S-asymptotic and the ordinary asymptotic, Univ. u Novom Sadu Zb. Rad. Prirod.-Mat. Fak., Ser. Mat. 18 (1988), 191–195.

[116] S. Pilipović, On the Behavior of Distribution at the Origin, Math. Nachr. 141 (1989), 27–32.

[117] S. Pilipović, Quasiasymptotic expansion of distributions from $\mathcal{S}'_+$ and the asymptotic expansion of the distributional Stieltjes transform, Internat. J. Math. Sci. 12 (1989) 673–684.

[118] S. Pilipović, Asymptotic expansion of Schwartz's distributions, Publ. Inst. Math. Beograd 45 (1989), 119–127.

[119] S. Pilipović, Generalized S-asymptotics, Tauberian Theorem, Publ. Inst. Math. Beograd 48 (1990), 129–132.

[120] S. Pilipović, The translation asymptotic and the quasiasymptotic behavior of distributions, Acta. Math. Hungarica 55 (1990), 239–243.

[121] S. Pilipović, On the behavior of distributions at infinity, Wiener-Tauberian type results, Publ. Inst. Mat. Beograd, 48 (1990), 129–132.

[122] S. Pilipović, Quasiasymptotic expansion and the Laplace transform, Appl. Anal. 35 (1991), 247–261.

[123] S. Pilipović, Characterization of bounded sets in spaces of ultradistributions, Proc. Amer. Math. Soc. 120 (1994), 1191–1206.

[124] S. Pilipović, Quasiasymptotics and S-asymptotics in S' and D', Publ. Inst. Math. Beograd 58 (1995), 13–20.

[125] S. Pilipović, B. Stanković, Abelian theorems for the distributional Stieltjes transform, Z. Anal. Anw. 6 (1987), 341–349.

[126] S. Pilipović, B. Stanković, Initial value Abelian theorems for the distributional Stieltjes transform, Studia Math. 86 (1987), 239–254.

[127] S. Pilipović, B. Stanković, S-asymptotics of a distribution, Pliska Studia Math. Bulgar. 10 (1989), 147–156.

[128] S. Pilipović, B. Stanković, Structural theorems for the S-asymptotics and quasi-asymptotics of distributions, Mathematica Panonica 1 (1993), 23–35.

[129] S. Pilipović, B. Stanković, Wiener-type Tauberian theorems for distributions, J. London Math. Soc. 47 (1993), 507–515.

[130] S. Pilipović, B. Stanković, Wiener-type Tauberian Theorems for ultradistributions, Rend. Sem. Math. Univ. Padova 92 (1994), 209–220.

[131] S. Pilipović, B. Stanković, Quasiasymptotics and S-asymptotics of ultradistributions, Bull. Acad. Serbe. Sci. 20 (1995), 61–74.

[132] S. Pilipović, B. Stanković, Properties of ultradistributions having the S-asymptotics, Bull. Acad. Serbe Sci. 21 (1996), 47–59.

[133] S. Pilipović, B. Stanković, Tauberian Theorems for integral transforms of distributions, Acta Math. Hungarica 74 (1997), 135–153.

[134] S. Pilipović, B. Stanković, Wiener Tauberian Theorems for Hyperfunctions, Z. Anal. Anw. 21 (2002), 1127–1042.

[135] S. Pilipović, B. Stanković, A. Takači, Asymptotic Behavior and Stieltjes Transformation of Distributions, Teubner, Leipzig BSB, 1990.

[136] S. Pilipović, M. Stojanović, B-transform and its quasi-asymptotics — Application to the convolution equation $(xu_{xx} + 2u_x + mu) * g = h$, $m \leq 0$, Publ. Math. Debrecen 44 (1994), 275–283.

[137] S. Pilipović, M. Stojanović, Generalized Solutions to Nonlinear Volterra Integral Equations with non-Lipshitz Nonlinearity, Nonlinear Analysis, TMA 37 (1999), 319–335.

[138] S. Pilipović, A. Takači, The quasiasymptotic behavior of some distributions, Univ. u Novom Sadu, Zb. Rad. Prirod. Mat. Fak., Ser. Math. 15 (1985), 37–46.

[139] S. Pilipović, J. Vindas, Multidimensional Tauberian theorems for wavelet and non-wavelet transforms, preprint.

[140] H. R. Pitt, Generalized Tauberian Theorems, Proc. London Math. Soc. 44 (1938), 243–288.

[141] H. R. Pitt, Tauberian Theorems, Oxford Univ. Press, Oxford, 1958.

[142] H. Reiter, Classical harmonic analysis and locally compact groups, Clarendon Press, Oxford, 1968.

[143] K. Saneva, J. Vindas, Wavelet expansions and asymptotic behavior of distributions, J. Math. Anal. Appl. 370 (2010), 543–554.

[144] M. Sato, On a generalization of the concept of functions, Proc. Japan Acad. 34 (1958), 126–130, and 604–608.

[145] M. Sato, Theory of hyperfunctions I, J. Fac. Sci. Univ. Tokyo IA 8 (1959), 139–193.

[146] L. Schwartz, Théorie des Distributions, Hermann, Paris, 1966.

[147] J. Sebastião e Silva, Axiomatic approach to the distribution theory, Integrals and orders of distributions, Lisbonne, Centro de Calculo Cientifico, 1964.

[148] E. Seneta, Regularly Varying Functions, Springer, Berlin, 1976.

[149] B. Stanković, Theorems of Tauberian Type for Measures, Glasnik Mat. 20 (1985), 383–390.

[150] B. Stanković, Abelian and Tauberian theorems for the Stieltjes transform of distributions, Usp. Mat. Nauk 40 (1985), 91–103. Russian Math. Surveys 40 (1985), 99–113.

[151] B. Stanković, Characterization of some subspaces of $\mathcal{D}'$ by S-asymptotics, Publ. Inst. Math. Beograd 41 (1987), 111–117.

[152] B. Stanković, S-asymptotic expansion of distributions, Internat. J. Math. Math. Sci. 11 (1988), 449–456.

[153] B. Stanković, S-Asymptotics of Elements Belonging to $\mathcal{D}_{L^p}$ and $\mathcal{D}'_{L^p}$, Univ. u Novom Sadu, Zb. Rad. Prir. Mat. Fak. Ser. Mat. 18 (1988), 169–175.

[154] B. Stanković, S-asymptotics of solutions of the elliptic partial differential equation, Univ. u Novom Sadu, Zb. Rad. Prir. Mat. Fak. Ser. Mat. 19 (1989), 65–72.

[155] B. Stanković, A Structural Theorem for the Distributions Having S-asymptotics, Publ. Inst. Math. Beograd 45 (1989) 129–132.

[156] B. Stanković, Regularly varying distributions, Publ. Inst. Math. Beograd 48 (1990), 119–128.

[157] B. Stanković, Abelian Theorem for the Stieltjes Transform of Distributions, Int. J. Math. Sci. 13 (1990), 677–686.

[158] B. Stanković, Abelian Theorems for integral transforms of distributions, Int. Trans. Spec. Func. 1 (1993), 61–73.

[159] B. Stanković, Relation between quasiasymptotics, equivalence at infinity and S-asymptotics, Univ. u Novom Sadu Zb. Rad. Prir. Mat. Fak. Ser. Mat. 21 (1993), 1–11.

[160] B. Stanković, Asymptotic almost periodic distributions and applications to PDE, Trudy Steklov Math. Inst. Moscow 203 (1994), 449–460.

[161] B. Stanković, Analyltic ultradistributions, Proc. Amer. Math. Soc. 123 (1995), 3365–3369.

[162] B. Stanković, Regularly varying ultradistributions, Publ Inst. Math. Beograd 58 (1995), 117–128.

[163] B. Stanković, Taylor Expansion for Generalized Functions, J. Math. Anal. Appl. 203 (1996), 31–37.

[164] B. Stanković, Asymptotic Taylor expansion for Fourier hyperfunctions, Proc. Royal Soc. London A 453 (1997), 913–918.

[165] B. Stanković, Convergence structures and S-asymptotic behaviour of Fourier hyperfunctions, Publ. Inst. Math. (Beograd) 64 (78), (1998), 98–106.

[166] H. Struwe, Semilinear Wave Equations, Bulletin Amer. Math. Soc. 26 (1992), 53–85.

[167] A. Takači, A note on the distributional Stieltjes transform, Math. Proc. Camb. Philos. Soc. 94 (1983), 523–527.

[168] A. Tauber, Ein Satz aus der Theorie der unendlichen Reihen, Monatsh. Math. Phys. 8 (1897), 273–277.

[169] E. C. Titchmarsh, The Theory of Functions, Second Edition, Oxford University Press, Oxford, 1939.

[170] E. C. Titchmarsh, Introduction to the theory of Fourier integrals, Second Edition, Clarendon Press, Oxford, 1948.

[171] J. Vindas, Structural theorems for quasiasymptotics of distributions at infinity, Publ. Inst. Math. (Beograd) (N.S.) 84(98) (2008), 159–174.

[172] J. Vindas, The structure of quasiasymptotics of Schwartz distributions, in: Linear and non-linear theory of generalized functions and its applications, 297–314, Banach Center Publ. 88, Polish Acad. Sc. Inst. Math., Warsaw, 2010.

[173] J. Vindas, Local behavior of distributions and applications, Dissertation, Louisiana State University, Baton Rouge, 2009.

[174] J. Vindas, Regularizations at the origin of distributions having prescribed asymptotic properties, Integral Transforms Spec. Funct. (2011), in press.

[175] J. Vindas, R. Estrada, Distributionally regulated functions, Studia Math. 181 (2007), 211–236.

[176] J. Vindas, R. Estrada, Distributional Point Values and Convergence of Fourier Series and Integrals, J. Fourier. Anal. Appl. 13 (2007), 551–576.

[177] J. Vindas, R. Estrada, A Tauberian theorem for distributional point values, Arch. Math. (Basel) 91 (2008), 247–253.

[178] J. Vindas, R. Estrada, On the jump behavior of distributions and logarithmic averages, J. Math. Anal. Appl. 347 (2008), 597–606.

[179] J. Vindas, R. Estrada, Measures and the distributional ϕ-transform, Integral Transforms Spec. Funct. 20 (2009), 325–332.

[180] J. Vindas, R. Estrada, A quick distributional way to the prime number theorem, Indag. Math. (N.S.) 20 (2009), 159–165.

[181] J. Vindas, R. Estrada, On the Order of Summability of the Fourier Inversion Formula, Anal. Theory Appl. 26 (2010), 13–42.

[182] J. Vindas, R. Estrada, On the support of tempered distributions, Proc. Edinb. Math. Soc. 53 (2010), 255–270.

[183] J. Vindas, R. Estrada, On the Cesàro behavior and some asymptotic notions for Schwartz distributions, preprint.

[184] J. Vindas, R. Estrada, On Tauber's second Tauberian theorem, preprint.

[185] J. Vindas, R. Estrada, Local boundary behavior of harmonic and analytic functions I: Abelian theorems, preprint.

[186] J. Vindas, S. Pilipović, Structural theorems for quasiasymptotics of distributions at the origin, Math. Nachr. 282 (2009), 1584–1599.

[187] J. Vindas, S. Pilipović, D. Rakić, Tauberian theorems for the wavelet transform, J. Fourier Anal. Appl. 17 (2011), 65–95.

[188] V. S. Vladimirov, Multidimensional Generalization of a Tauberian Theorem

of Hardy and Littlewood, Math. USSR, Izvestiya 10 (1976), 1032–1048.

[189] V. S. Vladimirov, Generalized Functions in Mathematical Physics, Mir, Moscow, 1979.

[190] V. S. Vladimirov, Equations in Mathematical Physics, Nauka, Moscow, 1988 (In Russian).

[191] V. S. Vladimirov, B. I. Zavyalov, Tauberian theorems in quantum field theory, Itogi Nauk. Tehn. 15 (1980), 95–130 (In Russian).

[192] V. S. Vladimirov, Yu. N. Drozhzhinov and B. I. Zavyalov, Tauberian Theorems for Generalized Functions, Kluwer, Dordrecht, 1988.

[193] V. S. Vladimirov, Yu. N. Drozhzhinov, B. I. Zavyalov, Tauberian theorems for generalized functions in the scale of regularly varying functions and functionals, Publ. Inst. Math. (Beograd) 71 (85) (2002), 123–132. (In Russian).

[194] P. Wagner, On the Quasi-asymptotic Expansion of the Causal Fundamental Solution of Hyperbolic Operators and Systems, Z. Anal. Anw. 10 (1991), 159–167.

[195] G. Walter, Pointwise convergence of distribution expansions, Studia Math. 26 (1966), 143–154.

[196] G. Walter, Fourier series and analytic representation of distributions, SIAM Review 12 (1970), 272–276.

[197] G. Walter, Local boundary behavior of harmonic functions, Illinois J. Math. 16 (1972), 491–501.

[198] N. Wiener, Tauberian Theorems, Ann. of Math. 33 (1932), 1–100.

[199] R. Wong, J. P. McClure, Generalized Mellin convolution and their asymptotic expansions, Canad. J. Math. 36 (1984), 924–960.

[200] B. I. Zavyalov, Automodel asymptotics of electromagnetic form-factors and the behavior of their Fourier transforms in the neighborhood of the light cone, Teoret. Mat. Fiz. 17 (1973), 178–188 (In Russian).

[201] B. I. Zavyalov, Asymptotic properties of functions that are holomorphic in tubular cones, Math. Sb. (N.S.) 136 (178) (1988), 97–114 (In Russian). English transl. Math. USSR, Sb. 64 (1989), 97–113.

[202] A. H. Zemanian, Generalized integral transforms, Interscience, John Wiley, New York, 1968.

[203] A. H. Zemanian, Some Abelian Theorems for the Distributional Hankel and K-Transformation, SIAM J. Appl. Anal 14 (1966), 1255–1265.

[204] V. V. Zharinov, Laplace transform of Fourier hyperfunctions and other related classes of analytical functionals, Teoret. Mat. Fiz. 33 (1977), 291–309 (In Russian).

[205] V. V. Zharinov, On the quasiasymptotics of Fourier hyperfunctions, Teoret. Mat. Fiz. 43 (1980), 32–38 (In Russian).

[206] A. Zygmund, Trigonometric Series, Vols. I & II, Second Edition, Cambridge University Press, Cambridge, 1959.

Index